Molecular Biology & Biotechnology

NIPA® GENX ELECTRONIC RESOURCES & SOLUTIONS P. LTD.
New Delhi-110 034

57	58	59	60	61	62	63	64	65	66	67	68	69	70	71
La	**Ce**	**Pr**	**Nd**	**Pm**	**Sm**	**Eu**	**Gd**	**Tb**	**Dy**	**Ho**	**Er**	**Tm**	**Yb**	**Lu**
138.91	140.12	140.91	144.24	(145)	150.36	151.96	157.25	158.93	162.50	164.93	167.26	168.93	173.04	174.97

89	90	91	92	93	94	95	96	97	98	99	100	101	102	103
Ac	**Th**	**Pa**	**U**	**Np**	**Pu**	**Am**	**Cm**	**Bk**	**Cf**	**Es**	**Fm**	**Md**	**No**	**Lr**
(227)	232.04	231.04	238.03	(237)	(244)	(243)	(247)	(247)	(251)	(252)	(257)	(258)	(259)	(262)

Actinium - **Ac**

Aluminum - **Al**

Americium - **Am**

Antimony - **Sb**

Argon - **Ar**

Arsenic - **As**

Astatine - **At**

Barium - **Ba**

Berkelium - **Bk**

Beryllium - **Be**

Bismuth - **Bi**

Boron - **B**

Bromine - **Br**

Cadmium - **Cd**

Caesium - **Cs**

Calcium – **Ca**

Californium - **Cf**

Carbon - **C**

Cerium - **Ce**

Chlorine - **Cl**

Chromium - **Cr**

Cobalt - **Co**

Copper - **Cu**

Curium - **Cm**

Dysprosium - **Dy**

Einsteinium - **Es**

Erbium - **Er**

Europium - **Eu**

Fermium - **Fm**

Fluorine - **F**

Francium - **Fr**

Gadolinium – **G**

Gallium - **Ga**

Germanium - **Ge**

Gold - **Au**

Hafnium - **Hf**

Helium - **He**

Holmium - **Ho**

Hydrogen - **H**

Indium - **In**

Iodine - **I**

Iridium - **Ir**

Iron - **Fe**

Krypton - **Kr**

Lanthanum - **La**

Lawrencium - **Lr**

Lead - **Pb**

Lithium – **Li**

Lutetium - **Lu**

Magnesium - **Mg**

Manganese - **Mn**

Meitnerium - **Mt**

Mendelevium -**Md**

Mercury - **Hg**

Molybdenum -**Mo**

Neodymium - **Nd**

Neon - **Ne**

Neptunium - **Np**

Nickel - **Ni**

Niobium - **Nb**

Nitrogen - **N**

Nobelium - **No**

Osmium - **Os**

Oxygen – **O**

Palladium - **Pd**

Phosphorus - **P**

Platinum - **Pt**

Plutonium - **Pu**

Polonium - **Po**

Potassium - **K**

Praseodymium -**Pr**

Promethium - **Pm**

Protactinium - **Pa**

Radium - **Ra**

Radon - **Rn**

Rhenium - **Re**

Rhodium - **Rh**

Rubidium - **Rb**

Ruthenium - **Ru**

Samarium - **Sm**

Scandium - **Sc**

Selenium - **Se**

Silicon - **Si**

Silver - **Ag**

Sodium - **Na**

Strontium - **Sr**

Sulphur - **S**

Tantalum - **Ta**

Technetium - **Tc**

Tellurium - **Te**

Terbium - **Tb**

Thallium - **Tl**

Thorium - **Th**

Thulium - **Tm**

Tin - **Sn**

Titanium - **Ti**

Tungsten - **W**

Unnilhexium -**Unh**

Unniloctium - **Uno**

Unnilpentium - **Unp**

Unnilquadium - **Unq**

Unnilseptium -**Uns**

Uranium - **U**

Vanadium - **V**

Xenon - **Xe**

Ytterbium - **Yb**

Yttrium - **Y**

Zinc - **Zn**

Periodic Table of the Element

SOLID | LIQUID | GAS

	1	2	3	4	5	6	7	8	9	10	11	12	13	14	15	16	17	18
1	1 **H** 1.0079																	2 **He** 4.0026
2	3 **Li** 6.941	4 **Be** 9.0122											5 **B** 10.811	6 **C** 12.011	7 **N** 14.007	8 **O** 15.999	9 **F** 18.998	10 **Ne** 20.180
3	11 **Na** 22.990	12 **Mg** 24.305											13 **Al** 26.982	14 **Si** 28.086	15 **P** 30.974	16 **S** 32.066	17 **Cl** 35.453	18 **Ar** 39.948
4	19 **K** 39.098	20 **Ca** 40.078	21 **Sc** 44.956	22 **Ti** 47.867	23 **V** 50.942	24 **Cr** 51.996	25 **Mn** 54.938	26 **Fe** 55.845	27 **Co** 58.933	28 **Ni** 58.693	29 **Cu** 63.546	30 **Zn** 65.409	31 **Ga** 69.723	32 **Ge** 72.64	33 **As** 74.922	34 **Se** 78.96	35 **Br** 79.904	36 **Kr** 83.798
5	37 **Rb** 85.468	38 **Sr** 87.62	39 **Y** 88.906	40 **Zr** 91.224	41 **Nb** 92.906	42 **Mo** 95.94	43 **Tc** (98)	44 **Ru** 101.07	45 **Rh** 102.91	46 **Pd** 106.42	47 **Ag** 107.87	48 **Cd** 112.41	49 **In** 114.82	50 **Sn** 118.71	51 **Sb** 121.76	52 **Te** 127.60	53 **I** 126.90	54 **Xe** 131.29
6	55 **Cs** 132.91	56 **Ba** 137.33	57 - 71 La-Lu	72 **Hf** 178.49	73 **Ta** 180.95	74 **W** 183.84	75 **Re** 186.21	76 **Os** 190.23	77 **Ir** 192.22	78 **Pt** 195.08	79 **Au** 196.97	80 **Hg** 200.59	81 **Tl** 204.38	82 **Pb** 207.2	83 **Bi** 208.98	84 **Po** (209)	85 **At** (210)	86 **Rn** (222)
7	87 **Fr** (223)	88 **Ra** (226)	89 - 103 Ac-Lr	104 **Rf** (261)	105 **Db** (262)	106 **Sg** (266)	107 **Bh** (264)	108 **Hs** (277)	109 **Mt** (268)	110 **Ds** (281)	111 **Rg** (272)							

Molecular Biology & Biotechnology

Microbial Methods

by:
Manoj V. Parakhia
Rukam S. Tomar
Sunil Patel
B.A. Golakiya

NIPA® GENX ELECTRONIC RESOURCES & SOLUTIONS P. LTD.
New Delhi-110 034

**NIPA® GENX ELECTRONIC
RESOURCES & SOLUTIONS P. LTD.**

101,103, Vikas Surya Plaza, CU Block
L.S.C. Market, Pitam Pura, New Delhi-110 034
Ph : +91 11 27341616, 27341717, 27341718
E-mail: newindiapublishingagency@gmail.com
web: www.nipabooks.com

For customer assistance, please contact
Phone: + 91-11-27 34 17 17
Fax: + 91-11- 27 34 16 16
E-Mail: feedbacks@nipabooks.com

ISBN: 978-81-19235-17-9

Composed and Designed by NIPA®.

Dr. N.C. Patel
Vice-Chancellor

Junagadh Agricultural University
Junagadh- 362 001

Message

Science knows no country because knowledge belongs to humanity and is the tourch which illuminates the world; science is the highest personification of the nation because that nation will remain the first which carries furthest the works of thought and intelligence.

Microbiology has great role to play in the field of agriculture. Agro ecosystems are teaming with microbes. We have not identified them sufficiently. This collection entitled "Molecular Biology and Biotechnology: Microbial Methods" will provide detailed guidelines, techniques and methods for identification and characterization of microbes at genomic level. This task will have long-lasting benefit in the eco-friendly agriculture.

The sincere efforts made by the authors are highly appreciated. I am sure, this book will be useful to the students, teachers and professionals related to the field of agricultural microbiology.

Junagadh
Date : 29-7-2009

(N.C. Patel)

Dr. D.B. Kuchhadiya
Director of Research &
Dean P.G. Studies

Junagadh Agricultural University
Junagadh- 362 001

Foreword

The invisible world of microbes is still a misty for us. We have yet to explore the synchrony of microbes with human life. The realm microbe is much more wast and intricate than the any other form of life on this planet. Since microbes have long history than any other living being on this earth, It should be understood closely and carefully to make ourselves wiser about the sustenance of their life in the biosphere Wisdom comes from knowledge. Knowledge is derived from the working on the object under investigation, working involves methods and techniques. The workings on microbial biotechnology consist of detail methodology. This book entitled "Molecular Biology and Biotechnology: Microbial Methods" embodied detailed methodology and techniques related to this subject. The manual of any branch of science are most important documents as it explain the tools and techniques of investigation. The life mechanism of microbes is time tested. Their adaptability, genetic diversity, genetic drift and survival are topics which attracted researchers and professionals. The molecular base of microbes made it much more interesting for gene cloning, silencing and gene transfer who is crossing the boundaries of genus and species. Thus one can insert any bacterial gene into plant or animals. These are possible through laboratory techniques. This manual will be good companion of students, teachers and researchers.

- Sd -

(D.B. Kuchhadiya)

Preface

Miniature world of microbes has fascinated mankind for centuries. Man has made miracles through microbes. The small pox vaccine is one example having collective effect on life of human being on this earth.

Microbes are still mysterious entities for us. Through their genome carries a few Mb genome we have not still understood the nature and properties of microbes exclusively. Microbes are assuming special importance in this era of nanobiotechnology. The diversities in the utilities of microbes interested the specialist of many branches of biological sciences. How to work with these microbes? Which are the fundamental techniques? What sort instrumentation it involves? How to work on microbial biotechnology? These are some of the aspects addressed in this collection entitled "Molecular Biology and Biotechnology: Microbial Methods".

The text consists of 370 pages divided into 36 chapters followed by detailed glossary. Most of the required protocols have been included. This is a huge subject and has many subsidiaries like food microbiology, textile microbiology, medical microbiology, and agriculture microbiology etc. Only the common modules of microbial biotechnology with broad generalization have been considered in this text. Working with the specific microbes is a trial & error game. It require lot of commonsense and tenacity. With that one can pave the way out in the methodological contest. And there will be no end of working with microbes. Open one gate and the several gate again will be in front of you. The reward for work well done is the opportunity to do more, and in the field of observation chance favours only the prepared mind. This text is just a guide line to set the hand. In actual working you will be doing much more beyond this text and that will be going to make us wiser. We hope that this text will prove as a good laboratory partner for those who set their hands on microbial biotechnology.

Authors

Abbreviations

A	adenine or adenosine; one-letter code for alanine
A260	absorbance at 260nm
ACES	N-(2-acetamido)-2-aminoethanesulfonic acid
Ad-2	adenovirus-2
ADA	N-(2-acetamido)-2-iminodiacetic acid
ADP	adenosine 5'-diphosphate
AEC	3-amino-9-ethylcarbazole
AMP	adenosine monophosphate
AP	alkaline phosphatase
ATP	adenosine 5'-triphosphate
BAP	bacterial alkaline phosphatase
BCIP	5-bromo-4-chloro-3-indolyl phosphate
BES	N,N-bis(2-hydroxyethyl)-2-aminoethanesulfonic acid
BICINE	N,N-bis (2-hydroxyethyl)glycine
Bio-dNTP	biotin-deoxynucleoside triphosphate
bp	base pair
Bq	Becquerel
BSA	bovine serum albumin
B/W	blue/white cloning
C	cytosine or cytidine; one-letter code for cysteine
CA	casamino-acids

cDNA	complementary deoxyribonucleic acid
Ci	Curie
CIAP	calf intestinal alkaline phosphatase
cpm	counts per minute
CTP	cytidine 5'-triphosphate
Da	Dalton
DAB	3,3'-diaminobenzidine tetrahydrochloride
dAMP	deoxyadenosine monophosphate
dATP	deoxyadenosine triphosphate
dCTP	deoxycytidine triphosphate
ddATP	dideoxyadenosine triphosphate
ddCTP	dideoxycytidine triphosphate
ddGTP	dideoxyguanosine triphosphate
ddNTP	dideoxythymidine triphosphate
DE-81	Whatman® diethylaminoethyl cellulose paper
DEPC	diethyl pyrocarbonate
DIG	digoxigenin
dGTP	deoxyguanosine triphosphate
DMSO	dimethyl sulfoxide
DNA	deoxyribonucleic acid
DNase	deoxyribonuclease
dNTP	deoxynucleoside triphosphate
dpm	disintegrations per minute
ds	double-stranded
DTE	dithioerythritol
DTT	dithiothreitol
dTTP	deoxythymidine triphosphate
dUTP	deoxyuridine triphosphate
EDTA	ethylenediaminetetraacetic acid
ELISA	enzyme-liked immunosorbent assay
EMBL	European Molecular Biology Laboratory

ENDO	endodeoxyribonuclease assay
exo	exonuclease
EXO	5' and 3'- exodeoxyribonuclease assay
FOA (5-FOA)	5-fluoroorotic acid
G	guanine or guanosine; one-letter code for glycine
Gal	D-galactose
GUS	beta-D-glucuronidase
HC	high concentration
HEPES	N-(2-hydroxyethyl) piperazine-N′-(2-ethanesulfonic acid)
HRP	horseradish peroxidase
HPLC	high-performance liquid chromatography
IEF	isoelectric focussing
IPTG	isopropyl-beta-D-thiogalactopyranoside
kb	kilobase
kDa	kiloDalton
LB	Luria Bertani media
LC	low concentration
LO	labeled oligonucleotide
MCS	multiple cloning site
MES	2-(N-morpholino) ethanesulfonic acid
M-MuLV	Moloney Murine Leukemia Virus
MOPS	3-(N-morpholino) propanesulfonic acid
mRNA	messenger ribonucleic acid
MSPU	mini power supply unit
MW	molecular weight
NAD	nicotinamide adenine dinucleotide
NADH	nicotinamide adenine dinucleotide, reduced form
NADP	nicotinamide adenine dinucleotide phosphate
NADPH	nicotinamide adenine dinucleotide phosphate, reduced form

NBT	nitro blue tetrazolium
nd	not determined
NR	not recommended
NP-40	Nonidet P-40 (detergent)
nt	nucleotide
NTP	nucleoside triphosphate
oligo(A)	oligoadenylic acid
oligo(dT)	oligodeoxythymidylic acid
OMP	orotidine monophosphate
OPD	1,2-phenylenediamine; ortho-phenylenediamine
Pi	inorganic phosphate
PAGE	polyacrylamide-gel electrophoresis
PBS	phosphate-buffered saline
PCR	polymerase chain reaction
PEI	polyethylenimine
PEG	polyethylene glycol
PIPES	piperazine-N,N′-bis(2-ethanesulfonic acid)
PNK	polynucleotide kinase
pNPP	4-nitrophenyl phosphate; para-nitrophenyl phosphate
poly(A)	polyadenylic acid
poly(A)$^+$	polyadenylated (mRNA)
poly(dA-dT)	poly (deoxyadenylic acid - deoxythymidylic acid)
poly(dT)	polydeoxythymidylic acid
PSU	power supply unit
QCA	quality control assay
RE	restriction enzyme
RNA	ribonucleic acid
RNase	ribonuclease
R-M	restriction-modification
rRNA	ribosomal ribonucleic acid
RT	reverse transcriptase

SAM	S-adenosylmethionine
SDS	sodium dodecyl sulfate
ss	single-stranded
SSC	sodium chloride/sodium citrate (buffer)
SSPE	sodium chloride/sodium phosphate/EDTA (buffer)
T	thymine or thymidine; one-letter code for threonine
TAE	Tris/acetate/EDTA (buffer)
Taq	*Thermus aquaticus*
TBE	Tris/borate/EDTA (buffer)
TdT	terminal deoxynucleotidyl transferase
TE	Tris/EDTA (buffer)
TEMED	N,N,N′,N′-tetramethylethylenediamine
TES	N-tris (hydroxymethyl)methyl-2-aminoethanesulfonic acid
TLC	thin layer chromatography
T_m	melting temperature
TMB	3,3', 5,5'-tetramethylbenzidine
TRICINE	N-tris(hydroxymethyl)methylglycine
Tris	tris(hydroxymethyl)aminomethane
tRNA	transfer ribonucleic acid
TPE	Tris/phosphate/EDTA (buffer)
TX-100	triton X-100
U	uracil or uridine
u	unit
UTP	uridine 5'-triphosphate
UV	ultraviolet
v/v	volume/volume
w/v	weight/volume
X-Gal	5-bromo-4-chloro-3-indolyl-beta-D-galactopyranoside
X-Gluc	5-bromo-4-chloro-3-indolyl-beta-D-glucuronic acid

Contents

CHAPTER 1

Laboratory Safety Rules

The Laboratory Worker is Surrounded by Many Dangers such as from..........

- Handling of infectious materials.
- Handling of broken glassware.
- Accidental spill of corrosive reagents.
- Swallowing of corrosive reagents such as concentrated Sulphuric acid,
- Hydrochloric acid, sodium hydroxide, trichloroacetic acid etc.
- Swallowing of infectious specimen.
- Inhalation of poisonous fumes.
- Infection in the conduct of necropsies.
- Potential hazards in the form of inflammable chemicals and gas leakages.

Important Practical Precautions

- The use of rubber gloves while handling corrosive substances such as strong acids or alkalies and also while handling poisonous chemicals such as potassium cyanide is essential.
- The use of laboratory coats is meant to protect the wearer from chemical splashes and infectious materials. Cotton is a better material for a laboratory coat since it has a greater absorptive capacity and is generally more resistant to chemical splashes.
- Safety spectacles or goggles should be used while carrying out any procedures where there is a risk to the eyes from reagent splashes.
- All chemicals should be considered as potentially dangerous, Contact with skin and clothing should be avoided. The solutions such as concentrated Sulphuric acid, concentrated sodium hydroxide and solutions containing sodium or potassium cyanide should never be pipette out by mouth.
- A gas cylinder and gas taps should be handled carefully, Accidents can happen through ignorance and incorrect use.
- All body fluids such as blood, serum, plasma, urine, CSF etc. should be handled with great care since they may be potential source of infections.
- Laboratory workers must know the meaning of safety signs for the following types of harmful substance.

The Laboratory Worker should take Following Precautions Regarding Fire and Explosions

- Open flames should not be left unattended. Leakages of gas from gas cylinder or from the gas taps should be promptly reported to the superior.
- Smoking should be strictly prohibited in the laboratory.
- Burning matchsticks should not be thrown in the wastebasket.
- The laboratory workers should know the location of following requirement:
- A charged fire Extinguisher.
- Bucket of sand with scoop and
- Fire Blankets.

Precautionary Measures Against Fire

- For a small blaze, water, sand and fire blanket should be used to put out the fire.
- A fire extinguisher is used for a longer blaze.
- Water should not be used on electrical fire. After putting off the main electrical switch, carbon dioxide fire extinguisher should be used.
- Water should not be used on a fire caused by organic solvents such as ether, alcohol, petrol, oil or grease. Sand should be used to smother the, fire.
- During the fire escape, it is safe to stay close to the floor and crawl by covering the mouth with a damp cloth. (This may help to filter out some of the flames and also may reduce the danger of inhaling the flames.)

Laboratory First Aid

- The laboratory first aid kit should contain (1) Cotton Wool and Guaze (2) Spirit (3) Roller Bandage (4) Medicinal Adhesive Tape (5) A pair of scissors (6) Tincture iodine (1g iodine in 95% alcohol) (7) A disinfectant solution (diluted dettol) (8)Sterile saline (9) 2% (W/V) Sodium carbonate (10) 5% (W/V) Acetic acid (11) 8% (W/V) Magnesium hydroxide (Milk of Magnesium) (12) 5% (W/V) Soap Solution.

The First Aid Measures given below must be Applied with common sense.

1) Contact with corrosive chemicals and reagents

Acid splashes on the skin

- Wash the affected skin with plenty of tap water.
- Bathe the affected skin with cotton wool soaked in 5% solution of sodium carbonate.
- Rinse the affected skin in mild detergent and then contact a physician.

Alkali Splashes on the Skin

- Wash the affected skin with plenty of tap water.
- Bathe the affected skin with cotton wool soaked in 5% acetic acid.
- Seek medical help.

Contact with Phenol

- Wash the affected skin with large volumes of the tap water.
- Use polyethylene glycol (PEG) mixed with water for further irrigation of affected skin.
- Consult a physician.

2) Eye Accidents

Burns of the eye by alkali or acid are among the most urgent ocular emergencies. Alkali burns are more disastrous than those of acid burns.

Acid or Alkali Splashes in the Eyes

- Immediately wash the eyes with plenty of tap water, by holding part of the eyelids manually.
- Rinse the eyes in sterile saline
- Consult an ophthalmologist immediately.

3) Accidental Swallowing of Poisonous Reagent

- Spit it out immediately.
- Rinse the mouth promptly with tap water and then induce vomiting by drinking warm salt water. (One tablespoon of common salt in glassful of water.)

4) Accidental Swallowing of Infectious Specimen

- Spit it out immediately.
- Wash the mouth with dilute antiseptic lotion.
- Rinse the mouth thoroughly with tap water.

5) Accident Swallowing of Corrosive Reagents

Acids

- Promptly rinse the mouth with tap water.
- Antidote such as 5% soap solution, 8% magnesium hydroxide, or white of egg mixed with about 500 ml of water can be used orally to neutralize the acid.

- Seek medical help immediately.

Alkalies

- After rinsing the mouth with tap water, antedate such as lemon juice or 5% acetic acid can be taken orally to neutralize the alkalies.
- Seek medical help immediately.

6) Contact of Lip and Tongue with Corrosive Reagents

Acids

- Immediately rinse in tap water.
- Bathe the affected part in 2% aqueous sodium Carbonate.

Alkalies

- Immediately rinse in tap water.
- Bathe the affected part in 5% acetic acid.

7) Injuries Caused by Broken Glass

- Wash the wound immediately with a disinfectant solution (diluted dettol).
- Cover with gauze and adhesive tape.

8) Bleeding

- Make the patient to lie down.
- Try to stop bleeding by applying direct pressure to the wound with a sterilized pad and by means of a firm bandage.
- Clean the affected area by using an antiseptic such as tincture iodine and apply sterile gauze and bandage.

9) Burns and Management

Minor Burns

- Immediately bathe the affected area in cold water.
- Cover the burnt area with sterile dressing as Early as possible.
- Seek medical help.

Severe Burns

- If the victim is on fire, put out the fire by using any material such as sheets, towels or coats.
- Pour cold water on the affected area.
- Remove smoldering clothes as quickly as possible.
- Call a physician, immediately.

CHAPTER 2

Milestones in DNA History and Biotechnology

BC

1750 The Sumerians brew beer.

500 The Chinese use moldy soybean curds as an antibiotic to treat boils.

250 The Greeks practice crop rotation to maximize soil fertility.

100 Powdered chrysanthemum is used in China as an insecticide.

AD: Before the 20th Century

1590 The microscope is invented by *Janssen*.

1663 Cells are first described by *Robert Hooke*.

1675 *Leeuwenhoek* discovers protozoa and bacteria.

1745 Maupertuis proposes an adaptationist account of organic design

1797 *Jenner* inoculates a child with a viral vaccine to protect him from smallpox.

1802 The word "biology" first appears.

1824 *Dutrochet* discovers that tissue is composed of living cells.

1830 Proteins are discovered.

1833 The cell nucleus is discovered

1839 The cell theory, or cell doctrine, states that all organisms are composed of similar units of organization, called cells. The concept was formally articulated in 1839 by *Schleiden & Schwann* and has remained as the foundation of modern biology.

1855 The *Escherichia coli* bacterium is discovered. It later becomes a major research, development and production tool for biotechnology. *Pasteur* begins working with yeast, eventually proving they are living organisms.

1859 Concepts of evolution was given by *Charles Darwin*

1863 *Mendel*, in his study of peas, discovers that traits were transmitted from parents to progeny by discrete, independent units, later called genes. His observations lay the groundwork for the field of genetics.

1865 *Gregor Mendel* laid down the rules of inheritance by distinct "factors" acting dominantly or recessively.

1866 *Mendel's* paper is published: units of inheritance in pairs; dominance and recessiveness; equal segregation; independent assortment. These ideas are not recognized for 34 years.

1869 *Miescher* discovers "nuclein" (DNA) in the cells from pus in open wounds- cells composed mostly of nuclear material. It became known as nucleic acid after 1874, when Miescher separated it into a protein and an acid molecule. The significance of DNA is not appreciated for over 70 years.

1877 A technique for staining and identifying bacteria is developed by *Koch*.

1878 The first centrifuge is developed by *Laval*.

The term "microbe" is first used.

1879 *Flemming* discovers chromatin, the rod-like structures inside the cell nucleus that later come to be called "chromosomes."

1883 *F. Galton* gave Quantitative aspects of heredity.

The first rabies vaccine is developed.

1888 The chromosome is discovered by *Waldyer.*

1889 *R. Altmann* first used the term "nucleic acid"

1892 *R. Ivanowski* first used the term "virus" introduced

1897 *Eduard Buchner,* discovered by accident that fermentation actually does not require the presence of living yeast cells. Buchner made an extract of yeast cells by grinding them and filtering off the remaining cell debris. Then he added a preservative—sugar—to the resulting cell-free solution to preserve it for future study. He observed that fermentation, the formation of alcohol from sugar, occurred. Buchner then realized that living cells were not required for carrying out metabolic processes such as fermentation. Instead, there must be some small entities capable of converting sugar to alcohol. These entities were enzymes. Buchner's accidental discovery won him the 1907 Nobel Prize in chemistry.

During Early 20th Century

1900 *H. de Vries, E.Tschermak and K. Correns,* independently recognized Mendel's discovery

Landsteiner identified AB0 blood group system

1902 Some diseases in man inherited according to Mendelian rules *(W. Bateson, A. Garrod)*

T. Boveri proved the individuality of chromosomes

W. Sutton stated Chromosomes and Mendel's factors are related

McClung identified Sex chromosomes

The term "immunology" first appears.

1906 *W. Bateson* proposed the term "genetics"

1907 The first in vivo culture of animal cells is reported.

1908 *Hardy, Weinberg* gave the principles of Population genetics

1909 Genes are linked with hereditary disorders.

British physician *Archibald Garrod* first proposes the relationship between genes and proteins. He hypothesizes that genes might be involved in creating the proteins that carry out the chemical reactions of metabolism.

W. Johannsen coined the terms "gene," "genotype," phenotype"

Janssens identified Chiasma formation during meiosis.

C. Little first inbred mouse strain DBA.

1910 *T. H. Morgan* begin work on *Drosophila* genetics
First *Drosophila* mutation *(white-eyed)*

1911 The first cancer-causing Sarcoma virus is discovered by *Peyton Rous*

1912 Crossing-over *(Morgan and Cattell)*
Genetic linkage *(Morgan and Lynch)*
First genetic map *(A. H. Sturtevant)*

1913 *A. Carrel* produced first cell culture

1914 *C. B. Bridges* identified Nondisjunction

Bacteria are used to treat sewage for the first time in Manchester, England.

1915 Genes located on chromosomes (chromosomal theory of inheritance) *(Morgan, Sturtevant, Muller, Bridges)*

Phages, or bacterial viruses, are discovered.

1919 The word "biotechnology" is first used by a Hungarian agricultural engineer.

1920 The human growth hormone is discovered by *Evans and Long.*

1922 *F. Blakeslee* characterized phenotypes of different trisomies in the plant *Datura stramonium.*

1924 Blood group genetics *(Bernstein)*

Fisher gave statistical analysis of genetic traits

1926 Enzymes are proteins *(J. Sumner)*

1927 *Muller* formulates the chief principles of spontaneous gene mutation as point effects of ultramicroscopic physico-chemical accidents; he induces such changes using X-rays

S.Wright gave the term Genetic drift

1928 Euchromatin/heterochromatin *(E. Heitz)*

Genetic transformation in pneumococci *(F. Griffith)*

Fleming discovers penicillin, the first antibiotic.

1933 Pedigree analysis *(Haldane, Hogben, Fisher, Lenz, Bernstein)*

Polytene chromosomes *(Heitz and Bauer, Painter)*

1935 *C. B. Bridges* gave first cytogenetic map in *Drosophila*

1937 *P. Gorer* identified mouse H2 gene locus

1938 The term "molecular biology" is coined.

1940 *E. B. Ford* explained the term Polymorphism.
Rhesus blood groups *(Landsteiner and Wiener).*

1941 *E. B. Lewis* gave the theory evolution through gene duplication.

Beadle and Tatum proved the genetic control of enzymatic biochemical reactions.

Auerbach identified mustard gas as Mutagen agent.

The term "genetic engineering" is first used by a Danish microbiologist.

1942 The electron microscope is used to identify and characterize a bacteriophage- a virus that infects bacteria.

1943 Avery demonstrates that DNA is the "transforming factor" and is the material of genes.

1944 *Oswald Avery* identifies nucleic acids as the active principle in bacterial transformation

DNA is shown to be the material substance of the gene.

1946 Genetic material can be transferred laterally between bacterial cells, as shown by *Lederberg and Tatum.* (Genetic recombination)

1947 Genetic recombination in viruses*(Delbrück and Bailey, Hershey)*

1949 Sickle cell anemia, a genetically determined molecular disease *(Neel, Pauling)*

Hemoglobin disorders prevalent in areas of malaria *(J. B. S. Haldane)*

X chromatin *(Barr and Bertram)*

1950 In DNA, there are equal amounts of A and T, and equal amounts of C and G, as shown by *Erwin Chargaff*. However, the A+T to C+G ratio can differ between organisms.

During 50's

1951 *B. McClintock* discovers transposable elements, or "jumping genes," in corn.

1952 *Alfred Hershey and Martha Chase* show that on infection of the host bacterium by a virus, at least 80% of the viral DNA enters the cell and at least 80% of the viral protein remains outside.

Plasmids *(Lederberg)*

Transduction by phages *(Zinder and Lederberg)*

Cori and Cori first identified enzyme defect in man.

Mohr identified first linkage group in man.

Colchicine and hypotonic treatment in chromosomal analysis *(Hsu and Pomerat).*

J. Warkany identified exogenous factors as a cause of congenital malformations.

1953 determine that deoxyribonucleic acid (DNA) is a double-strand helix of nucleotides. Each nucleotide consists of a deoxyribose sugar molecule to which is attached a phosphate group and one of four nitrogenous bases: two purines (adenine and guanine) and two pyrimidines (cytosine and thymine). The nucleotides are joined together by covalent bonds between the phosphate of one nucleotide and the sugar of the next, forming a phosphate-sugar backbone from which the nitrogenous bases protrude. The two strands are linked by selective hydrogen bonds: the purine adenine bonds only with the pyrimidine thymine, and the purine cytosine only with the pyrimidine guanine.

Nonmendelian inheritance *(Ephrussi).*

Howard and Pelc gave cell cycle.

Dietary treatment of phenylketonuria *(Bickel)*

1954 *Muller* gave the mechanism of DNA repair.

Leukocyte drumsticks *(Davidson and Smith)*

Polani identified cells in Turner syndrome as X-chromatin negative.

Cell-culturing techniques are developed.

1955 *F. Sanger* gave amino acid sequence of insulin.

Lysosomes *(C. de Duve)*

Buccal smear *(Moore, Barr, Marberger)*

A. Pardee and R. Litman identified 5-Bromouracil, an analogue of thymine, to be an inducer of mutations in phages.

An enzyme involved in the synthesis of a nucleic acid is isolated for the first time.

1956 46 Chromosomes in man *(Tijo and Levan, Ford and Hamerton)*

The fermentation process is perfected in Japan.

Kornberg discovers the enzyme DNA polymerase I, leading to an understanding of how DNA is replicated.

Genetic heterogeneity *(Harris, Fraser)*

1957 During a dysentery epidemic in Japan, biologists discover that some strains of bacterium are resistant to antibiotics. Later scientists will find that this resistance is transferred by olasmids.

Amino acid sequence of hemoglobin molecule *(Ingram)*

Benzer identified cistron, as the smallest nonrecombinant unit of a gene

Genetic complementation *(Fincham)*

DNA replication is semi-conservative, as shown by *Meselson and Stahl* using equilibrium density gradient centrifugation.

Genetic analysis of radiation effects in man *(Neel and Schull)*

Sickle cell anemia is shown to occur due to a change of a single amino acid.

1958 Somatic cell genetics *(Pontocorvo)*
Ribosomes *(Roberts, Dintzis)*
Human HLA antigens *(Dausset)*
Cloning of single cells *(Sanford, Puck)*
Synaptonemal complex, the area of synapse in meiosis *(Moses)*

1959 Messenger RNA is the intermediate between DNA and protein.

First chromosomal aberrations described in man: trisomy 21 *(Lejeune, Gautier, Turpin),*

Turner syndrome: 45,XO *(Jacobs).*

Klinefelter syndrome: 47 XXY *(Ford)*

Isoenzymes *(Vesell, Markert)*

Pharmacogenetics *(Motulsky, Vogel)*

1960 Phytohemagglutinin-stimulated lymphocytecultures *(Nowell, Moorehead,Hungerford)*

Exploiting base pairing, hybrid DNA-RNA molecules are created.

Messenger RNA is discovered.

During 60's

1961 *Sidney Brenner and Francis Crick* establish that groups of three nucleotide bases, or codons, are used to specify individual amino acids.

Nirenberg, Mathaei, and Ochoa determined the genetic code.

X-chromosome inactivation *(M. F. Lyon),* confirmed by *Beutler, Russell, Ohno)*

Jacob and Monod gave the gene regulation, concept of operon

Galactosemia in cell culture *(Krooth)*

Cell hybridization *(Barski, Ephrussi)*

Thalidomide embryopathy *(Lenz, McBride)*

1962 *Nowell and Hungerford* identified Philadelphia chromosome.

Xg, the first X-linked human blood group *(Mann, Race, Sanger).*

Screening for phenylketonuria *(Guthrie, Bickel).*

Edelman and Franklin made the molecular characterization of immunoglobulins.

Identification of individual human chromosomes by 3H-autoradiography *(German,Miller).*

Jacob and Brenner gave the concept of Replicon.

S. Brenner gave the term "codon" for a triplet of (sequential) bases.

1963 Lysosomal storage diseases *(C. de Duve)*

First autosomal deletion syndrome (cridu-chat syndrome) *(J. Lejeune*)

1964 Excision repair *(Setlow).*

Bach and Hirschhorn, Bain and Lowenstein gave the concept of MLC test.

Terasaki and McClelland gave the concept of Microlymphotoxicity test.

Selective cell culture medium HAT *(Littlefield)*

German, Schroeder gave the concept of Spontaneous chromosomal instability.

Cell fusion with Sendai virus *(Harris and Watkins)*

Cell culture from amniotic fluid cells *(H. P. Klinger)*

Hereditary diseases studied in cell cultures *(Danes, Bearn, Krooth, Mellman)*

Population cytogenetics *(Court Brown)*

Fetal chromosomal aberrations in spontaneous abortions *(Carr, Benirschke)*

The existence of reverse transcriptase (RT) is predicted.

1965 Limited life span of cultured fibroblasts *(Hayflick, Moorehead)*

1966 The genetic code is cracked by a number of researchers (including *Nirenberg, Matthaei, Leder, and Khorana*) using RNA homopolymer and heteropolymer experiments as well as tRNA labeling experiments.

The genetic code is deciphered when biochemical analysis reveals which codons determine which amino acids.

Catalog of Mendelian phenotypes in man *(McKusick)*

1967 The first automatic protein sequencer is perfected.

1968 HLA-D the strongest histocompatibility system *(Ceppellini, Amos)*

Repetitive DNA *(Britten and Kohne)*

Biochemical basis of the AB0 blood group substances *(Watkins)*

DNA excision repair defect in xeroderma pigmentosum *(Cleaver)*

Restriction endonucleases *(H. O. Smith, Linn and Arber, Meselson and Yuan)*

First assignment of an autosomal gene locus in man *(Donahue, McKusick)*

Khorana synthesied gene in vitro.

1969 An enzyme is synthesized in vitro for the first time.

1970 *Hamilton Smith,* at Johns Hopkins Medical School, isolates the first restriction enzyme, an enzyme that cuts DNA at a very specific nucleotide sequence. Over the next few years, several more restriction enzymes will be isolated.

D. Baltimore, H. Temin gave the concept of Reverse transcriptase, independently.

Renwick gave the new term Synteny, that refers to all gene loci on the same chromosome.

Enzyme defects in lysosomal storage diseases *(Neufeld, Dorfman)*

Individual chromosomal identification by specific banding stains *(Zech, Casperson, Lubs, Drets and Shaw, Schnedl, Evans)*

Y-chromatin *(Pearson, Bobrow, Vosa)*

Thymus transplantation for immune deficiency *(van Bekkum)*

Specific restriction nucleases are identified, opening the way for gene cloning.

RT is discovered independently in murine and avian retroviruses.

During 70's

1971 Two-hit theory in retinoblastoma *(A. G. Knudson)*

RT is shown to have ribonuclease H (Rnase H) activity.

1972 *Stanley Cohen and Herbert Boyer* combine their efforts to create recombinant DNA. This technology will be the beginning of the biotechnology industry.

High average heterozygosity *(Harris and Hopkinson, Lewontin)*

Association of HLA antigens and diseases

The DNA composition of humans is discovered to be 99% similar to that of chimpanzees and gorillas.

Purified RT is first used to synthesize cDNA from purified mRNA in vitro.

1973 Receptor defects in the etiology of genetic defects, genetic hyperlipidemia *(Brown, Goldstein, Motulsky)*

S. A. Latt gave demonstration of sister chromatid exchanges with BrdU.

Philadelphia chromosome as translocation *(J. D. Rowley)*

Cohen and Boyer perform the first successful recombinant DNA experiment, using bacterial genes.

1974 Chromatin structure, nucleosome *(Kornberg, Olins and Olins)*

P. C. Doherty and R. M. Zinkernagel gave the concept of dual recognition of foreign antigen and HLA antigen by T lymphocytes.

Clone of a eukaryotic DNA segment mapped to a specific chromosomal location *(D. S. Hogness)*

The National Institute of Health forms a Recombinant DNA Advisory Committee to oversee recombinant genetic research.

1975 Asilomar conference First protein-signal sequence identified *(G. Blobel)*

Southern blot hybridization *(E. Southern).*

Monoclonal antibodies *(Köhler and Milstein).*

Colony hybridization and Southern blotting are developed for detecting specific DNA sequences.

The first monoclonal antibodies are produced.

1976 *Herbert Boyer* cofounds Genentech, the first firm founded in the United States to apply recombinant DNA technology

Overlapping genes in phage ! X174 *(Barell, Air, Hutchinson)*

R. Jaenisch made first transgenic mouse.

Loci for structural genes on each human chromosome known at Baltimore Conference on Human Gene Mapping.

The tools of recombinant DNA are first applied to a human inherited disorder.

Molecular hybridization is used for the prenatal diagnosis of alpha thalassemia.

Yeast genes are expressed in *E. coli* bacteria.

1977 *R. J. Roberts, P. A. Sharp,* independently gave the concept that genes contain coding and noncoding DNA segments ("split genes") (exon/intron structure).

First recombinant DNA molecule that contains mammalian DNA *F. Sanger; Maxam and Gilbert* gave the methods to sequence DNA.

Finch et al. gave X-ray diffraction analysis of nucleosomes.

Genetically engineered bacteria are used to synthesize human growth protein.

1978 Somatostatin, which regulates human growth hormones, is the first human protein made using recombinant technology.

Leder, Weissmann, Tilghman and others gave Globulin gene structure.

North Carolina scientists *Hutchinson and Edgell* show it is possible to introduce specific mutations at specific sites in a DNA molecule.

1979 The first monoclonal antibodies are produced.

1980 The U.S. Supreme Court, in the landmark case *Diamond v. Chakrabarty,* approves the principle of patenting genetically engineered life forms.

The U.S. patent for gene cloning is awarded to *Cohen and Boyer*.

Genes for embryonic development in *Drosophila* studied by mutational screen *(C. Nüsslein-Volhard and others)*

During 80's

1981 The North Carolina Biotechnology Center is created by the state's General Assembly as the nation's first state-sponsored initiative to develop biotechnology. Thirty-five other states follow with biotechnology centers of various kinds.

The first gene-synthesizing machines are developed.

The first genetically engineered plant is reported.

Mice are successfully cloned.

S. Anderson, S. G. Barrell, A. T. Bankier did the sequencing of a mitochondrial genome.

1982 Humulin, Genentech's human insulin drug produced by genetically engineered bacteria for the treatment of diabetes, is the first biotech drug to be approved by the Food and Drug Administration.

Tumor suppressor genes *(H. P. Klinger)*

Prions (proteinaceous infectious particles) proposed as cause of some chronic progressive central nervous system diseases (kuru, scrapie, Creutzfeldt Jakob disease) *(S. B. Prusiner)*

1983 The Polymerase Chain Reaction (PCR) technique is conceived. PCR, which uses heat and enzymes to make unlimited copies of genes and gene fragments, later becomes a major tool in biotech research and product development worldwide.

The first genetic transformation of plant cells by TI plasmids is performed.

The first artificial chromosome is synthesized.

The first genetic markers for specific inherited diseases are found.

Efficient methods are developed to synthesize double-stranded DNA from first-strand cDNA involving minimal loss of sequence information.

L. Montagnier, R. Gallo reported HIV virus.

1984 The DNA fingerprinting technique is developed.

The first genetically engineered vaccine is developed.

Chiron clones and sequences the entire genome of the HIV virus.

Gusella Localizes the gene for Huntington disease.

Tonegawa Identifies the T-cell receptor.

Variable DNA sequences as "genetic fingerprints" *(A. Jeffreys)*

McGinnis discovers homeotic (Hox) regulatory genes, responsible for the basic body plan of most animals. In subsequent work, his team demonstrates that a single mutation in a Hox gene suffices to suppress all limb development in the thoracic region of fruit flies.

1985 Fully active murine RT is cloned and overexpressed in *E. coli.*

*Gitschie*Characterizes the gene for clotting factor VIII *(r).*

Sequencing of the AIDS virus Localization of the gene for cystic fibrosis Hypervariable DNA segments Genomic imprinting in the mouse *(B. Cattanach)*

1986 PCR is developed by *Kary Mullis*. The first field tests of genetically engineered plants (tobacco) are conducted.

The first biotech-derived interferon drugs for the treatment of cancer, Biogen's Intron A and Genentech's Roferon A, are approved by the FDA. In 1988, the drugs are used to treat Kaposi's sarcoma, a complication of AIDS.

The first genetically engineered human vaccine, Chiron's Recombivax HB, is approved for the prevention of hepatitis B.

First cloning of human genes. First identification of a human gene based on its chromosomal location (positional cloning) *(Royer-Pokora et al.)*

RNA as catalytic enzyme *(T. Cech)*

1987 Humatrope is developed for treating human growth hormone deficiency.

Advanced Genetic Sciences' Frostban, a genetically altered bacterium that inhibits frost formation on crop plants, is field tested on strawberry and potato plants in California, the first authorized outdoor tests of an engineered bacterium.

Genentech's tissue plasminogen activator (tPA), sold as Activase, is approved as a treatment for heart attacks.

Reverse transcription and PCR are combined to amplify mRNA sequences.

Cloned murine RT is engineered to maintain polymerase and eliminate Rnase H activity.

Fine structure of an HLA molecule *(Björkman, Strominger et al.)*

Cloning of the gene for Duchenne muscular dystrophy *(Kunkel)*

Knockout mouse *(M. Capecchi)*

A genetic map of the human genome *(H. Donis-Keller et al.)*

Mitochondrial DNA and human evolution *(R. L. Cann, M. Stoneking, A. C. Wilson)*

1988 Congress funds the Human Genome Project, a massive effort to map and sequence the human genetic code as well as the genomes of other species.

Molecular structure of telomeres at the ends of chromosomes *(E. Blackburn and others)*

1989 Amgen's Epogen is approved for the treatment of renal disease anemia.

Microorganisms are used to clean up the Exxon Valdez oil spill. The gene responsible for cystic fibrosis is discovered.

Cloning of a defined region of a human chromosome obtained by microdissection *(Lüdecke, Senger, Claussen, Horsthemke)*

1990 The first federally approved gene therapy treatment is performed successfully on a 4-yearold girl suffering from an immune disorder.

Evidence for a defective gene causing inherited breast cancer *(Mary-Claire King)*

During 90's

1991 *Amgen* develops Neupogen, the first of a new class of drugs called colony stimulating factors, for the treatment of low white blood cells in chemotherapy patients.

Complete sequence of a yeast chromosome.

Increasing use of microsatellites as polymorphic DNA markers

1992 The three-dimensional structure of HIV RT is elucidated. Recombinate, developed by Genetics Institute and used in the treatment of hemophilia A, becomes the first genetically engineered blood clotting factor approved in the U.S.

Trinucleotide repeat expansion as a new class of human pathogenic mutations.

High density map of DNA markers on human chromosomes

O. Smithies identifies X chromosome inactivation center

p53 knockout mouse *(O. Smithies)*

1993 Chiron's Betaseron is approved as the first treatment for multiple sclerosis in 20 years.

The FDA declares that genetically engineered foods are "not inherently dangerous" and do not require special regulation.

The Biotechnology Industry Organization (BIO) is created by merging two smaller trade associations.

Gene for Huntington disease cloned

1994 Genentech's Nutropin is approved for the treatment of growth hormone deficiency.

The first breast cancer gene is discovered.

Calgene's Flavr Savr tomato, engineered to resist rotting, is approved for sale.

Physical map of the human genome in high resolution.

Mutations in fibroblast growth factor receptor genes as cause of achondroplasia and other human diseases.

1995 The first baboon-to-human bone marrow transplant is performed on an AIDS patient.

The first full gene sequence of a living organism other than a virus is completed for the bacterium *Hemophilus influenzae*.

The three-dimensional structure of a catalytically active fragment of murine RT is elucidated.

Master gene of the vertebrate eye, *sey* (small-eye) *(W. J. Gehring)*

STS-band map of the human genome *(T. J. Hudson et al.)*

1996 Scottish scientists clone identical lambs from early embryonic sheep.

Yeast genome sequenced.

Mouse genome map with more than 7000 markers *(E. S. Lander)*

1997 A group of Oregon researchers claims to have cloned two Rhesus monkeys.

A new DNA technique combines PCR, DNA chips, and a computer program, providing a new tool in the search for disease-causing genes.

F. R. Blattner et al. Sequences *E. coli.*

Wilmut cloned Mammal by transfer of an adult cell nucleus into an enucleated oocyte.

Embryonic stem cells

1998 University of Hawaii scientists clone three generations of mice from nuclei of adult ovarian cumulus cells.

Human skin is produced in vitro.

Embryonic stem cells are used to regenerate tissue and create disorders mimicking diseases.

The first complete animal genome for the elegans worm is sequenced. A rough draft of the human genome map is produced, showing the locations of more than 30,000 genes.

Cloned vain RT with fully active polymerase and minimized Rnase H activity is engineered.

The Biotechnology Institute is founded by BIO as an independent national, 501(c)(3) education organization with an independent Board of Trustees.

Nematode *C. elegans* genome sequenced

1999 The complete genetic code of the human chromosome is first deciphered. The rising tide of public opinion in Europe brings biotech food into the spotlight.

First human chromosome (22) sequenced Ribosome crystal structure

2000 A rough draft of the human genome is completed by Celera Genomics and the Human Genome Project.

Pigs are the next animal cloned by researchers, hopefully to help produce organs for human transplant.

"Golden Rice," modified to make vitamin A, promises to help third-world countries alleviate blindness.

The 2.18 million base pairs of the commonest cause of bacterial meningitis, Neisseria meningitidis, are identified.

Drosophila genome sequenced First draft of the complete sequence of the human genome First complete plant pathogen *(Xylella fastidiosa)* genome sequence *Arabidopsis thaliana*, the first plant genome sequenced

During 21's Century

2001 The sequence of the human genome is published in Science and Nature, making it possible for researchers all over the world to begin developing treatments.

2002 Scientists complete the draft sequence of the most important pathogen of rice, a fungus that destroys enough rice to feed 60 million people annually. By combining an understanding of the genomes of the fungus and rice, scientists will elucidate the molecular basis of the interactions between the plant and pathogen.

2003 Dolly, the cloned sheep that made headlines in 1997, is euthanized after developing progressive lung disease. Dolly was the first successful clone of a mammal.

2007 Controversies continue over human and animal cloning, research on stem cells, and genetic modification of crops.

❑❑❑

CHAPTER 3

Contribution in Development of DNA and Biotechnology

Search for genetic material – nucleic acid or protein/DNA or RNA?

- Griffith's Transformation Experiment
- Avery's Transformation Experiment
- Hershey-Chase Bacteriophage Experiment
- Tobacco Mosaic Virus (TMV) Experiment

Timeline of Events

- 1890 Weismann - substance in the cell nuclei controls development.
- 1900 Chromosomes shown to contain hereditary information, later shown to be composed of protein & nucleic acids.
- 1928 Griffith's Transformation Experiment
- 1944 Avery's Transformation Experiment
- 1953 Hershey-Chase Bacteriophage Experiment
- 1953 Watson & Crick propose double-helix model of DNA
- 1956 Gierer & Schramm/Fraenkel-Conrat & Singer Demonstrate RNA is viral genetic material

Frederick Griffith's Transformation Experiment – 1928

Transforming principle" demonstrated with *Streptococcus pneumoniae*

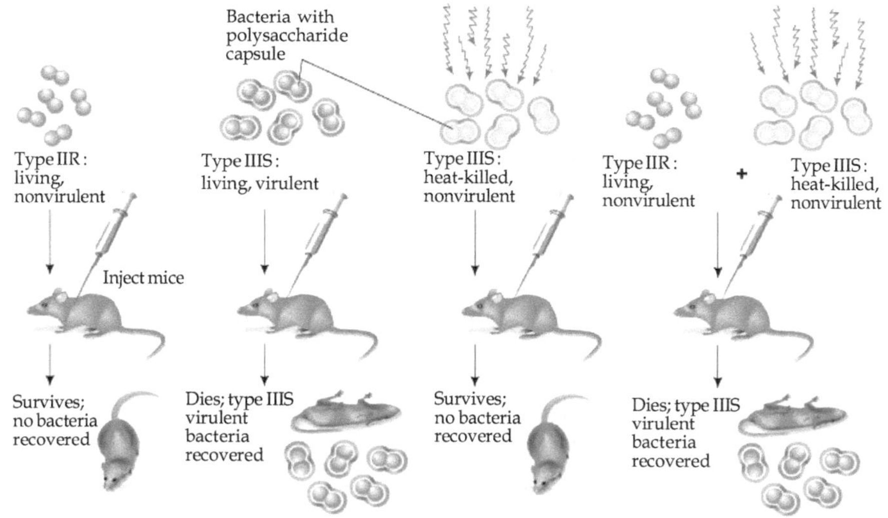

Oswald T. Avery's Transformation Experiment - 1944

Determined that "IIIS" DNA was the genetic material responsible for Griffith's results (not RNA).

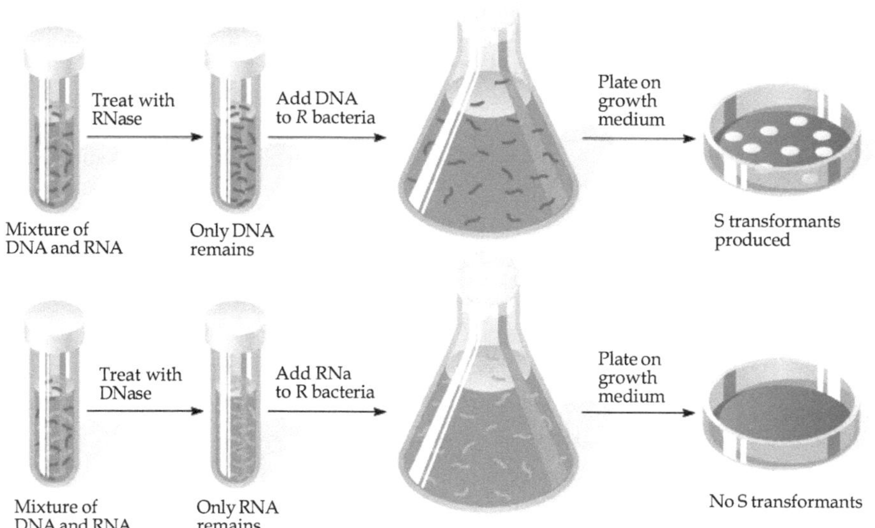

Hershey-Chase Bacteriophage Experiment - 1953

Bacteriophage = Virus that attacks bacteria and replicates by invading a living cell and using the cell's molecular machinery.

Structure of T_2 phage

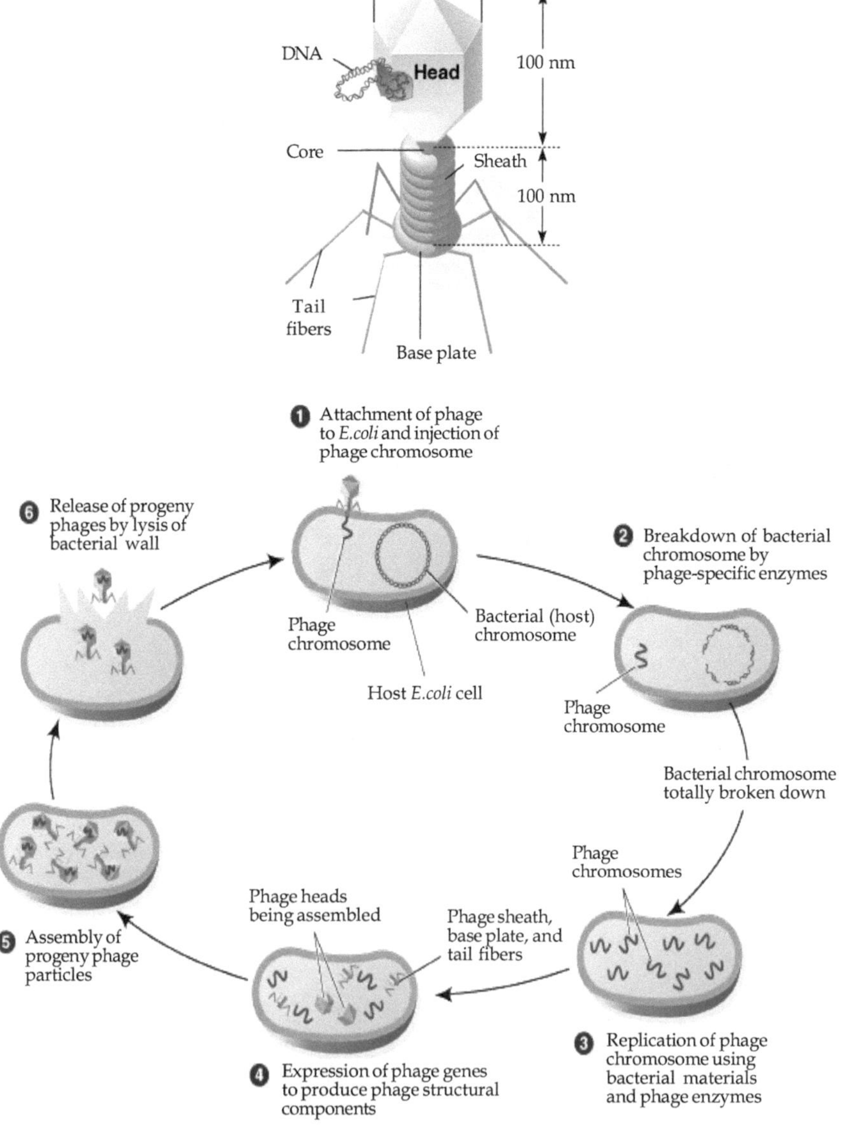

Life Cycle of Virulent T_2 Phage

Hershey-Chase Bacteriophage Experiment - 1953

1. T_2 bacteriophage is composed of DNA and proteins:
2. Set-up two replicates:
   - Label DNA with ^{32}P
   - Label Protein with ^{35}S

3. Infected *E. coli* bacteria with two types of labeled T_2
4. ^{32}P is discovered within the bacteria and progeny phages, whereas ^{35}S is not found within the bacteria but released with phage ghosts.

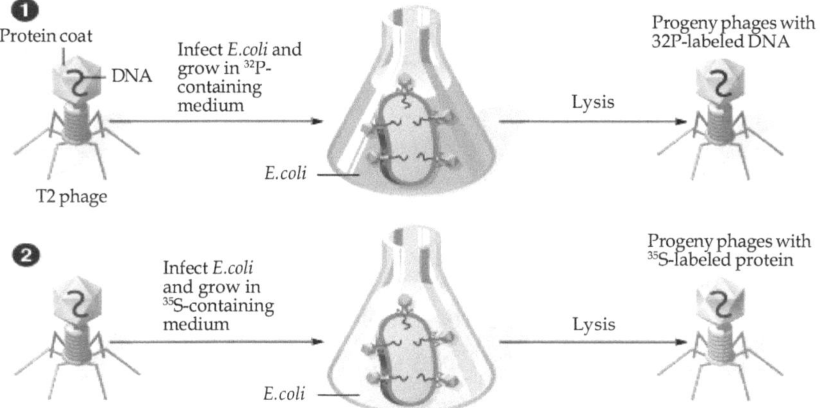

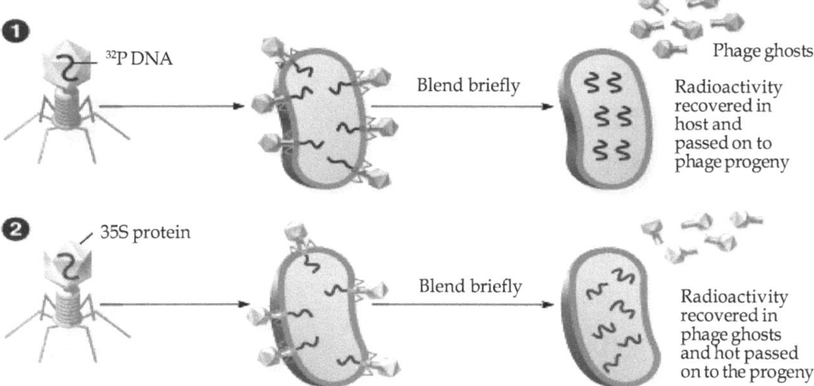

Fig. Hershey -- Chase Experiment

Gierer & Schramn Tobacco Mosaic Virus (TMV) Experiment – 1956 Fraenkel-Conrat & Singer - 1957

Used 2 viral strains to demonstrate RNA is the genetic material of TMV

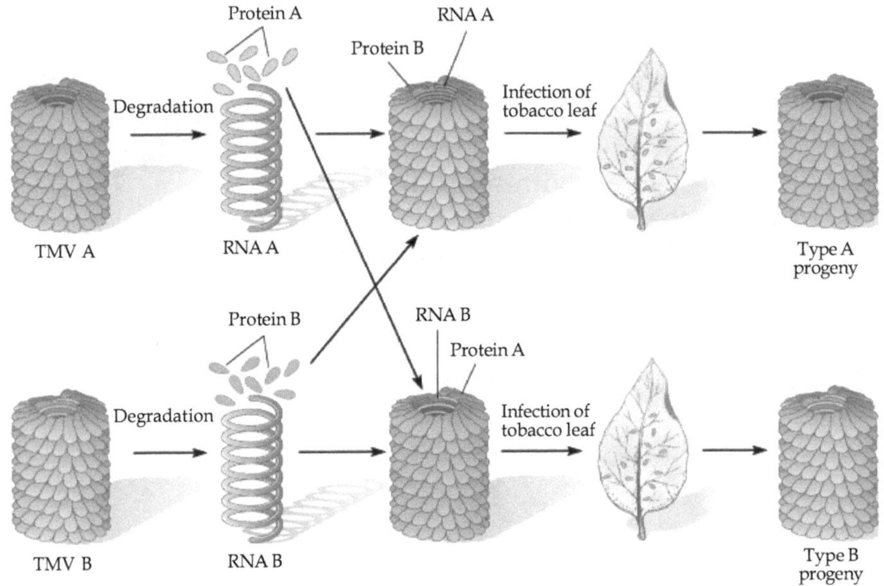

Fig. Gierer & Schramn experiment

Conclusions about these early experiments:

Griffith 1928 & Avery 1944: DNA (not RNA) is transforming agent.

Hershey-Chase 1953: DNA (not protein) is the genetic material.

Gierer & Schramm 1956/Fraenkel-Conrat & Singer 1957: RNA (not protein) is genetic material of some viruses

James D. Watson & Francis H. Crick - 1953

Double Helix Model of DNA

Two sources of information:

1. Base composition studies of Erwin Chargaff

Indicated double-stranded DNA consists of ~50% purines (A,G) and ~50% pyrimidines (T, C)

The amount of A = amount of T and amount of G = amount of C (Chargraff's rules)

%GC content varies from organism to organism

Examples	%A	%T	%G	%C	%GC
Homo sapiens	31.0	31.5	19.1	18.4	37.5
Zea mays	25.6	25.3	24.5	24.6	49.1
Drosophila	27.3	27.6	22.5	22.5	45.0
Aythya americana	25.8	25.8	24.2	24.2	48.4

James D. Watson & Francis H. Crick - 1953

Double Helix Model of DNA

Two sources of information:

X-ray diffraction studies - Rosalind Franklin & Maurice Wilkins

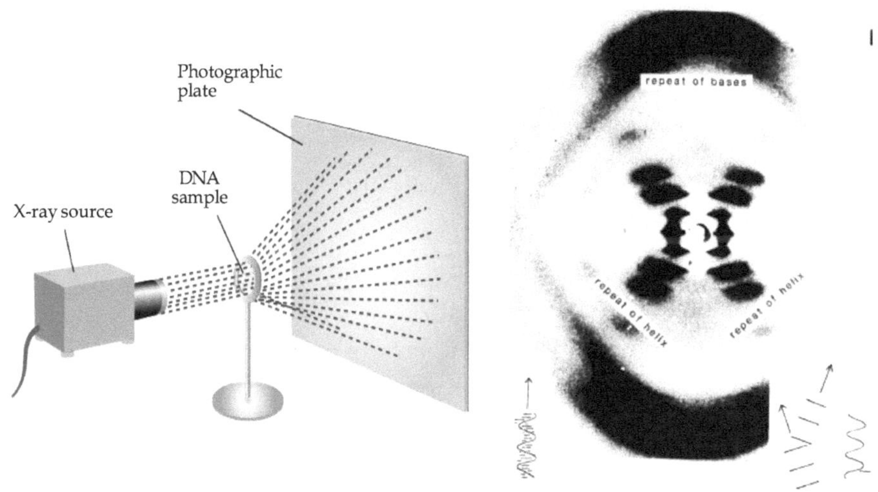

Conclusion-DNA is a Double Helix Model of DNA: Six main features

1. Two polynucleotide chains wound in a right-handed (clockwise) double-helix.
2. Nucleotide chains are anti-parallel: 5′ → 3′
   3′ ¬ 5′
3. Sugar-phosphate backbones are on the outside of the double helix, and the bases are oriented towards the central axis.
4. Complementary base pairs from opposite strands are bound together by weak hydrogen bonds.

A pairs with T (2 H-bonds), and G pairs with C (3 H-bonds).

e.g., 5′-TATTCCGA-3′

3′-ATAAGGCT-5′

5. Base pairs are 0.34 nm apart. One complete turn of the helix requires 3.4 nm (10 bases/turn).
6. Sugar-phosphate backbones are not equally-spaced, resulting in major and minor grooves.

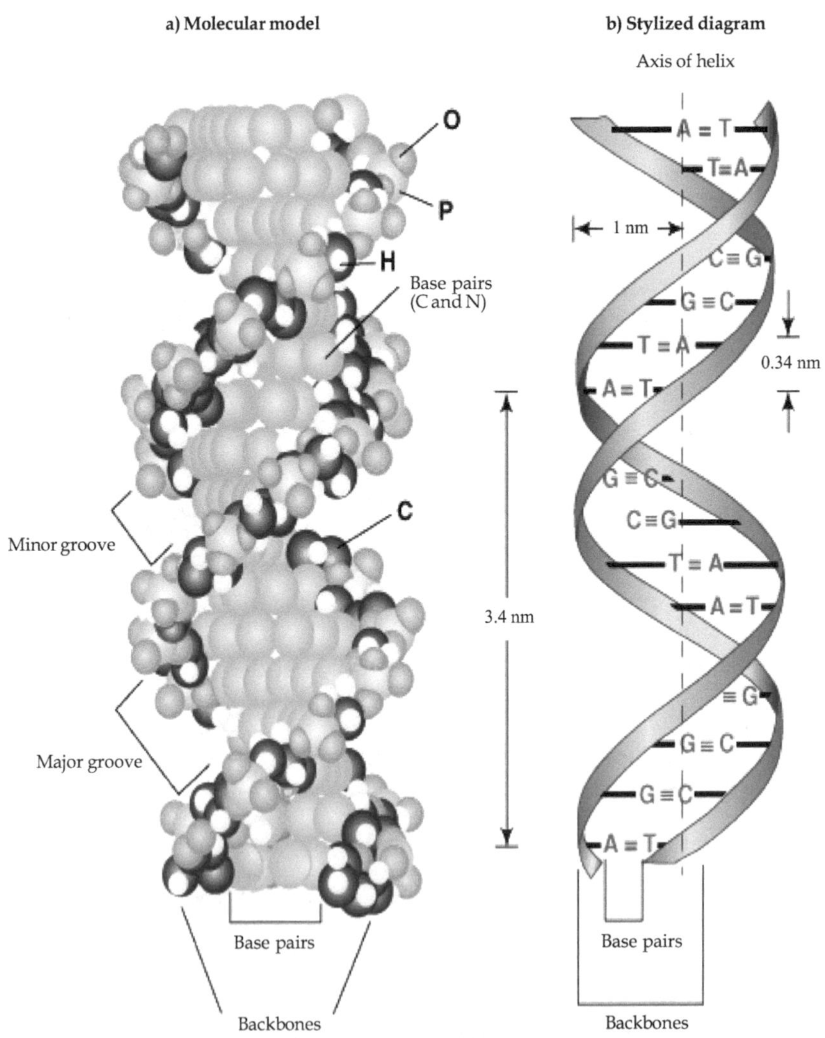

Fig. Watson & Crick DNA double helix model

1962: Nobel Prize in Physiology and Medicine

James D. Watson

Francis H. Crick James Watson

Maurice H. F. Wilkins

Rosalind Franklin

Structure of DNA, Gene, Chromosome and Genome

What is DNA?

DNA is a long, double-stranded, helical molecule composed of building blocks called *deoxyribonucleotides*. A deoxyribonucleotide is composed of 3 parts: a molecule of the 5-carbon sugar deoxyribose, a nitrogenous base, and a phosphate group.

A deoxyribose Deoxyribose is a ringed 5-carbon sugar. The 5 carbons are numbered sequentially clockwise around the sugar. The first 4 carbons actually form the ring of the sugar with the 5' carbon coming off of the 4' carbon in the ring. The nitrogenous base of the nucleotide is attached to the 1' carbon of the sugar and the phosphate group is bound to the 5' carbon. During DNA synthesis, the phosphate group of a new deoxyribonucleotide is covalently attached by the enzyme DNA polymerase to the 3' carbon of a nucleotide already in the chain.

The 5-Carbon Sugar Deoxyribose

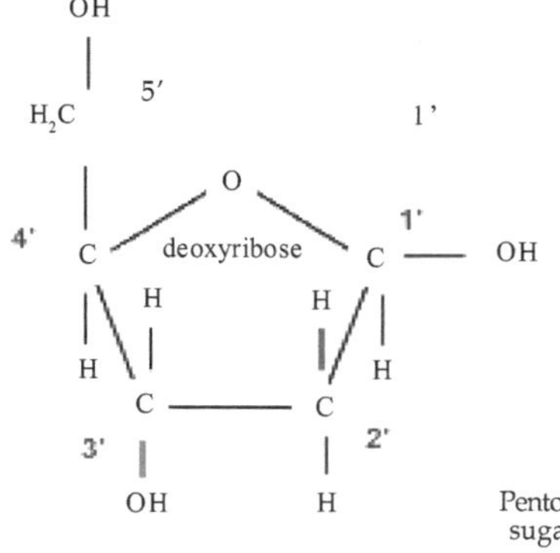

During nucleotide production, the nitrogenous base will attach to the 1' carbon and the phosphate group will attach to the 5' carbon. The first 4 carbons shown form the actual ring of the sugar. The 5' carbon comes off of the ring.

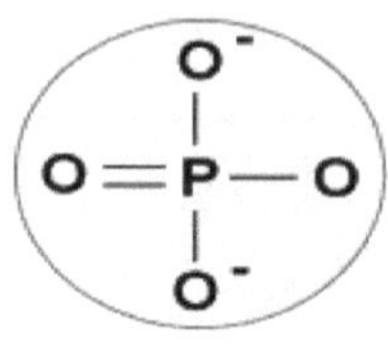

Phosphate group

B. a phosphate group.

A. Phosphate group.

C. a nitrogenous base: There are four nitrogenous bases found in DNA: adenine, guanine, cytosine, or thymine. Adenine and guanine are known as purine bases while cytosine and thymine are known as pyrimidine bases

Pyrimidines

Thymine

Cytosine

Nitrogenous Bases of DNA

Adenine

Guanine

Purines

Adenine and guanine are also known as purine bases; cytosine and thymine are also called pyrimidine bases. Each deoxyribonucleotide will contain one of these four bases.

Note *the phosphate group attached to the 5' carbon of the deoxyribose and the nitrogenous base, in this case thymine, attached to the 1' carbon.*

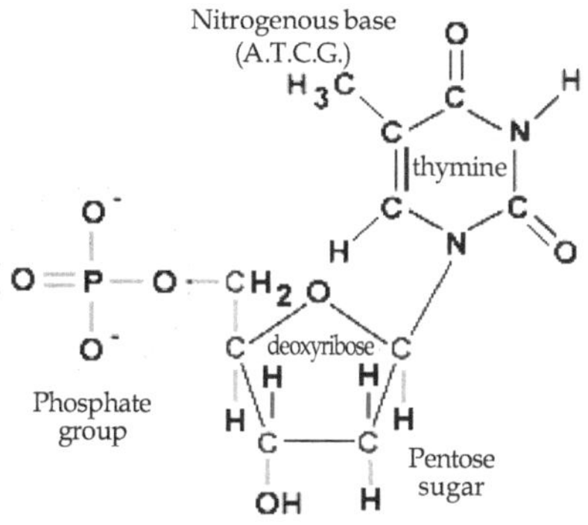

The phosphate of one deoxyribonucleotide binding to the 3' carbon of the deoxyribose of other forms the sugar-phosphate backbone of the DNA (the sides of the "ladder"). The hydrogen bonds between the complementary nucleotide bases (adenine-thymine; guanine-cytosine) form the rungs. Note the antiparallel nature of the DNA. One strand ends in a 5' phosphate and the other ends in a 3' hydroxyl.

To synthesize the two chains of deoxyribonucleotides during DNA replication, the DNA polymerase enzymes involved are only able to join the phosphate group at the 5' carbon of a new nucleotide to the hydroxyl (OH) group of the 3' carbon of a nucleotide already in the chain. The covalent bond that joins the nucleotides is called a phosphodiester bond. Each DNA strand has what is called a 5' end and a 3' end. This means that one end of each DNA strand, called the 5' end, will always have a phosphate group attached to the 5' carbon of its terminal deoxyribonucleotide. The other end of that strand, called the 3' end, will always have a hydroxyl (OH) on the 3' carbon of its terminal deoxyribonulceotide.

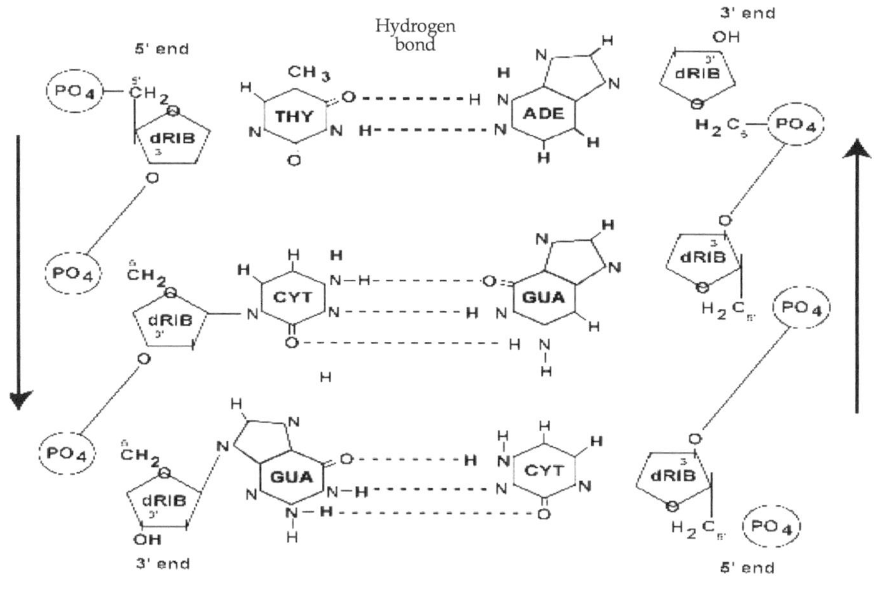

Complementary base pairing refers to nucleotides with the base adenine hydrogen bonding only with nucleotides having the base thymine (A-T). Likewise, nucleotides with the base guanine can hydrogen bond only with nucleotides having the base cytosine (G-C). (As a result of this bonding, the DNA assumes its helical shape.) Therefore, the two strands of DNA are said to be complementary. Wherever one strand has an adenine-containing nucleotide, the opposite strand will always have a thymine nucleotide; wherever there is a guanine-containing nucleotide, the opposite strand will always have a cytosine nucleotide.

While the two strands of DNA are complementary, they are *oriented in opposite directions* to each other. One strand is said to run 5' to 3'; the opposite DNA strand runs *antiparallel* or 3' to 5'.

DNA, or deoxyribonucleic acid, is the hereditary material in humans and almost all other organisms. Nearly every cell in a person's body has the same DNA. Most DNA is located in the cell nucleus (where it is called nuclear DNA), but a small amount of DNA can also be found in the mitochondria (where it is called mitochondrial DNA or mtDNA).

The information in DNA is stored as a code made up of four chemical bases: adenine (A), guanine (G), cytosine (C), and thymine (T).

Human DNA consists of about 3 billion bases, and more than 99 percent of those bases are the same in all people. The order, or sequence, of these bases determines the information available for building and maintaining an organism, similar to the way in which letters of the alphabet appear in a certain order to form words and sentences.

Each strand of DNA in the double helix can serve as a pattern for duplicating the sequence of bases. This is critical when cells divide because each new cell needs to have an exact copy of the DNA present in the old cell. DNA is a double helix formed by base pairs attached to a sugar-phosphate Backbone

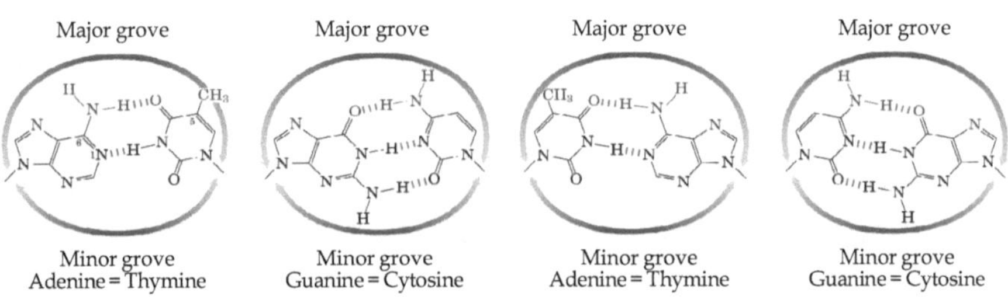

Central Dogma & New Dogma of Genetic Material

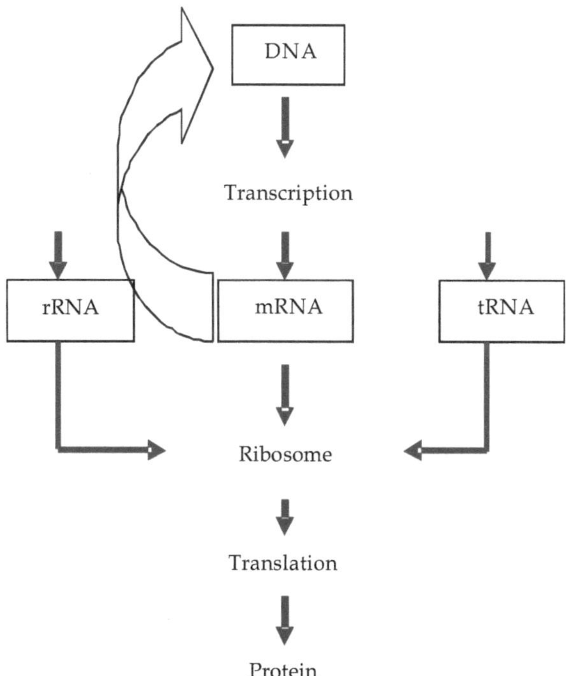

What is a Gene?

GENE : a specific sequence of nucleotides in DNA or RNA that controls the transmission and expression of one or more traits by specifying the structure of a protein or RNA

A gene is the basic physical and functional unit of heredity. Genes, who are made up of DNA, act as instructions to make molecules called proteins. In humans, genes vary in size from a few hundred DNA bases to more than 2 million bases. The Human Genome Project has estimated that humans have between 20,000 and 25,000 genes. Every person has two copies of each gene, one inherited from each parent. Most genes are the same in all people, but a small number of genes (less than 1 percent of the total) are slightly different between people. Alleles are forms of the same gene with small differences in their sequence of DNA bases. These small differences contribute to each person's unique physical features. Genes are made up of DNA. Each chromosome contains many genes.

Gene = The Basic Unit of Heredity

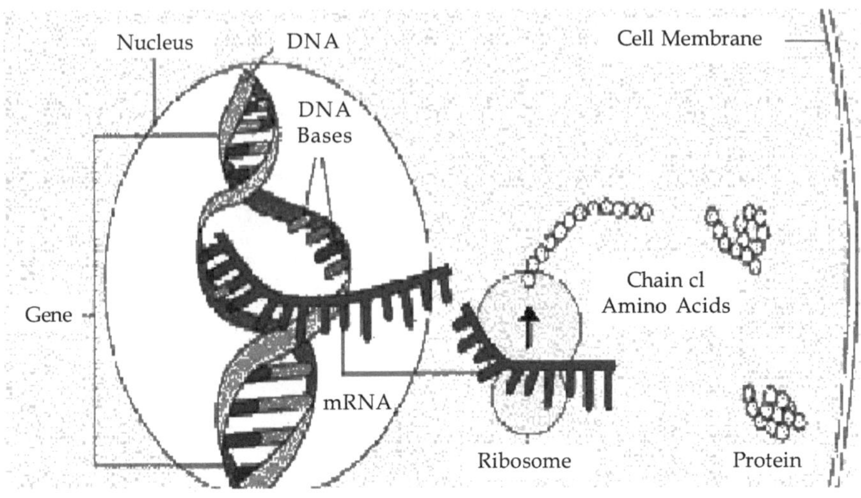

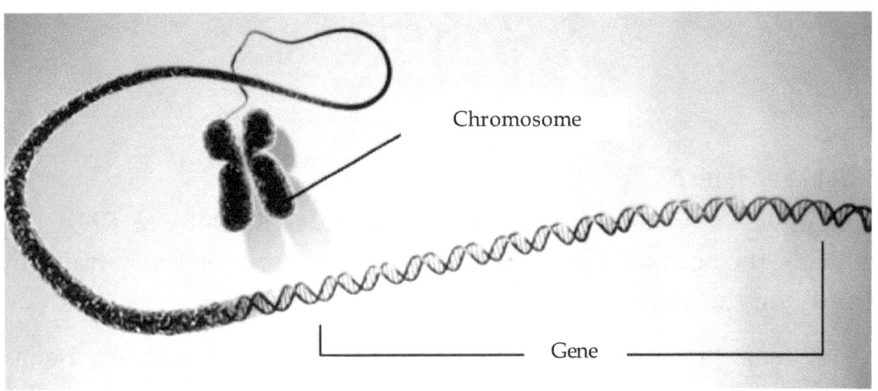

Prokaryotic Gene

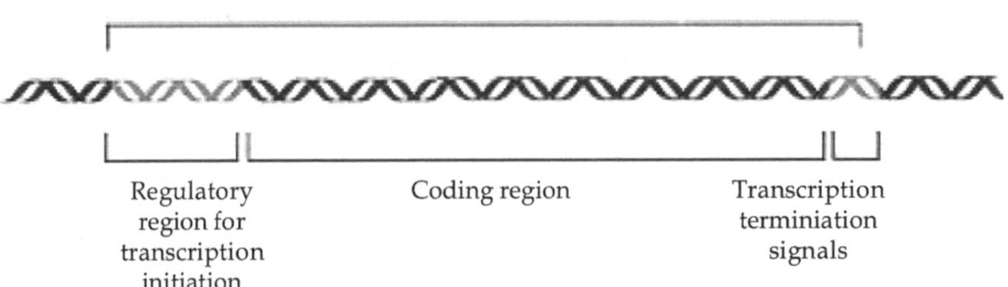

Eukaryotic Gene

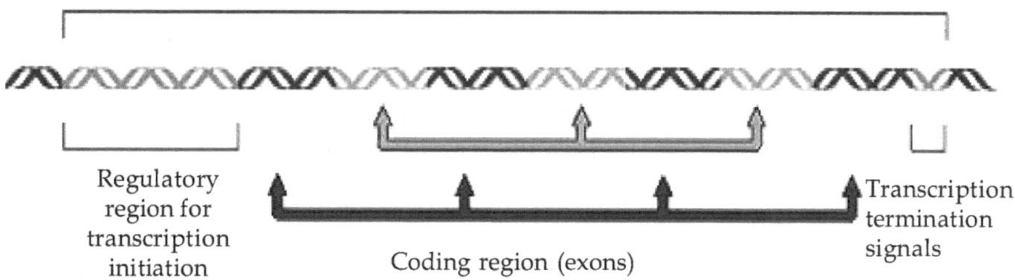

Coding region : encodes the amino acid sequence of a polypeptide

Gene Expression

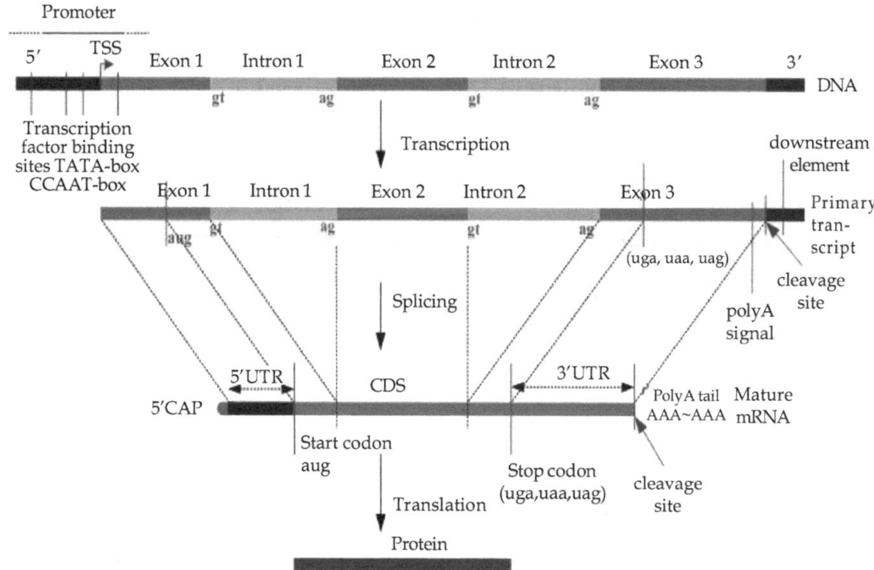

What is a Chromosome?

In the nucleus of each cell, the DNA molecule is packaged into thread-like structures called chromosomes. Each chromosome is made up of DNA tightly coiled many times around proteins called histones that support its structure.

Chromosomes are not visible in the cell's nucleus—not even under a microscope—when the cell is not dividing. However, the DNA that makes up

Chromosomes become more tightly packed during cell division and are then visible under a microscope. Most of what researchers know about chromosomes was learned by observing chromosomes during cell division.

Each chromosome has a constriction point called the centromere, which divides the chromosome into two sections, or "arms." The short arm of the chromosome is labeled the "p arm." The long arm of the chromosome is labeled the "q arm." The location of the centromere on each chromosome gives the chromosome its characteristic shape, and can be used to help describe the location of specific genes.

Chromosome Structure

Level 1: Nucleoside - the most fundamental unit of packaging

- Eukaryotic DNA is associated with DNA histone proteins
- Histones are small (102 to 135 amino acids) proteins that contain a very high proportion of positively charged amino acids such as lysine and arginine. Thus, they have high affinity for DNA (negatively charged molecules).

Level 1: Nucleosome - the most fundamental unit of packaging Nucleosome (200 bp): Nucleosome core particle (146 bp) + linker DNA Nucleosome core particle: is consisted of histone core octamer (Two subunit of H2A, H2B, H3, and H4) and bp of DNA wrapped 1.75 turns around the core.

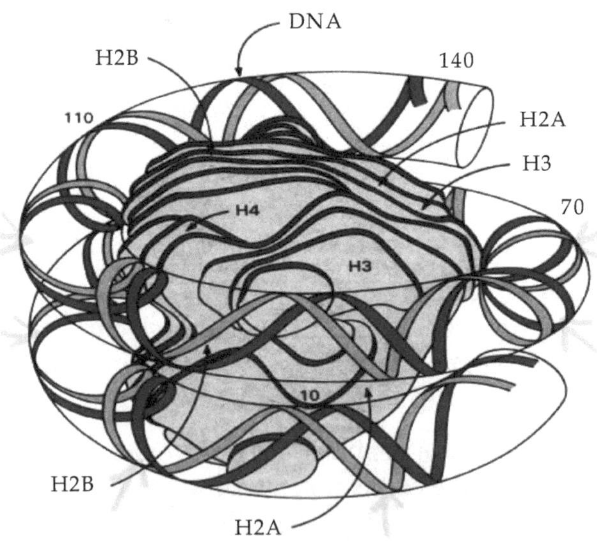

Level 2: 30-nm chromatin fiber

- Histone H1 brings nucleosomes together
- DNA is 40-fold more compact

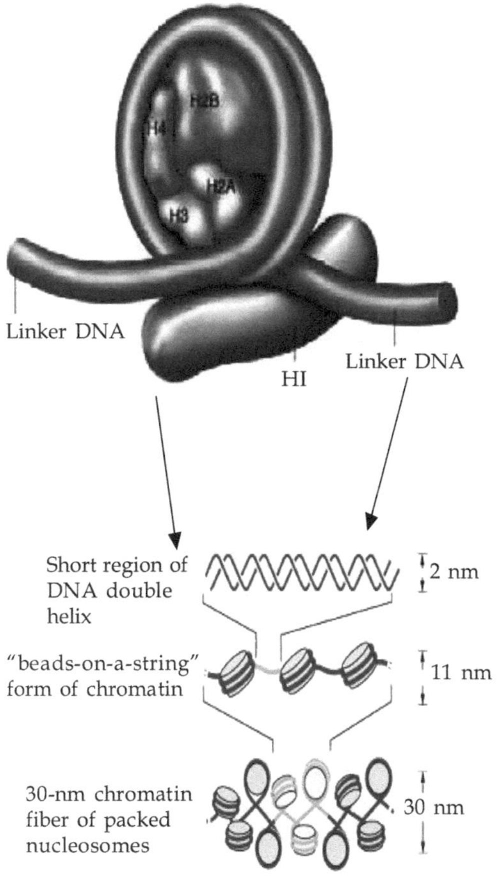

H2B
H4
H2A
H3
Linker DNA
HI
Linker DNA

Short region of DNA double helix
2 nm
"beads-on-a-string" form of chromatin
11 nm
30-nm chromatin fiber of packed nucleosomes
30 nm

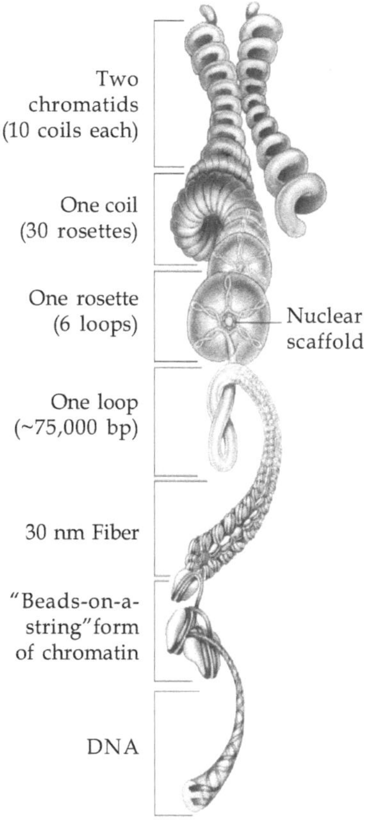

Two chromatids (10 coils each)
One coil (30 rosettes)
One rosette (6 loops)
Nuclear scaffold
One loop (~75,000 bp)
30 nm Fiber
"Beads-on-a-string" form of chromatin
DNA

Level 3: Radial loop scaffold

- Scaffold proteins loop the 30-nm fiber
- Specific, repeated DNA sequences interact With the scaffold proteins

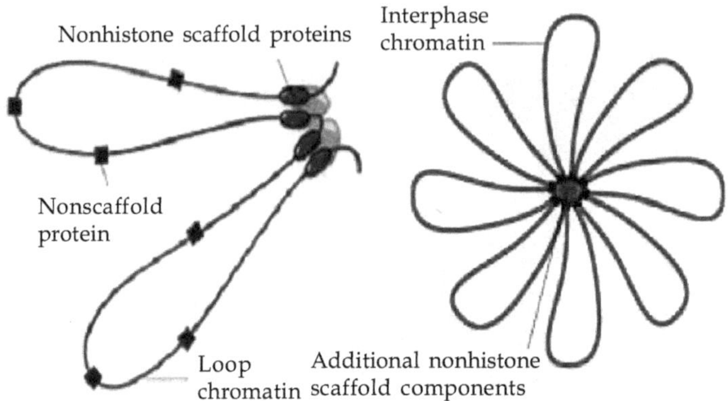

Level 4: Radial loop scaffold

- Additional looping and gathering of loops
- 10000-fold more compact at metaphase

Controlling of Gene Expression by Altering Chromatin Structures

- Affinity of transcription factor for its binding site on DNA is decreased when the DNA is reconstituted into nucleosomes.
- The packaging of DNA into nucleosomes is generally regarded as a block to transcription, presumably because the nucleosome interferes with binding of activators.

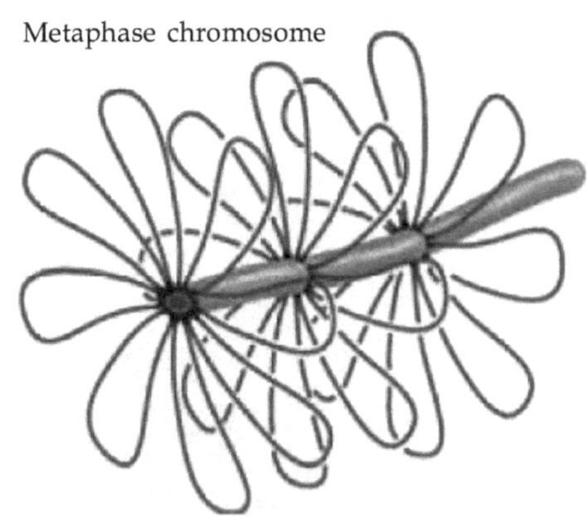

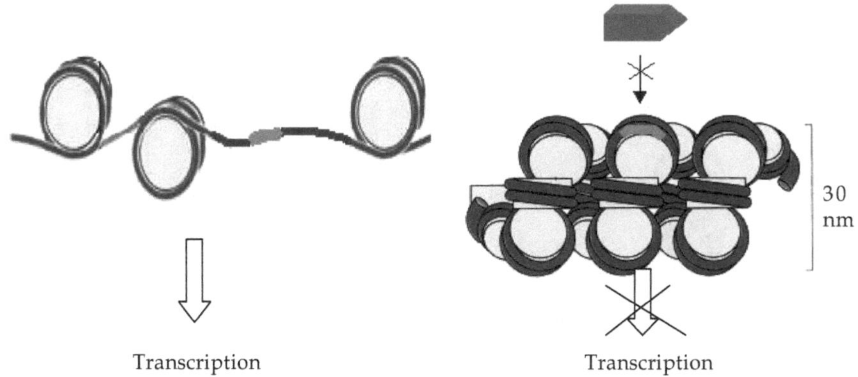

Heterochromatin = a portion of the chromatin in he interphase which remains relatively compacted and is transcriptionally inactive. Probably consists of closely packed region of 30-nm chromatin fiber.

Euchromatin= the more diffuse region of the interphase chromosome consisting of less dense chromatin.

Important Features of Chromosome

- Centromere is required to attach to spindle at mitosis's chromosome segregate into new cells
- Telomeres protect the ends of chromosomes
- Replication origins are where DNA replication state

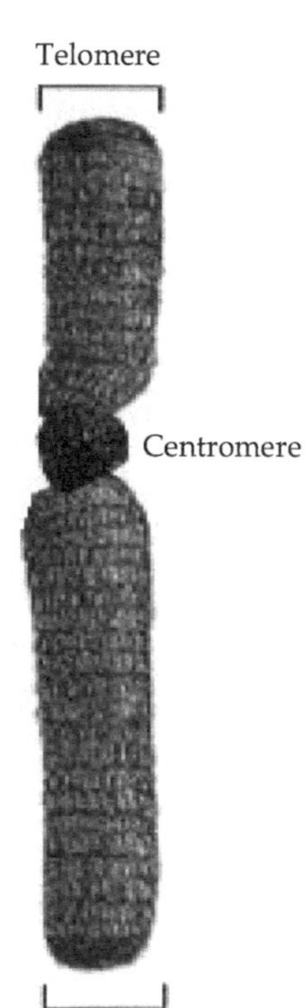

Centromere

- In mammals, it consists of Blocks of **satellite**
- Tightly condensed chromatin Structure (heterochromatin)
- Hold sister chromatid together
- Bind spindle fiber, allowing Segregation

Telomere : consists of 10-15 kb TTAGGG sequence telomeric family of minisatellite DNAs

- Protect the ends of chromosomes from degradation and loss of DNA sequence

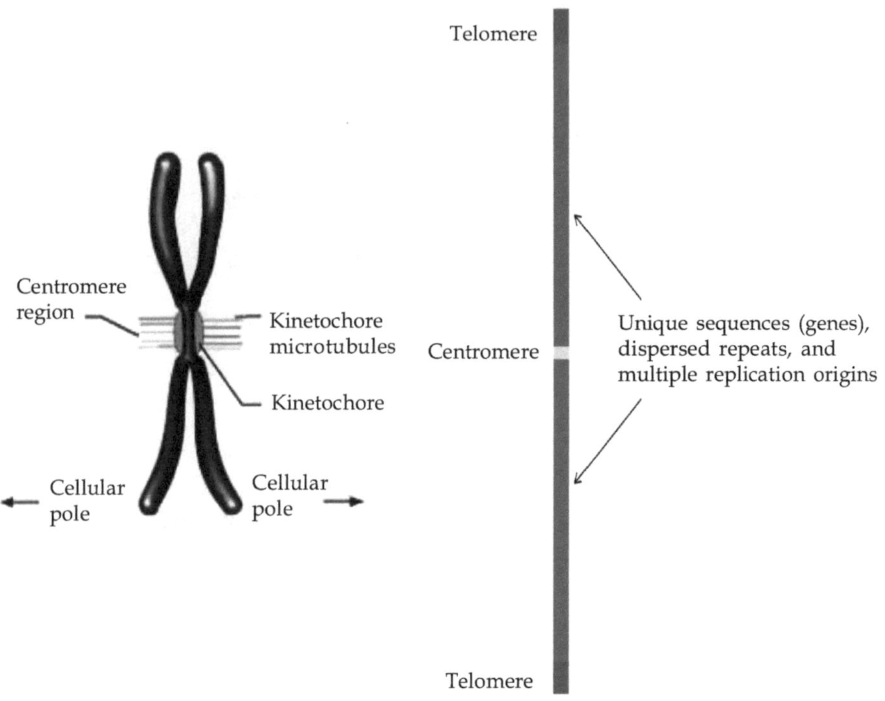
Centromere region
Kinetochore microtubules
Kinetochore
Cellular pole
Cellular pole
Telomere
Centromere
Telomere
Unique sequences (genes), dispersed repeats, and multiple replication origins

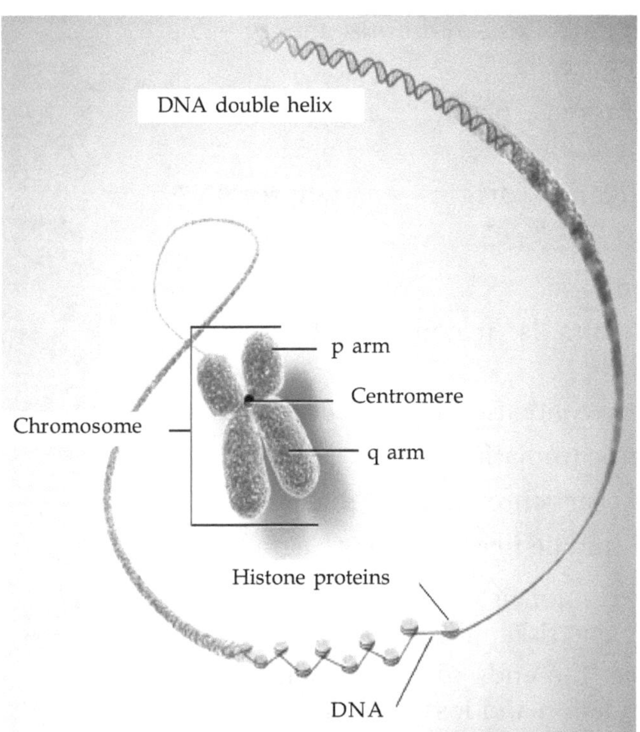
DNA double helix
p arm
Centromere
Chromosome
q arm
Histone proteins
DNA

What is a Genome?

A genome is an organism's complete set of DNA, including all of its genes. Each genome contains all of the information needed to build and maintain that organism.

Gene and Genome

• The term *genome* refers to the complete complement of DNA for a given species.

Organism	Genome Size Bases	Estimated Genes
Human (Homo sapiens)	3 billion	25,000
Laboratory mouse (M. musculus)	2.6 billion	30,000
Mustard weed (A. thaliana)	100 million	25,000
Roundworm (C. elegans)	97 million	19,000
Fruit fly (D. melanogaster)	137 million	13,000
Yeast (S. *cerevisiae*)	12.1 million	6,000
Bacterium (*E. coli*)	4.6 million	3,200
Human immunodeficiency virus (HIV)	9700	9

1. The Prokaryotic (Bacterial) Genome

The bacterial nuclear body is called a *nucleoid* (The prokaryotic (bacterial) nuclear body, usually composed of a single molecule of circular, chromosomal DNA)

a. The nucleoid of prokaryotes is one long, single molecule of double stranded, helical, supercoiled DNA which forms a physical and genetic circle. The chromosome is generally around 1000 µm long and frequently contains around 4000 genes. *E. coli*, which is 2-3 µm in length has a chromosome approximately 1400 µm long. To enable a macromolecule this large to fit within the bacterium, histone-like proteins bind to the DNA, seggregating the DNA molecule into around 50 chromosomal domains and making it more compact. Then an enzyme called *DNA gyrase* supercoils each domain around itself forming a compacted, supercoiled mass of DNA approximately 0.2 µm in diameter. Bacterial enzymes called *DNA gyrase* and *DNA topoisomerases* are essential in the unwinding, replication, and rewinding of the circular, supercoiled bacterial DNA.

b. The prokaryotic nucleoid has no nuclear membrane surrounding the DNA.

c. The nuclear body does not divide by mitosis. The cytoplasmic membrane plays a role in DNA separation during bacterial replication. Since bacteria are haploid (have only one chromosome), there is also no meiosis.

2. The Eukaryotic Genome

Prokaryotic and eukaryotic cells differ a great detail in both the amount and the organization of their molecules of DNA.

a. Eukaryotic cells contain much more DNA than do bacteria, and this DNA is organized as multiple chromosomes located within a nucleus.

b. The nucleus in eukaryotic cells is surrounded by a nuclear membrane and contains linear chromosomes composed of negatively charged DNA associated with positively charged basic proteins called histones to form structures known as nucleosomes. The nucleosomes are part of what is called chromatin, the DNA and proteins that make up the chromosomes.

c. The nucleus divides by mitosis and haploid sex cells are produced from diploid cells by meiosis.

The DNA in eukaryotic cells is packaged in a highly organized way. It consists of a basic unit called a nucleosome, a bead like structure 11 nm in diameter that consists of 146 base pairs of DNA wrapped around eight histone molecules. The nucleosomes are linked to one another by a segment of DNA approximately 60 base pairs long called linker DNA. Another histone associated with the linker DNA then packages adjacent nucleotides together to form a nucleosome thread 30nm in diameter. Finally, these packaged nucleosome threads form large coiled loops that are held together by non histone scaffolding proteins. These coiled loops on the scaffolding proteins interact to form the condensed chromatin seen in chromosomes during mitosis.

In recent years its been found that the structural nature of the deoxyribonucleoprtein contributes to whether or not DNA is transcribed into RNA. For example, chemical changes to the chromatin can enable portions of it to condense or relax. When a region is condensed, genes cannot be transcribed. In addition, chemical can attach to or be removed from the histone proteins around which the DNA wraps. The attachment or removal of these chemical groups to the histones determine whether nearby gene expression is amplified or repressed.

❑❑❑

CHAPTER 4

Equipments Required in a Molecular Marker Laboratory

1. pH Meter

pH meter/electrode systems are the most common method of measuring pH in the biology laboratory. Although more expensive and more complicated to use than chemical indicators, pH meters provide greater accuracy, sensitivity, and flexibility. When used properly, pH meters can measure the pH of a solution to the nearest 0.1 pH unit or better, and they can be used with a variety of samples.

2. UV Spectrophotometer

Perhaps there is no other technique except UV-visible spectroscopy, which has been so extensively employed and has such a profound impact on progress in biochemical investigations. UV-visible spectroscopy has undoubtedly become an indispensable tool and is routinely used for identification and quantitative estimation of most of the biomolecules. Besides other applications, most of the enzyme assays and the screening of biomolecules recovered in different fractions by separation processes such as column chromatography, density gradient centrifugation etc, rely heavily on spectrophotometer or colorimetric technique.

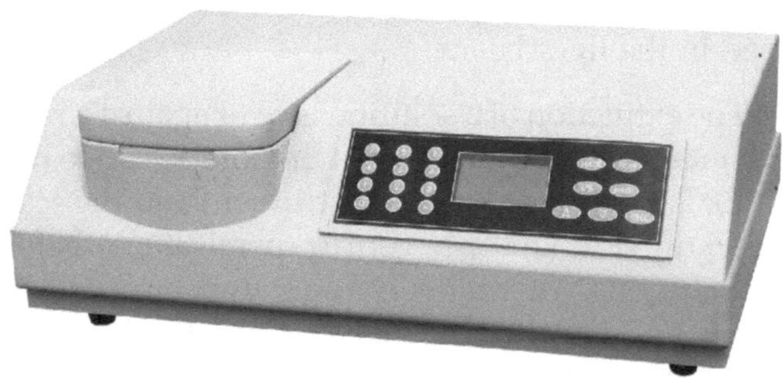

Points to Remember while Using UV Spectrophotometer

The detailed operation of a particular instrument must, of course, be obtained by carefully reading the instruction manual, but a few general points concerning the use and care of calorimeters and spectrophotometers are given below.

1. *Cleaning cuvettes:* Cuvettes are cleaned by soaking in 50 per cent v/v nitric acid and then thoroughly rinsed in distilled water.
2. *Using the cuvettes:* First of all, fill the cuvettes with distilled water and check them against each other to correct for any small differences in optical properties. Always wipe the outside of the cuvettes with soft tissue paper before placing in the cell holder and do not handle them by the optical faces. When all the measurements have been taken, wash them with distilled water and leave in the inverted position to drain.
3. *Absorption of radiation by cuvettes:* All cuvettes absorb radiation and the wavelengths at which significant absorption occurs

depend on the material from which the cuvette is made. Silica cuvettes are the most transparent to UV. Glass cuvettes are expensive so than silica and so they are used whenever possible and invariably in the visible region of the spectrum. However, they do absorb UV and cannot be used below 360 nm. Plastic disposable cuvettes also absorb in the UV but much less than glass and they can be conveniently used at 340 nm for the assay of dehydrogenease enzymes that require NAD or NADP as coenzymes.

4. *Light source:* A tungsten lamp produces a broad range of radiant energy down to about 360 nm. To obtain the ultraviolet reaction of the spectrum a deuterium is used as the light source. If the tungsten lamp is used in the range 360-400 nm, then a blue filter is placed in the light beam.
5. *Blanks:* The extinction of a solution is read against a reagent blank, which contains everything, except the compound to be measured. The blank is first placed in the instrument and the scale adjusted to zero extinction (100 per cent transmittance) before reading any test solutions. Alternatively, the extinction can be read against distilled water and the absorbance of the blank subtracted from that of the test solution.
6. *Replicates:* It is essential to prepare all blanks and standard solutions in duplicate so that an accurate standard curve can be constructed. In addition, the test solutions should also be prepared in duplicate wherever possible.

3. Centrifuge

Biologists seldom need to separate gravel and sand from one another; rather, they are interested in such material as cells, organelles, bacteria, and viruses. These biological materials, called particles in the context of centrifugation, are much smaller than sand and would take a long time to sediment due to the force of gravity alone.

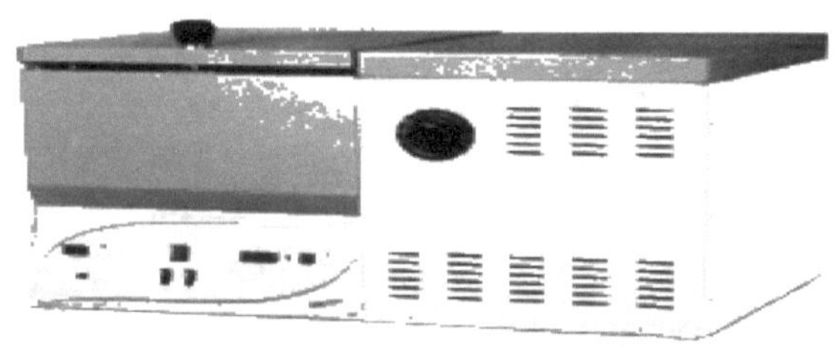

In a simple centrifuge there is a central drive shaft that rotates during centrifugation. A rotor sits on top of the drive shaft and holds tubes, bottles, or other sample containers. As the drive shaft rotates, the sample containers spin rapidly, creating the force that facilitates the sedimentation of particles.

$$RCF = 11.2 \times r\left(\frac{RPM}{1{,}000}\right)^2$$

Where,

RCF = the relative centrifugal field in units of X g

RPM = the speed of rotation in revolutions per minute

r = the radius of rotation in centimeters

You may also see this equation rearranged as:

$$RCF = 1.12 \times 10^{-5}\ (RPM)^2\ (r)$$

Examples of Applications of Preparative Centrifugation

1. Separation of intact, single cell suspension such as bacteria, viruses, and blood cells from a liquid medium.
2. Separation of organelles, such as mitochondria, and, ribosomes from disrupted cells.
3. Separation of biological macromolecules, including DNA, RNA, and protein from a solution.
4. Separation of plasma (the portion of blood) from blood cells.
5. Separation of immiscible liquids from one another.
6. Separation of cells from broth after fermentation.

Analytical centrifugation is used in specialized research settings to determine the molecular weight, purity, shape, and other physical characteristics of proteins and other macromolecules Analytical centrifugation involves specially designed, computer controlled centrifuges that allow the operator to monitor the movement of particles.

4. Autoclave

The laboratory apparatus designed to use steam under regulated pressure to achieve sterilization is called autoclave. It used for sterilization of nutrient media and glass wares.

Principle

Water boils at 100°C depending on the vapour pressure of the atmosphere. If the atmospheric pressure is raised the boiling temperature for water will also rise. So if the steam pressure inside a close vessel is increased to 15 lb/inch2. The temperature can go up to 121.6°C. An autoclave is equipped with device that can maintain saturated steam at a designated temperature and pressure for any period of time.

Description

It is essentially a double jacketed vertical cylinder made up of a strong metal and is separated by a case made up of iron sheet. The lid is very heavy and made up of gun metal and is screwed by buffer files screws. It is rendered air tight with interposition of an asbestos washer. The cylinder contains water up to certain level. The articles to be sterilized are placed on a perforated diaphragm serving as a platform situated above the water level. Heating can be done either by gas burner or by electricity. The chamber is furnished with a steam tap at the top and a pressure gauge with a safety valve. The pressure employed is 15 lb/inch for 15-20 minutes under this pressure H_2O boils at 121°C under operation it is necessary that air in the chamber is replaced by saturated steam otherwise required by temperature will not be obtained. It is not the pressure that kills the organism but the temperature of steam. The time of operation to achieve sterility depends upon the nature of the material being sterilized, the type of the containers and the volumes.

Sterilization by autoclaving all organisms including spores. Death of cells and spores is due to coagulation of proteins.

Uses : Sterilization of most type of solid or liquid media, distilled water, normal saline, discarded cultures, contaminated media, aprons, rubber tubes, gloves etc.

5. Hot Air Oven

Dry heat or hot air sterilization is recommended where it is either undesirable or unlikely that steam under pressure will make direct and complete contact. With the materials to be sterilized. The apparatus employed for this type of sterilization may be special electric or gas oven or even the kitchen stove oven.

Principle

Sterilization is carried out by dry heat at high temperature. Bacterial cells and spore die due to dehydration.

Description

It consists of a chamber having triple wall. The inner two walls are made up of copper sheet and outer of asbestos. The chamber is filled with several adjustable shelves and a thermometer is interested to record temperature. There are adjustable holes which are kept open during rising the temperature and are again open partially when sterilization completes clean glassware, like petridishes, test tube, flasks, pipettes etc. are place inside and the door is closed. Electric heated is put ON and the temperature of chamber is opened only after the temperature comes down to normal.

Uses : It is useful device for sterilization of laboratory glassware such as petridishes and pipettes. It is also used to sterilize substances which can not b e sterilized by moist heat like oils, fats and powders.

6. Deep Freeze

It is observed that in case of most of the bacteria, their preservation needs low temperature. This is because their metabolic activities are retarded. It is used for routine preservation of laboratory cultures. It is also used for storage of vaccines, sera, blood, plasma, antibiotics, media and several heat sensitive substances and many chemical. In DNA isolation it used for given a low (-20 °C) temperature during precipitation.

It is a cupboard like chamber with storage facility for wide variety of material. The cooling in done by pheron gas and it is circulated in the chamber by a compressor.

7. BOD Incubator

Specially designed to meet the requirements for incubation of bacteria to decompose organic matter in waste water. It is used in Biology, Botany Virology, and Oceanography. Water pollution, sewage, agriculture, food and research departments also use it. In addition to this, it helps in preservation of vaccines. It provides a wide range of conditions for the study of synthesizing organisms. Temperature range, 5°C to 50°C, is attainable within the cabinet to the study the growth of organisms under tropical or temperate and summer or winter conditions. Temperature is controlled with an Electronic Digital Temperature Controller cum Indicator with an accuracy of ± 0.5°C or better.

8. Microwave Oven

It is use for maintaining the specific temperature during DNA isolation as the lysis of the cell require high temperature.

9. Gel-Documentation

In gel-documentation system uses UV lights to make ethidium bromide or other dye fluorescence. With fluorescent staining of nucleic acids, a fluorescent substance that has bound to nucleic acids is excited by ultraviolet irradiation and emits fluorescent light. The fluorescent substance EtBr binds specifically to nucleic acid and the amount of bonding depends on the molecular weight and concentration of the nucleic acid. In other words, a band for a large molecular weight or large amount will shine brighter; conversely, fluorescence will be weaker for a band for a small molecular weight or small amount. This instrument is very useful in analyzing the band in gel.

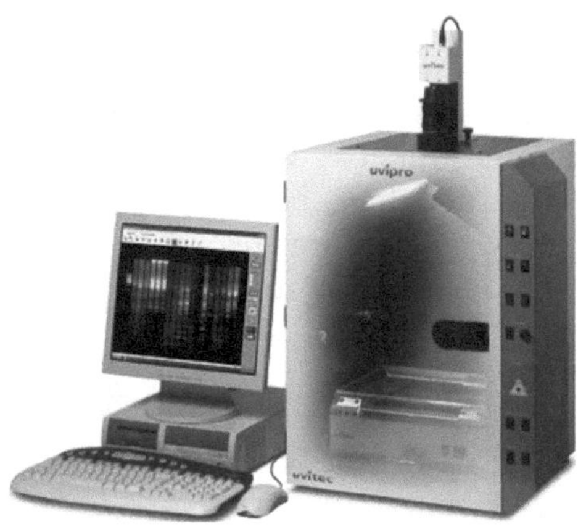

10. Deep Freeze Centrifuge

A *centrifuge* is a piece of equipment that accelerates the rate of sedimentation by rapidly spinning the samples, thus, creating a centrifugal force many times that of gravity.

It is use for the separation of cells, organelles, bacteria, and viruses. In separation of DNA &RNA from rest of the material by using high speed centrifugation.

11. Micropipette

In molecular biology practical the most important thing is that to take a reagent /chemical a specific volume by using automated pipette we can take a micro liter volume of any chemical which can not take sing conventional pipette.

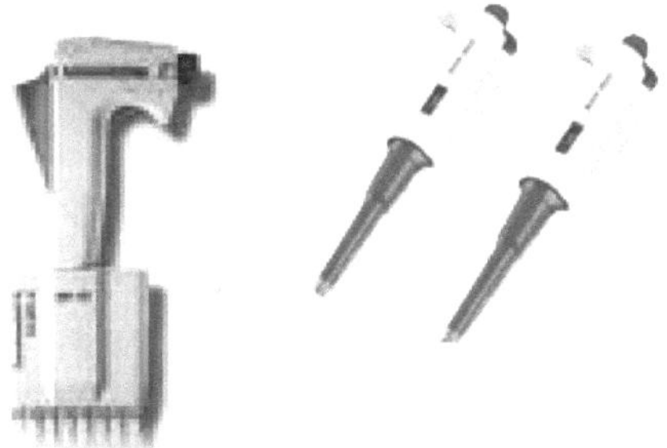

12. PCR

The purpose of a PCR (Polymerase Chain Reaction) is to make a huge number of copies of a gene. This is necessary to have enough starting template for sequencing.

13. Mortar & Pestle

Mortar and pestle is used to crush/grind the leaf tissues / plant tissues to isolate DNA/RNA/Isoenzyme etc. for molecular marker study.

14. Digital Balance

Digital balance is used to precisely weigh chemicals, plant material and other things used in molecular marker study and plant biotechnology.

15. Liquid Nitrogen Cylinder

Isolation of DNA and storage of plant tissues. Helps in grinding of tissues to the powder form and avoid enzymatic activities that harm/degrade DNA/isoenzyme etc. while working with them.

16. Water Bath

Maintains water temperature to incubate DNA (or any other) reagent / samples/ restriction digestion mixture etc. at a constant temperature.

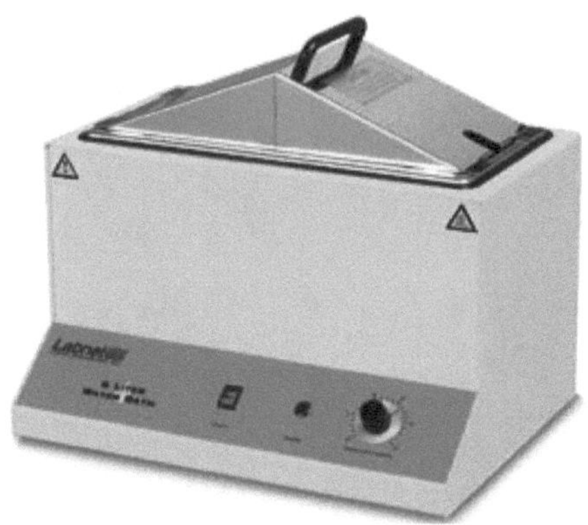

17. Ice Bucket

Ice bucket is used to keep the working samples of DNA / isoenzymes etc. while working with any molecular biology protocol.

18. Ice Flaker

Ice flaker is used to prepare ice flakes, which is further used in different place while working with DNA/RNA/Protein, like it is put in the ice buckets for storage of working DNA / Isoenzyme samples.

19. Electrophoresis Apparatus

Electrophoresis is a method of separating out large molecule (DNA or protein). An electric current is passed through a medium containing the molecules, and each molecule travels at a different rate depending on its electrical charge, size and shape. Separation by electrophoresis is based on these differences. In electrophoresis, agarose and acrylamide gels are used for electrophoresis of nucleic acids and proteins.

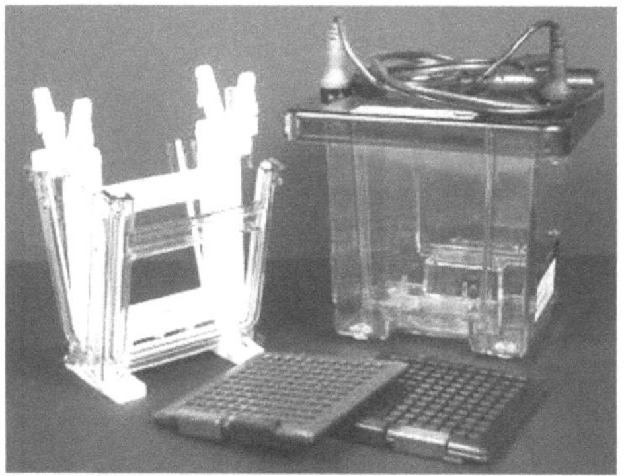

20. Sequencing Gel Apparatus

Sequence gel apparatus is to run PAGE (Polyacrylamide gel electrophoresis). SDS-PAGE is the most widely used method for qualitatively analyzing any protein mixtures. It is particularly useful for monitoring protein purity and determines their relative mass.

21. UV transilluminator

UV transilluminator is used for gel documentation. In gel-documentation system uses UV lights to make ethidium bromide or other dye fluorescence.

22. Genome sequencer

The purpose of genome sequencer is to determine the order of the nucleotides of a gene.

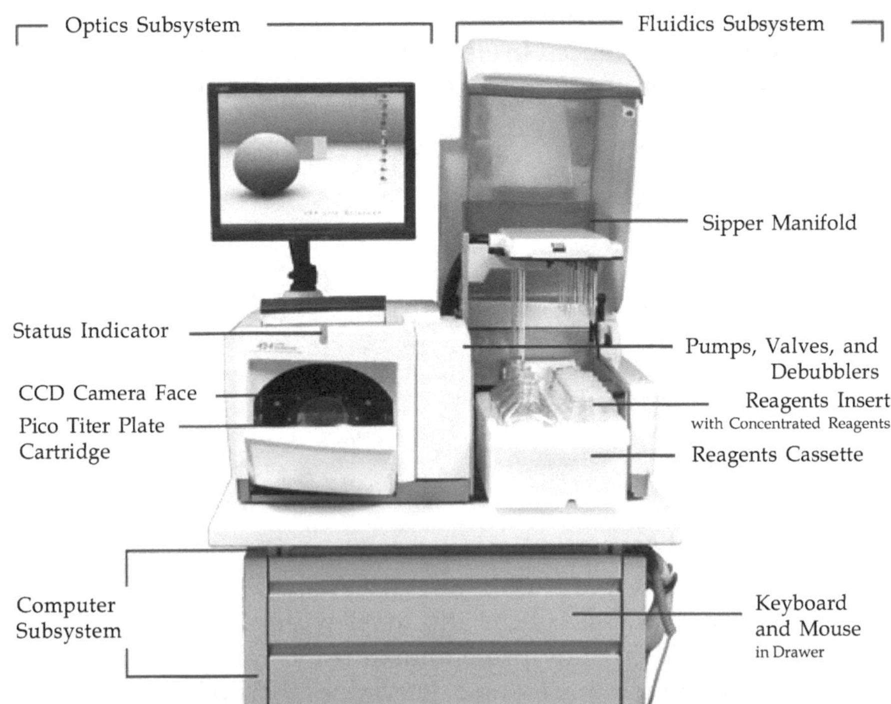

23. Aminoacid Sequencer

The purpose of amino acid sequencer is to determine the order of the nucleotides of proteins. Sequence information can be used to identify a protein or homologous proteins through searches in databases; the minimal number of residues for a succesful search is ten, but often more residues are required. The N-terminal sequence is often used for confirmation of the identity of a protein; in that case a few residues are enough. The sequence also gives information about post translational cleavage points. In addition, the sequence results give informations about the purity of a preparation; limits of detectable contamination depend on the sequences of the analysed proteins.

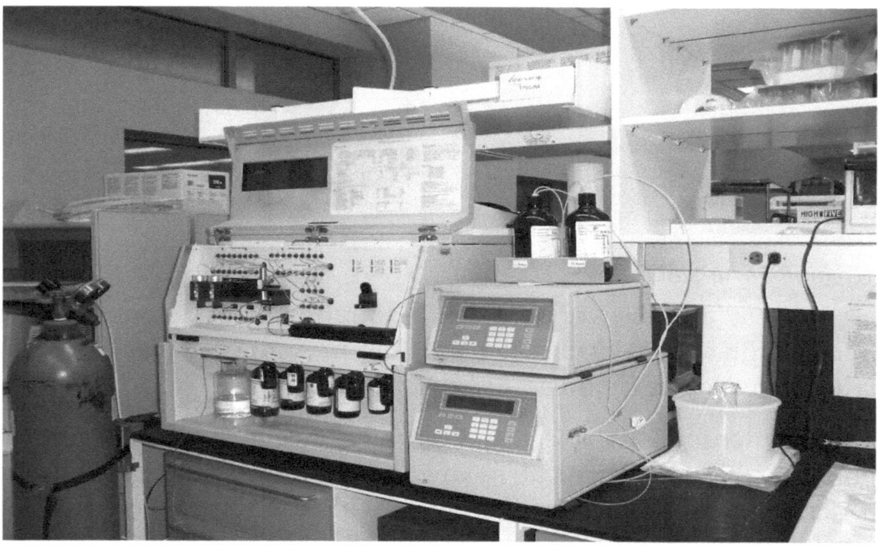

24. Western Blotting Apparatus

The Western Blotting Apparatus is made with a view to blotting a maximum of three gels at one time. It is used to transfer DNA to the nylon membrane from the gel.

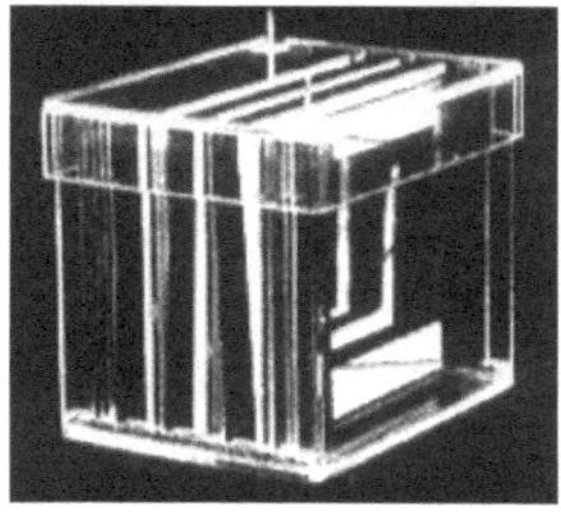

25. Picodrop

Picodrop uses UV/Vis measurements to be taken directly through the pipette tip for finding out purity and concentration of DNA. It requires only 2µl of sample for measurement. It can be further used for RNA and Protein estimation.

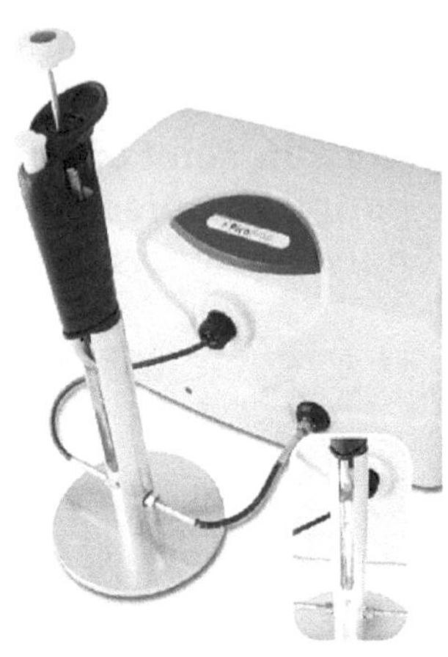

26. Laminar Air Flow

Laminar air flow work on basis of filtration in this HEPA filter are used for filter the contaminates air flow prevent the enter the other bacteria inside the chamber.

In microbiology one major problem is contamination and in molecular biology practical /plant tissue culture practical to need sterile condition. Laminar air flow provide the sterile condition for specific region in this chamber is totally sterile.

CHAPTER 5

Basic Calculations used in Microbial Biotechnology

Mole (mol)

1 mole of any substance contains 6.02 x 10^{23} units. The number of moles in a given weight of substance can be calculated using following equation.

$$mol = \frac{wt\,(g)}{mol\ wt}$$

The weight of a given mount of moles can be calculated using equation given below

wt (g) = mol x mol wt (g/mol)

Equivalent Weight (equiv wt)

This is the weight of an acid or a base containing 1 mole of replaceable H^+ or OH^-. Gram equivalent weight is used when equivalent weight is expressed in grams, as calculated in equation given below

equiv wt = mol wt/n

Equivalent (equiv)

One equivalent of a compound contain 1 g-equiv wt of the compound, as calculated with following equation

$$equiv = \frac{wt\,(g)}{equiv\ wt}\ or\ equiv = mol\ x\ n$$

Molarity (M)

This defines the number of moles of a substance per liter (L) of solution, as calculated with following.

$$M = \frac{mol}{L}\ or\ M = mol\ x\ n$$

1M = 1GMW/L
(g/mole) x (moles/L) = g/L
(g/mole) x (moles/L) x L = g

Example

How many grams of H_3BO_3 (Boric acid, MW=61.83g/M) are required to make 2 liters of a 2M solution?

61.83g/M x 2M/L x 2L = Zg
Z = 247.32g

Example

What is the molarity of a solution that has 60g of NaOH (sodium hydroxide, MW=40g/M) in a total volume of 2 liters?

40g/M x XM/L x 2L = 60g
XM/L = 60g ¸ (40g/M x 2L)
XM/L = (60/80)M/L
X = 0.75M/L = 0.75M

Example

How many liters of 1M NaOH can be made with 80g of substance?

40g/M x 1M/L x XL = 80g
XL = 80g ¸ (40g/M x 1M/L)
XL = (80/40)L
X = 2L

Molality (M)

Moles of solute/Kg of solvent, this is a W/W expression. Molality is not a practical expression of concentration for routine laboratory use.

Normality (N)

"Normality is the number of "equivalent weights" of the solute per litre of solution".

- Biologists are most likely to encounter normality in reference to acids and bases.
- *For an acid,* 1 equivalent weight is equal to the number of grams of that acid that react to yield. 1 mole of H+ ions i.e. 1 mole of reactive group.
- *For bases,* 1 equivalent weight is equal to the number of grams of that base that supplies 1 mole of OH^- i.e.1 mole of reactive group.

$$NaOH \rightarrow Na^+ + OH^-$$

$$H_2SO_4 \rightarrow SO_4^{-2} + 2H^+$$

1 equivalent weight of NaOH is the number of grams that will produce 1 mole of OH^- ions, so equivalent weight of NaOH is 40 gms (equal to molecular weight).

1 mole of H_2SO_4 dissociates to form 2 molecules of H^+ ions. Therefore molecular weight of H_2SO_4 is 98.1 gm. & equivalent weight = 98.1 / 2 = 49 gms

$$N = \frac{equiv}{L} \; or \; N = \frac{wt}{mol\,wt\,x\,L}$$

N = nM

Example

NaCl has a valence of +1 in solution; for every molecule of NaCl, there is one molecule of Na^+ that can replace a molecule of hydrogen ions. For H_2SO_4, there is the equivalent of two molecules of H^+ for every molecule of H_2SO_4. Therefore the gram equivalent weight of H_2SO_4 is equal to the gram molecular weight (98) divided by 2; 98/2 = 49.

Calculating Concentrations Based on Weight Percent (Weight/Weight) or % (w/w)

The percent weight of a pure analyte in a sample is calculated using following equation (e.g the weight in g of an analyte in a 100g sample).

$$\%(w/w) = \frac{g\ of\ analyte\ in\ sample}{g\ of\ sample} x100\%$$

Percent (Weight/Volume) or %(w/v)

The percent weight of a pure analyte in a solution is calculated using equation (e.g. the weight in g of an analyte in 100ml sample)

Part dilutions : "Part solutions tell us how many parts of each component to mix together". The parts may have any units but must be the same for all the components of the mixture.

For example : A Solution that is 3:2:1 ethylene: chloroform: isoamylalcohol is:

1 parts ethylene, 2 parts Chloroform, 1 part isoamylalcohol.

Parts Per Million and Parts Per Billion

"Parts per million" (ppm) is the number of parts of solute per 1 million parts of total solution. Any units may be used but must be the same for the solute and the total Solution.

"Parts per billion" (ppb) is the number of parts of solute per billion parts of solution. Concentration is most often expressed in terms of ppm (or ppb) in environmental applications.

For example : an environmental scientist might speak of a pollutant in a lake as being present at a concentration of 5 ppm.

For any solute,

$$1\ ppm\ in\ water \frac{1\,\mu g}{ml}$$

Also,

$$1\ ppb\ in\ water \frac{1\,ng}{ml}$$

Parts Per Million (Weight/Weight), or ppm(w/w)

$$ppm(w/w) = \frac{g\ of\ analyte\ in\ sample}{g\ of\ sample} x 10^6\ ppm$$

Part Per Million (Weight/Volume) or ppm (w/v)

The grams of pure analyte in 10^6 mL of sample is calculated using equation.

$$ppm(w/v) = \frac{g\ of\ analyte\ in\ sample}{mL\ of\ sample} x 10^6\ ppm$$

Density (p) and Specific Gravity (sp gr)

Density is the mass per volume ration of a substance and is calculated using equation.

$$\rho = \frac{wt}{vol}$$

Specific gravity is the density of a fluid or other material relative to that of H_2O as expressed in equation.

$$sp\ gr = \frac{\rho_{sample}}{\rho^{H_2O}} = \rho_{....}\ at\ 4^{\circ} C$$

Equations for Calculating the Weight or Volume of a Reagent Needed to Prepare a Given Solution

To calculate the number of gram of a substance to create a pre-assigned molar concentration use equation

$$wt(g) = mol\ wt(g)\ x\ M\ x\ L$$

When correcting for % purity use equation.

$$wt(g) = mol\ wt(g)\ x\ M\ x\ L\ x \frac{100}{purity\ of\ reagent(\%)}$$

For % and ppm concentration use equation

$$wt(g) = \frac{\%(w/v)\ x\ mL\ of\ solution}{100\%}$$

$$wt(g) = \frac{ppm(w/v)\ x\ mL\ of\ solution}{10^{-6}\ ppm}$$

Dilution Factor in Concentration and Volume Calculations

To calculate the volume of a stock reagent necessary to add to a known volume of diluent to yield a desired concentration use equation.

$$V_1 = \frac{C_2 V_2}{C_1}$$

where V_1 is the volume of the stock reagent which, when diluted to a final volume V_2, yields a desired concentration C_2; C_1 is the concentration of the stock reagent. The units of C_1 and C_2 must be identical, and those of V_1 and V_2 must also be identical.

Dilutions

General Concepts

Many laboratory procedures, especially those in microbiology, Immunology / Serology, require that a series of dilutions be performed on serum or other body fluids. These dilutions are used to establish a relative measure of the quantity of antibody or antigen present in the fluid. This section will review the steps necessary to make simple and serial dilutions.

A dilution is expressed as the parts of solute relative to parts total volume. This is written using a slash (/). The colon (:) is used to express a ratio; the parts of one substance relative to parts of a second substance. Therefore, dilution is an expression of *relative* concentration. The dilution of a substance can be determined by dividing the volume of the solute by the volume of the solute + solvent or the total volume.

Example

What is the dilution when 1ml of Sample is mixed with 9ml of saline?

dilution = 1ml serum ÷ (1ml Sample+ 9ml saline)
dilution =1/10

Therefore the Sample has been diluted 1 to 10 with saline.

Example

What is the dilution achieved when 3ml of Sample are mixed with 22ml of saline?

dilution = 3ml serum ÷ (3ml of Sample + 22ml saline)
dilution = 3/25

Express this dilution as 1 to X dilution.

3/25 = 1/X

X = 25/3 = 8.33

Therefore 3/25 is the same as 1/8.33.

The volume of solute and solvent required to make a given volume of a specific dilution can be determined using ratio and proportion. The specified dilution represents one ratio and the amount of solute required becomes the unknown in the second ratio.

Example

How many milliliters of Sample are required to make 300ml of a 1/10 dilution?

1 part Sample /10 parts total volume = Xml Sample /300ml total volume

1/10 = X/300

X = 300/10ml

X = 30ml

This solution is made by diluting 30ml of Sample up to 300ml with saline.

The amount of undiluted solute present in a specified volume of a given dilution can also be calculated using ratio and proportion.

Example

How many milliliters of Sample are present in 25ml of a 1/8 dilution.

1 part Sample /8 parts tot. vol. = Xml Sample /25ml tot. vol.

1/8 = X/25

X = 25/8 = 3.125ml

Therefore there are approximately 3.1ml of undiluted Sample in 25ml of a 1/8 dilution.

In the preceding examples, the concentration of Sample is implied to be 100%, that is a 1/1 dilution. When diluting solutions with a specified concentration the final concentration can be calculated by multiplying the original concentration by the dilution.

Example

What is the final concentration of a solution created by diluting 5ml of 5N NaCl to a final volume of 500ml?

First calculate the dilution:

5/500 = 1/100

Multiply the original concentration by the dilution:

5N X 1/100 = 0.05N

Therefore the final concentration of the solution is 0.05N ($5x10^{-2}$N)

Dilution Series

In some laboratory situations it becomes necessary to perform a series of dilutions on a given solution. This usually done for two reasons. First, the desired dilution may be technically impractical. For example it is difficult to make a 1/10,000 dilution in a single step. This would require accurately diluting 1ml of starting solution to a final volume of 10,000ml. In this situation, the volume of the solute used is smaller that the inherent inaccuracy of the measuring device for the solvent (the error is usually + 1%; in this case + 10ml). When performing this type of dilution procedure, it is necessary to keep track of two dilutions: the tube dilution-the dilution factor of that particular tube or container; the solution dilution-the actual dilution of the starting substance in that particular tube. In both situations the tube dilution is calculated by dividing the #parts solute by parts total volume. The solution dilution of a given tube is determined by multiplying the tube dilution of the specified tube by the solution dilution of the preceding tube.

Note *When Sample is diluted, undiluted serum is assumed to be 1/1.*

Example

One ml of serum is added to 4ml of saline (tube1). One ml of this dilution is added to 9ml of saline (tube2). One ml of this dilution is added to 99ml of saline (tube3). Determine the tube and solution dilution in each tube.

Tube	1	2	3
Tube dilution	1/(1=4) = 1/5	1/(1+9) = 1/10	1/(1+99) = 1/100
Solution dilution	1/1 X 1/5 = 1/5	1/5 X 1/10 = 1/50	1/50 X 1/100 = 1/5000

Therefore the final dilution of serum in Tube 3 is 1/5000.

Example

What would be the concentration of HCl present in Tubes 1-3 if 10N HCl was diluted in the manner described above?

For each tube, the concentration of HCl is determined by multiplying the initial concentration of HCl by the solution dilution for that tube.

Tube 1 = 1/5X10N = 2N HCl

Tube 2 = 1/50X10N = 0.2N HCl

Tube 3 = 1/5000X10N = 10 ¸ 5X10^{3} = 2X10^{-3}N HCl

Preparing Dilutions of Specified Volume

It is often necessary to produce a specified volume of a dilution or to produce a series of dilutions of specified volume from a limited amount of starting material. It then becomes necessary to determine how the make each dilution.

Example

Prepare a series of dilutions of Sample (1/10, 1/100, 1/500) so that the total volume of the final container is 100ml and starting with only 1.5ml of Sample.

The easiest way to solve this problem is to establish a series of dilutions where and only 1ml of serum is used for the first dilution and the final tube has a volume of 100ml.

Tube 1 = 1/10=>1ml of Sample + 9ml of saline

Tube 2 = 1/100 = 1/10 x X => X = 1/10

Tube 3 = 1/500 = 1/100 x Y => Y = 1/5

1/5 = Z/100

Z = 20ml

To prepare this dilution, first dilute 1ml of serum 1/10. Take 2ml of Tube 1 and dilute it 1/10. To prepare 100ml of the final dilution take 20ml of tube 2 and dilute 1/5 to yield a final volume of 100ml.

Remember the following general rule: final conc. = init. Conc. x dil tube1x diltube 2x.

Serial Dilutions

Serial dilutions represent a specialized application of dilution procedures. Serial dilutions are performed when it is necessary to dilute a substance in a series of equal dilutions.

This is most commonly done when trying to determine the relative concentration of antibody or antigen in a serum specimen. Samples are generally diluted twofold; a series of ½ dilutions. The same concepts used above are employed when making serial dilutions. Again, the dilution of serum is set as 1/1. Consequently the dilution factor increases by powers of two.

Example

A sample specimen is diluted in the following manner: 1ml of undiluted sample is mixed with 1ml of saline (tube1); 1ml of this tube is added to a second tube containing 1ml of saline (tube2); this same procedure is repeated twice more (tubes3-4). What is the tube dilution and solution dilution for each tube?

Tube	**1**	**2**	**3**	**4**
Tube dilution	1/(1+1) = 1/2	1/(1+1) = 1/2	1/(1+1) = 1/2	1/(1+1) = 1/2
Solution dilution	1/1 X 1/2 = 1/2	1/2 x 1/2 = 1/4	1/4 X 1/2 = 1/8	1/8 X 1/2 = 1/16

Note the pattern developing. The solution dilution is equal to ½ raised to the power corresponding to the tube number:

tube1 = $(1/2)^1$= 1/2 tube2 = $(1/2)^2$=1/4 tube 3=$(1/2)^3$=1/8 tube4=$(1/2)^4$=1/16

This holds for the original serum if it is labeled tube 0: $(1/2)^0$= 1=1/1

Therefore in any serial twofold (2X) dilution series, the tube dilution is always 1/2 and the solution dilution is (1/2)

IN SUMMARY: The dilution series requested is prepared by adding 0.5ml of saline to each of 10 tubes. 0.5ml of undiluted serum is added to tube#1 and mixed well. 0.5ml of tube 1 is then added to tube 2. This is repeated out to tube 10. 0.5ml of sample is removed from tube 10 and discarded to yield a final volume in tube 10 of 0.5ml.

Tube Number	0	1	2
Tube dilution 1/(1+9)	1/1	1/(1+9)	1/(1+9)
1/10	1/1	1/10	1/10
Sample dil. 0.05N X 1/10	5N X 1/1	5N X 1/10	0.5N X 1/10
0.005N	5N	0.5N	0.05N
Tube Number	0	2	3
Tube dilution 1/(1+9)	1/1	1/(1+9)	1/(1+9)
1/10	1/1	1/10	1/10
Sample dil. 1/100 X 1/10	1/1 X 1/1	1/1 X 1/10	1/10 X 1/10
1/1000	1/1	1/10	1/100

Twofold Dilutions

The standard dilution technique most frequently used in the serology laboratory is the twofold dilution. The guiding principle for this dilution type is the fact that each tube contains half serum and half diluent (usually saline). The amount of antibody is directly proportional to the quantity of serum. The first tube of the twofold dilution series usually contains specified quantities of undiluted patient's serum - the same undiluted serum originally found to cause agglutination in the Qualitative Test. The second tube contains half the amount of serum and therefore half the amount of antibody; the third tube contains a quarter of the amount of antibody, the fourth tube contains an eighth, the fifth contains a sixteenth, the sixth a thirty-second, the seventh tube a sixty-fourth, etc.

The Twofold Dilution Procedure should be Performed in the Following Manner

1. Place 10 clean test tubes in a test-tube rack.
2. Add 1.0 ml of 0.9% saline to tubes 2 through 10, using a 5.0 ml pipet.

3. Add 1 ml of patient's serum to tubes 1 and 2.
4. With the same pipet, mix the contents of tube 2 by drawing the contents up into the pipet. The process of drawing up and blowing out is considered one mixing. To be thorough, mix four times. Note that tube 1 contains only undiluted serum and tube 2 contains half the amount of the serum because 1.0 ml of undiluted serum was diluted with 1.0 ml saline. When serum is diluted to half the original quantity, it is said to be diluted 1:2. This means that there is one part of undiluted serum and one part of diluent, which equals two parts or two volumes.
5. With the same pipet that was used in step 4, transfer 1.0 ml of serum from tube 2 to tube 3 which also contains 1.0 ml of saline. Mix four times as described above. The half quantity of serum in tube 2 has been diluted to one-quarter in tube 3. When serum is diluted to one-quarter the original quantity, it is said to be diluted 1:4. This means that 1 volume of a 1:2 serum diluted with 1 volume of saline is equal to 2 volumes. Examine these dilutions carefully. If a serum is diluted 1:4 directly from undiluted serum, this may be accomplished by adding 1 ml of serum to 3 ml of saline for a total of 4 volumes.

   1 + 3 = 4 or 1 volume of serum diluted to 4 volumes or 1:4 dilution

   But a 1:4 dilution has been prepared from a 1:2 dilution, not from undiluted serum. This is done by diluting 1 volume of a 1:2 dilution with 1 volume of saline to make a 1:4 dilution.
6. With the same pipet used in step 5, transfer 1.0 ml of serum from tube 3 (1:4) to tube 4 to prepare a 1:8 dilution. Mix four times as described above.

   If one were to prepare a 1:8 dilution from undiluted serum, one could do so by adding 1 volume of serum to 7 volumes of saline to get 8 volumes.

   1 + 7 = 8 or 1 volume of serum diluted to 8 volumes or 1:8

   But because a twofold dilution is being prepared, one merely dilutes 1 ml of a 1:4 with 1 volume of saline to get a 1:8 dilution, or 1:4 diluted 1:2 = 1:8
7. Continue to transfer 1 ml of each serum dilution from tube 4 to tube 5, mix well; take 1 ml from tube 5 and place it in tube 6, mix well; take 1 ml from tube 6 and place it in tube 7, etc., to tube 10.

8. Mix tube 10 and discard 1 ml from that tube. Identical volumes in all tubes have been maintained. Each contains exactly 1 ml total volume.

Example

Given 0.6 ml of serum, diagrammatically prepare a twofold dilution, using 0.3 ml of sample in tubes 1 and 2.

Answers

1 + 1 = twofold dilution

X 0.3 ml (multiply both sides by 0.3 ml) 0.3 ml sample + 0.3 ml saline = 0.6 ml of a 1:2 dilution

Fivefold Serum Dilution

Rarely is a five or tenfold dilution required, except perhaps for problem-solving or research purposes. The standard method previously described is for the preparation of a twofold dilution. Note the following facts about these ranges:

The second test tube in the five and tenfold series contains less serum than that in the twofold series.

The dilution in the fourth tube of the twofold dilution (1:8) is between the first and second tubes of the ten fold (1:10) and between the second and third tubes of the fivefold (1:5).

The Densities of Liquids

	Density [g/ml]	Additional information
Acetic acid glacial	1.049	T_b=118°C
Acetic acid 36%	1.049	
Acetone	0.788	T_b=56.5°C
AcOK, 5.0M	1.218	
1.0M	1.046	
AcONa, 3.0M	1.121	
2.5M	1.147	
$AcONH_4$, 10M	1.073	
Acrilamide:BIS, 40%	1.044	
Ammonium hydroxide, NH_4OH, 28%	0.89	
Blocking reag., 10%	1.038	

Contd...

Butanol (n)	0.81	T_b=117-118°C
$CaCl_2$, 1M	1.086	
Chlorophorm	1.48	T_b=61-62°C
Denhardt sol., 50X	1.004	
DMFA	0.945	
DMSO	1.100	
EDTA, 0.5M	1.096	
Ethanol, anhydr. 96% 70%	0.79 0.80 0.868	 T_b=78.5°C
FGRB, 50X	1.104	
Formaldehyde, 37%	1.09	
Formamid	1.133	
Glucose, 10% (w/v) 2M	1.040 1.13	
Glycerol	1.25	
GTGB, 20X	1.094	
HCl, 37% (12.07M)	1.19	
Isoamil alchogol	0.813	T_b=132°C
Isobutanol	0.871	T_b=108°C
Isopropanol	0.785	T_b=82.5°C
KCl, 0.25M	1.044	
KOH, 5.0M	1.217	
LB	1.024	
LiCl, 9M 8M	1.194 1.172	
Mercaptoethanol (b)	1.114	14.192M, Mw 78.13
$MgCl_2$, 1M	1.073	
$MgSO_4$, 1M	1.112	
$MnCl_2$, 1M	1.099	
Nitric acid, $HNO_3$71%	1.42	
NaCl, 5M	1.186	
NaOH, 50%	1.53	
NaOH, 10M	1.329	
$NaxHPO_4$, 1M,pH~7	1.104	
Nonident P-40	1.064	
Octane	0.71	T_b=125.6°C
PEG 6000, 20% 40%	1.041 1.075	

Contd...

PEG/NaCl, (20%/2.5M)	1.145	
PEG/TAE, (15%/1X)	1.032	
Phosphoric acid H_3PO_4, 85%	1.70	
Phenol	1.054	T_b=182°C, T_m=43°C
Sulfuric acid H_2SO_4, 96%	1.84	
Sarcosile, 10% (w/v) 30% (w/v)	1.017 1.027	
SDS, 10%	1.018	
SSC, 20X 30X	1.163 1.245	
SSPE, 20X	1.136	
TAE, 50X	1.088	
TBE, 10X	1.070	
Tris.Cl, 1M, pH 7-8 2M, pH7.9	1.041 1.10	
TritonX-100, 10%	1.016	
TritonX-100, 100%	1.06	
Tween 20, 100%	1.108 (1.101)	
$ZnCl_2$, 1M	1.111	

Buffer Preparation

Preparation of 0.1 M **Sodium and Potassium Acetate buffers**.

Solution A : 11.55ml glacial acetic acid/liter (0.2M).

Solution B : 0.2M Sodium/Potassium Acetate.

Desired pH	Solution A (ml)	Solution B (ml)	5 ml	
			Solution A (ml)	Solution B (ml)
3.6	46.3ml	3.7ml	4.63	0.37
3.8	44.0ml	6.0ml	4.4	0.6
4.0	41.0ml	9.0ml	4.1	0.9
4.2	36.8ml	13.2ml	3.68	1.32
4.4	30.5ml	19.5ml	3.05	1.95
4.6	25.5ml	24.5ml	2.55	2.45
4.8	20.0ml	30.0ml	2	3
5.0	14.8ml	35.2ml	1.48	3.52
5.2	10.5ml	39.5ml	1.05	3.95
5.4	8.8ml	41.2ml	0.88	4.12
5.6	4.8ml	45.2ml	0.48	4.52

Preparation of 1 M Potassium **Phosphate buffer.**

Potassium Phosphate/Monobasic, KH_2PO_4, Mw=136.1g/M; 1M=136.1g/L

Desired pH	1M K_2HP_4	1M KH_2P_4	5 ml	
			1M K_2HP_4	1M KH_2P_4
5.8	8.5ml	91.5ml	0.43	4.58
6.0	13.2ml	86.8ml	0.66	4.34
6.2	19.2ml	80.8ml	0.96	4.04
6.4	27.8ml	72.2ml	1.39	3.61
6.6	38.1ml	61.9ml	1.91	3.1
6.8	49.7ml	50.3ml	2.49	2.52
7.0	61.5ml	38.5ml	3.08	1.93
7.2	71.7ml	28.3ml	3.59	1.42
7.4	80.2ml	19.8ml	4.01	0.99
7.6	86.6ml	13.4ml	4.33	0.67
7.8	90.8ml	9.2ml	4.54	0.46
8.0	94.0ml	6.0ml	4.7	0.3

Preparation of 1 M Sodium Phosphate buffer

NaH_2PO_4 Sodium Phosphate/Monobasic, NaH_2PO_4, Mw = 120.0 g/M;

Na_2HPO_4 Sodium Phosphate/Dibasic, Na_2HPO_4, Mw=142.0g/M;

Desired pH	1M Na_2HP_4	1M NaH_2P_4	5ml	
			1M Na_2HP_4	1M NaH_2P_4
5.8	7.9ml	92.1ml	0.4	4.61
6.0	12.0ml	88.0ml	0.6	4.4
6.2	17.8ml	82.2ml	0.89	4.11
6.4	25.5ml	74.5ml	1.28	3.73
6.6	35.2ml	64.8ml	1.76	3.24
6.8	46.3ml	53.7ml	2.32	2.69
7.0	57.7ml	42.3ml	2.89	2.12
7.2	68.4ml	31.6ml	3.42	1.58
7.4	77.4ml	22.6ml	3.87	1.13
7.6	84.5ml	15.5ml	4.23	0.78
7.8	89.6ml	10.4ml	4.48	0.52
8.0	93.2ml	6.8ml	4.66	0.34

Solution Conversions

1) Percent solution to molarity
   M = (% X 10)/Molecular wt.
2) Percent solution to normality
   N = (% X 10)/Equivalent wt
3) Molarity to Normality
   M = Normality/Valence
4) Normality to Molarity
   N = Molarity X Valence

Unit Conversion

Ratios and proportion can be used to convert among other units of measure; i.e. English to Metric, Metric to English. With enough time and determination it is possible for the student to derive all the necessary conversion factors in a manner similar to that used to convert among the different temperature scales. The information listed below can be used to develop factors for converting among commonly used units of measure.

Length	:	1 meter (m)	=	39.37 inches
		2.54 centimeters (cm)	=	1 inch
		30.48 cm	=	1 foot
		1.61 kilometers (km)	=	1 mile
Mass	:	1 kilogram (kg)	=	2.205 pounds
		453.6 grams (g)	=	1 pound
		28.35 g	=	1 ounce
Volume	:	1 liter (L)	=	1.057 quarts
		29.57 milliliters (ml)	=	1 fluid ounce
		3.78 L	=	1 gallon (U.S.)

In addition to being used to convert among units of measure, factors can be developed to relate or convert among other types of units or to simplify calculations.

Example

Convert amount of nitrogen assayed in a blood test to amount of urea present. This needs to be done since some tests assay for nitrogen not urea. The test assumes that all the nitrogen detected is in the form of urea.

Molecular Weight (MW) of nitrogen = 14

Urea consists of two atoms of nitrogen, one carbon four hydrogen. Thus, the molecular weight of urea is 60.

Factor: 60 parts urea/28 pts N = X pts urea/1 pt N

Parts urea nitrogen x 2.14 = parts urea

Example

Determine how much NaCl is needed to yield a given concentration of Cl^-?

MW Na = 23 MW Cl = 35.5

58.5/35.5 = 1.65 = X parts NaCl/1 pt Cl

To yield 100 mg Cl, 165 mg NaCl is needed.

Converting Among Metric Units

The use of the base 10 system makes converting among different metric units easy. In general, subtract the exponent of the desired unit from the exponent of the given unit to determine the exponent for the conversion factor.

Example

Convert 7.5 milligrams (mg) to nanograms (ng). First convert each unit to the base unit. In this example convert mg and ng to grams.

$$1\ mg = 10^{-3}g$$
$$1\ ng = 10^{-9}g$$

Subtract exponents

$$-3 - (-9) = 6$$
$$7.5mg = (7.5 \times 10^6) = 7.5\times10^6 ng$$

Alternate Method

$$(10^{-3}g/mg) \div (10^{-9}g/ng) = (10^{-3}/10^{-9})ng/mg = 10^6 ng/mg$$
$$7.5mg \times 10^6 ng/mg = 7.5\times10^6 ng$$

Correction Factor

At times it is necessary to perform a test or assay with other than the specified amount of material. In general correction for variation in procedure quantities follows the general rule:

Correction Factor = amount specified/amount used

Example

Need 5 ml of urine for sodium determination; received only 3.5 ml

Take 3 ml and dilute to 5 ml with H_2O

5/3 = 1.67

1.67 x Test result = corrected answer

Test result 40 mEq Na

40 x 1.67 = 66.8 mEq Na in original specimen

Converting Molarity to Percent Solution

MW x M = g/L Therefore M = (g/L)/MW

M = (g/L)/MW = (g/100ml x 10)/MW

M= ($\%^{w/v}$ x 10)/MW

Example

Convert a $30\%^{w/v}$ solution of NaCl (MW = 58.5) to a molar solution.

M = (% x 10)/MW

M = (30 x 10)/58.5

M = 5.13

Therefore 30% solution of NaCl is a 5.13M solution.

Example

Convert a 6M solution of NaOH (MW = 40g) to a $\%^{w/v}$ solution.

Rearrange the formula from above: M x MW = (% x 10)

(M x MW)/10 = %

6x40/10 = 24%

A 6M solution of NaOH is equivalent to a $24\%^{w/v}$ solution.

Converting A Normal Solution to Percent Solution

g/EqW x N = g/L

N = (g/L) ÷ EqW

N = (% x 10) ÷ EqW

Example

Convert a 3N H_2SO_4 (MW = 98, % assay = 96, SG = 1.84) solution to a $\%^{w/v}$ solution.

$$N = (\% \times 10)/EqW$$
$$\% = N \times EqW/10$$
$$\% = 3 \times 49/10$$
$$\% = 14.7\ \%^{w/v}$$

Note *that for this problem the information on % assay and SG was not needed since the original normality concentration accounted for this.*

Conversion of mg/dl to mEq/L

The old unit for reporting out the amount of sodium, potassium or chloride ions in body fluids was mg/dl. The new standard for reporting is mEq/L (milliequivalents/L or "millinormals"/L). Depending on the procedure used to measure these ions in solution, it may be necessary to convert from the old to the new units. It may also be helpful to some physicians to have the values converted to the older, more familiar units of concentration.

mEq/L = mgEqW/L (milligram equivalent weight/L)

mEq/L = (mg/L) ¸ EqW = (mg/dl x 10)/EqW.

Example

Convert 300mg/dl of Cl^- to mEq/L of Cl^-.

$$mEq/L = (300 \times 10)/35.5 = 3000/35.5$$
$$mEq/L = 84.5$$

Therefore 300mg/dl of Cl^- is equivalent to 84.5mEq/L.

Example

Convert 150 mEq/L NaCl to mg/dl of NaCl.

$$mEq/L = (mg/dl \times 10)/EqW$$
$$mg/dl = (mEq/L \times Eq)/10$$
$$mg/dl = (150 \times 58)/10$$
$$870\ mg/dl\ NaCl$$

Conversion of Complex Metric Units

Conversion among complex metric units of concentration is an extension of conversion among metric units of measure. The complex unit of concentration, for example mg/dl is broken into its component parts, mg and dl. Each of these is then converted to the corresponding desired unit of measure, for example Tg and ml.

Convert mg/dl to Tg/ml

$$Tg = 10^{-3}mg,\ mg \times 10^{3} = Tg,\ ml = 10^{-2}dl,\ dl \times 10^{2} = ml$$
$$mg/dl \times 10^{3}/10^{2} = mg/dl \times 10^{1} = Tg/ml$$

Example

Convert 21g/L to mg/Tl.

$$g \times 10^{3} = mg\ L \times 10^{6} = Tl$$
$$21 \times 10^{3}/10^{6} = 21 \times 10^{-3}$$
$$21g/L = 2.1 \times 10^{-2} = 0.021mg/Tl$$

Example

Convert 316 Tg/ml to mg/dl.

$$Tg \times 10^{-3} = mg$$
$$ml \times 10^{-2} = dl$$
$$316Tg/ml \times 10^{-3}/10^{-2} = 316 \times 10^{-1} = 31.6mg/dl$$

1. If you have 20 grams of a substance in 100ml then how many grams in 20 ml?

Known ratio 20g/100ml

Unknown ratio Xg/20ml

(1) 20g/100ml = Xg/20ml
(2) (100ml) x (Xg) = 20gx20ml
(3) Xg = (20gx20ml)/100ml
(4) Xg = (400g/ml)/100ml
(5) X = 4g

Therefore 20g/100ml = 4g/20ml

2. How would you set up a 1:1000 dilution from undiluted sample?

Answer : Format : 1 volume undiluted serum + 999 volumes of diluent = 1000 volumes or 1:1000 dilution

Obviously, if one actually had to prepare this dilution, 1000 ml of a 1:1000 dilution would be too great a volume to be practical. It can be decreased while maintaining the 1:1000 dilution as follows:

1 + 999 = 1000 or 1:1000

X 0.1 ml

0.1 ml + 99.9 = ml of 1:1000 (100 ml is still too large a quantity for practicality)

Or

1 + 999

X 0.01 ml

0.01 ml sample + 9.99 ml saline = 10 ml of 1:1000

This is a practical volume of serum with which to work. Plan to measure small quantities such as 0.01 ml of undiluted serum with a 0.01 ml serologic pipet to insure accuracy.

Note *When the decimal point is moved to the left on the left side of the formula (0.1 to 0.01), it must also be moved to the left on the right side (99.0 to 9.99) in order to maintain the 1:1000 dilution. The decimal must be moved the same distance on both sides.*

3. Prepare a 1:80 dilution from an undiluted sample

Answer : Format 1 + 79 = 1:80

1 volume of undiluted sample + 79 volumes of saline = 80 volumes of a 1:80 dilution. Since 80 ml is too great a volume with which to work, use 0.1 ml undiluted sample + 7.9 ml saline = 8.0 ml of a 1:80 dilution.

4. Prepare a 1:8 dilution directly from 1:1 or undiluted sample

Answer : 1 volume of serum + 7 volumes of diluent = 8 volumes

Or

1 volume of serum is diluted to 8 volumes *Or* the serum is diluted 1:8

The above example indicates how a 1:8 dilution is prepared. Now, how can a 1:8 dilution be made if there is only 0.1 ml of undiluted serum with which to make that dilution?

First : multiply each side of the basic formula (1 and 7) by 0.1 ml serum.

1+ 7 = 8
X 0.1 ml
0.1 ml serum + 0.7 ml diluent = 0.8 ml total volume

Thus, by adding 0.1 ml of undiluted serum to 0.7 ml saline, 0.8 ml of a 1:8 dilution has been prepared.

5. Prepare a dilution of 1:64 using 0.5 ml of serum

Answer : 1 volume of serum + 63 volumes of diluent = 64 volumes or 1:64

Multiply each side of the formula (1 and 63) by 0.5 ml serum

1 + 63 = 64 or 1:64
X 0.5 ml
0.5 ml serum + 31.5 ml diluent = 32 ml of a 1:64 dilution

Note that one can also prepare a dilution of 1:512 on paper similarly:

1 volume of serum + 511 volumes of saline = 512 volumes of 1:512 dilution

Or if 0.1 ml of serum is used,

0.1 ml serum + 51.1 ml saline = 51.2 ml of a 1:512 dilution.

6. Prepare a twofold dilution - any volume

When a twofold dilution was described above in STEP 7, the final volume in each tube was 1.0 ml; however, any final volume can be prepared. If one used 0.2 ml of each serum dilution instead of 0.1 ml, the final volume in each tube would be 0.2 ml. This is how a twofold dilution with a final volume of 0.2 ml would be prepared:

1 + 1 = 2 (1:2) twofold
X 0.2 ml
0.2 ml serum + 0.2 ml saline = 0.4 ml

If each side is multiplied (1 and 1) by 0.2 ml, a result of 0.2 ml serum + 0.2 ml of diluent is obtained, which equals 0.4 ml. the result is 0.4 ml of a twofold dilution. Now if one wishes to set up a twofold dilution using 0.2 ml of each serum dilution, after mixing, 0.2 ml must be removed, leaving 0.2 ml of the dilution in the tube. In this way, all test tubes have a final volume of 0.2 ml. Note that to set up a twofold dilution using 0.2 ml of serum in tube 2, a total volume of 0.4 ml undiluted serum is needed (0.2 ml for tube 1 and 0.2 ml for tube 2). The less serum required, the better, since remaining serum may be used for other tests, if requested.

7. Prepare a twofold dilution-minimum serum volume

To economize without adversely affecting accuracy, perform steps 1-7 by using 0.1 ml of serum in tubes 1 and 2, stipulating the format on paper, and diagrammatically prepare the twofold dilution to determine how one would actually perform this procedure using this quantity of serum.

1 + 1 = 2 (1:2)

X 0.1 ml

0.1 ml of serum + 0.1 ml saline = 0.2 ml of a 1:2 dilution

8. You have a frozen 1:10 dilution of serum and you need to prepare from this a 1:35 dilution for an antibody test. Show how you would do this.

Answer : Step 1. Find the working dilution to be prepared:

1:35 = 1:3.5

1:10

A dilution of 1:35 is 3.5 times more dilute than the 1:10. Note that a 1:3.5 of a 1:1 gives a final dilution of a 1:35.

Step 2. Write the format: 1:3.5 = 1 + 2.5 = 3.5 ml of a 1:3.5 dilution

Step 3. Write the directions: Add 1 ml of a 1:10 dilution to 2.5 ml saline, to prepare a 1:3.5 dilution.

9. You have just begun your Senior Honor's Thesis. Your advisor asks you to make a series of stock solutions. She also explains that you may use a pH meter to adjust the pH. Calculate how much of the solid reagent you would add to make the following:

A. 500 ml 1 M Tris, pH 8.0 (MW=121.4 g/mole)

B. 1.0 L 5 M NaCl (MW= 58.44 g/mole)

C. 10 ml 100 mg/ml ampicillin (MW=371.4 g/mole)

D. 500 ml 1 M $MgCl_2$

Note *$MgCl_2$ is sold by the chemical company as co-crystallized with H_2O. Thus, the MW of MgCl2 $6H_20$ is 203.30 g/mole*

a. $\frac{1 \text{ mole of Tris}}{L} \times \frac{121.4 \text{ g} \times 0.5 \text{ L}}{\text{mole}} = 60.7 \text{ g of Tris}$

Dissolve 60.7 g of Tris in about 300 ml of H_2O. Once dissolved adjust the pH to 8.0, and then make up to 500 ml mark with H_2O in a graduated cylinder.

b. $\frac{5 \text{ moles of NaCl}}{L} \times \frac{58.44 \text{ g} \times 1 \text{ L}}{\text{mole}} = 292.2 \text{ g of NaCl}$

Dissolve 292.2 g of NaCl in about 0.7 L of H_2O. Once dissolved, dilute to 1 L with H_2O in a graduated cylinder.

c. $\frac{100 \text{ mg}}{\text{ml}}$ of ampicillin x 10 ml = 1 g of ampicillin

Dissolve 1 g of ampicillin in about 5 ml of H_2O. Once dissolved, dilute to 10 ml with H_2O in a graduated cylinder.

d. $\frac{1 \text{ mole of } MgCl_2}{L} \times \frac{203.3 \text{ g} \times 0.5 \text{ L}}{\text{mole}} = 101.65 \text{ g of } MgCl_2$

Dissolve 101.65 g of $MgCl_2$ in about 250 ml of H_2O. Once dissolved, dilute to 500 ml with H_2O in a graduated cylinder.

10. For your first experiment, you need to make a solution that is commonly called TE and stands for Tris-EDTA. It is comprised of 10 mM Tris, 1 mM EDTA, pH 8.0. Calculate and describe how you would make 100 ml of TE using the stock solutions (stock EDTA: 0.5 M EDTA pH 8.0) or that you have already made above in

For dilutions, use the formula $C_1V_1=C_2V_2$.

This solution has 2 components:

Tris-dilute 1M to 10 mM $C_1V_1=C_2V_2$

1000 mM x V_1 = 10 mM x 100 ml

$$V_1 = \frac{10 \text{ mM} \times 100 \text{ ml}}{1000 \text{ mM}} = 1 \text{ ml of } 1 \text{ M Tris}$$

EDTA-dilute 0.5 M to 1 mM $C_1V_1=C_2V_2$

500 mM x V_1 = 1 mM x 100 ml

$$V_1 = \frac{1 \text{ mM} \times 100 \text{ ml}}{500 \text{ mM}}$$

= 0.2 ml or 200 mL of 0.5 M

EDTA

Using a 100 ml graduated cylinder, add 1 ml of 1M Tris and 200 mL of 0.5 M EDTA. Add water to dilute to a final volume of 100 ml.

11. Your advisor asks you to make two concentrated stock solutions for the whole laboratory to use. She explains that it is common to make a concentrated solution to dilute for use. Thus, a 10x solution refers to one that is 10-fold concentrated. Calculate how you would make 1.0 L of each of the following:

A. 10x TBE (0.9 M Tris, 0.9 M Boric Acid, 20 mM EDTA)
   Tris MW=121.4 g/mole
   Boric Acid MW=61.84 g/mole
   EDTA MW=292.2 g/mole

B. 50x TAE (2 M Tris, 2 M Acetic acid, 0.5 M EDTA)
   Tris MW=121.4 g/mole
   Concentrated Acetic acid (glacial) comes as a solution and is 17.4 M
   EDTA MW=292.2 g/mole

1. A

Tris

$$\frac{0.9 \text{ moles of Tris}}{\text{L}} \times \frac{121.4 \text{ g} \times 1.0 \text{ L}}{\text{mole}} = 109.3 \text{ g of Tris}$$

Boric Acid

$$\frac{0.9 \text{ moles of Boric Acid}}{\text{L}} \times \frac{61.84 \text{ g} \times 1.0 \text{ L}}{\text{mole}} = 55.6 \text{ g of Boric Acid}$$

EDTA

$$\frac{20 \text{ mM of EDTA}}{\text{L}} \times \frac{292.2 \text{ g} \times 1.0 \text{ L}}{\text{mole}} \times \frac{1 \text{ mole}}{1000 \text{ mmoles}} = 5.84 \text{ g of EDTA}$$

Dissolve 109.3 g of Tris, 55.6 g of Boric Acid and 5.84 g of EDTA in about 0.5 L of H_2O. Once dissolved, dilute the stock solution to 1.0 L with H_2O in a graduated cylinder to form 10x TBE.

B

Tris

$$\frac{2 \text{ moles of Tris}}{\text{L}} \times \frac{121.4 \text{ g} \times 1.0 \text{ L}}{\text{mole}} = 242.8 \text{ g of Tris}$$

EDTA

$$\frac{0.5 \text{ moles of EDTA}}{\text{L}} \times \frac{292.2 \text{ g} \times 1.0 \text{ L}}{\text{mole}} = 146.1 \text{ g of EDTA}$$

Acetic Acid 17.4M to 2M $C_1V_1 = C_2V_2$

$17.4 \text{ M} \times V_1 = 2 \text{ M} \times 1.0 \text{ L}$

$$V_1 = \frac{2 \text{ M} \times 1.0 \text{ L}}{17.4 \text{ M}}$$

= 0.115 L or 115 ml of 17.4 M Acetic Acid

Dissolve 242.8 g of Tris, 146.1 g of EDTA and also dilute 115 ml of concentrated acetic acid in about 0.5 L of H_2O. Once the solids are dissolved, dilute the stock solution to 1.0 L with H_2O in a graduated cylinder to make 50x TAE.

12. Our polyacrylamide gel electrophoresis experiment requires 1.0 L of 0.5x TBE. Calculate and describe how you would make this solution using your stock solutions.

TBE-dilute 10x to 0.5x $C_1V_1 = C_2V_2$

$(10x) \times V_1 = (0.5x) \times 1.0 \text{ L}$

$$V_1 = \frac{(0.5x) \times 1.0\ L}{(10x)}$$

= 0.05 L or 50 ml of 10x TBE

Dilute 50 ml of 10x TBE to 1.0 L with H_2O in a graduated cylinder to make 0.5x TBE. (Also, you could say 50 ml of 10x TBE plus 950 ml of H_2O.)

13. Your agarose gel electrophoresis experiment requires 500 ml of 1x TAE. Calculate and describe how you would make this solution using your stock solutions.

2. TAE-dilute 50x to 1x $C_1V_1 = C_2V_2$

$(50x) \times V_1 = (1x) \times 500$ ml

$$V_1 = \frac{(1x) \times 500\ ml}{(50x)}$$

= 10 ml of 50x TAE

Dilute 10 ml of 50x TAE to 500 ml with H_2O in a graduated cylinder to make 1x TAE. (Also, you could say 10 ml of 50x TAE and 490 ml of H_2O.)

14. Your experiment requires 100 ml 20 mM Tris, pH 8.0 which also contains 0.5 mM EDTA.

Calculate and describe how you would make this solution using a stock of 1 M Tris, pH 8.0 and 0.5M EDTA, pH 8.0.

1. 20 mM Tris from 1M:

$C_1V_1 = C_2V_2$ 1 M=1000 mM

$(1000\ mM) \times V_1 = (20\ mM) \times 100$ ml

$$V_1 = \frac{(20\ mM) \times 100\ mL}{(1000 mM)}$$

= 2.0 ml of 1 M Tris

0.5 mM EDTA from 0.5 M

$C_1V_1 = C_2V_2$ 0.5 M=500 mM

$(500 \text{ mM}) \times V_1 = (0.5 \text{ mM}) \times 100 \text{ ml}$

$$V_1 = \frac{(0.5 \text{ mM}) \times 100 \text{ mL}}{(500 \text{ mM})}$$

= 0.1 ml or 100 ml of 0.5 M EDTA

Add 2.0 ml of 1 M Tris and 0.1 ml of 0.5 M EDTA in a graduated cylinder and fill it up with H_2O to 100 ml.

15. The enzyme you need is EcoRI and it has a stock concentration of 1 U/ml. Your DNA concentration is 5mg/ml. You need to set up a 50 ml reaction which incorporates 1 mg DNA, 1x buffer, 1x BSA, and 0.5 U EcoRI. Calculate and describe how much of each stock solution, DNA, enzyme and water you would add to make up your restriction digest.

1. 50 ml Reaction: Reaction Mixture:

1 mg DNA	0.2 ml 5 mg/ml DNA
1x Buffer	5.0 ml 10x Buffer
1x BSA	0.5 ml 100x BSA
0.5 U EcoR I (U=units)	0.5 ml 1 U/ml EcoR I
	43.8 ml H_2O
	50.0 ml total volume

DNA

$$1 \text{ mg X } \frac{1 \text{ µl}}{5 \text{ mg}} = 0.2 \text{ ml of } 5 \text{ mg/ml DNA}$$

Buffer

$C_1V_1 = C_2V_2$

$(10x) \times V_1 = (1x) \times 50 \text{ ml}$

$$V_1 = \frac{(1x) \times 50 \text{ ml}}{(10x)}$$

= 5 µl of 10x Buffer

BSA

$C_1V_1 = C_2V_2$

$(100x) \times V_1 = (1x) \times 50$ ml

$$V_1 = \frac{(1x) \times 50 \text{ ml}}{(100x)}$$

= 0.5 ml of 100x BSA

EcoR I Enzyme

$$0.5 \text{ U} \times \frac{1 \text{ µl}}{1 \text{ U}} = 0.5 \text{ ml of } 0.5 \text{ U/ml } EcoR \text{ I}$$

16. You need to make 100 ml of a 1% solution of agarose in 1x TAE. You remember that you made up a 50x TAE stock solution about a month ago. Calculate and describe how you would make your agarose gel.

2. 100 ml of 1% Agarose, 1x TAE

Agarose 1% = 1 g/100 ml = 1 g of Agarose

TAE

$C_1V_1 = C_2V_2$

$(50x) \times V_1 = (1x) \times 100$ ml

$$V_1 = \frac{(1x) \times 100\text{ml}}{(50x)}$$

= 2 ml of 50x TAE

Combine 2.0 ml of 50x TAE and 1 g of Agarose. Once dissolved, dilute to 100 ml with H_2O.

17. Given a 1:3 dilution, prepare a 1:60 dilution.

Step 1. Working dilution: $\frac{1:60}{1:3} = 1:20$

A dilution of 1:60 is 20 times more dilute than the 1:3. Note that a 1:20 of a 1:3 gives a final dilution of 1:60.

Step 2. Format: 1:20 = 1 + 19.0 = 20 ml of a 1:20

Step 3. Directions

Add 1 ml of a 1:3 to 19 ml of saline, to prepare a 1:60 dilution or 0.1 ml of a 1:3 to 1:9 ml of saline = 2.0 ml final volume.

18. Given a 1:60 dilution, prepare a 1:100 dilution.

Step 1. 1:100
-------- = 1:1.66
1:60

Step 2. 1:1.66 = 1 ml + 0.66 ml of diluent = 1.66 ml

Step 3. Add 1 ml of the 1:60 dilution to 0.66 ml of saline to prepare a 1:100 dilution.

CHAPTER 6

Basic Principles in Preparation of Reagents Required in Microbial Biotechnology

DNA Migration in Agarose and Polyacrylamide Gels

Recommended Gel Percentages for Separation of Linear DNA

Agarose gel, %	Range of separation, bp	Polyacrylamide gel, %	Range of separation, bp
0.5	1,000-30,000	3.5	100-1,000
0.7	800-12,000	5.0	80-500
1.0	500-10,000	8.0	60-400
1.2	400-7,000	12.0	40-200
1.4	200-4,000	20.0	5-100
2.0	50-2,000		

Migration Rates of the Marker Dyes through Polyacrylamide Gels (1)

Polyacrylamide gel, %	Bromophenol blue*	Xylene cyanol FF*
Non-denaturing gels		
3.5	100 bp	460 bp
5.0	65 bp	260 bp

Contd...

8.0	45 bp	160 bp
12.0	20 bp	70 bp
15.0	15 bp	60 bp
20.0	12 bp	45 bp
Denaturing gels		
5.0	35 bases	130 bases
6.0	29 bases	106 bases
8.0	26 bases	76 bases
10.0	12 bases	55 bases
20.0	8 bases	28 bases

DNA Size Migration with Sample Loading Dyes

Agarose concentration, %	Xylene cyanol FF	Bromophenol blue	Orange G
0.7-1.7	~4000bp	~300bp	~50bp
2.5-3.0	~800bp	~100bp	~30bp

Common Conversions of Nucleic Acids

Molar Conversions

1 µg of 1000 bp DNA	= 1.52 pmol
1 µg of pUC18/19 DNA (2686 bp)	= 0.57 pmol
1 µg of pBR322 DNA (4361 bp)	= 0.35 pmol
1 µg of SV40 DNA (5243 bp)	= 0.29 pmol
1 µg of PhiX174 DNA (5386 bp)	= 0.28 pmol
1 µg of M13mp18/19 DNA (7250 bp)	= 0.21 pmol
1 µg of lambda phage DNA (48502 bp)	= 0.03 pmol
1 pmol of 1000 bp DNA	= 0.66 µg
1 pmol of pUC18/19 DNA (2686 bp)	= 1.77 µg
1 pmol of pBR322 DNA (4361 bp)	= 2.88 µg
1 pmol of SV40 DNA (5243 bp)	= 3.46 µg
1 pmol of PhiX174 DNA (5386 bp)	= 3.54 µg
1 pmol of M13mp18/19 DNA (7250 bp)	= 4.78 µg
1 pmol of lambda phage DNA (48502 bp)	= 32.01 µg

Spectrophotometric Conversions

1 A_{260} of dsDNA = 50 µg/ml	=	0.15 mM (in nucleotides)
1 A_{260} of ssDNA = 33 µg/ml	=	0.1 mM (in nucleotides)
1 A_{260} of ssRNA = 40 µg/ml	=	0.12 mM (in nucleotides)
1 mM (in nucleotides) of dsDNA	=	6.7 A_{260} units
1 mM (in nucleotides) of ssDNA	=	10.0 A_{260} units
1 mM (in nucleotides) of ssRNA	=	8.3 A_{260} units

The average MW of a deoxyribonucleotide base = 333 Daltons The average MW of a ribonucleotide base = 340 Daltons.

Estimation of Ends (3' or 5') Concentration

Circular DNA

pmol ends = pmol DNA x number of cuts x 2

Linear DNA

pmol ends = pmol DNA x (number of cuts x 2 + 2)

1 µg of 1000 bp DNA	=	3.04 pmol ends
1 µg of linear pUC18/19 DNA	=	1.14 pmol ends
1 µg of linear pBR322 DNA	=	0.7 pmol ends
1 µg of linear SV40 DNA	=	0.58 pmol ends
1 µg of linear PhiX174 DNA	=	0.56 pmol ends
1 µg of linear M13mp18/19 DNA	=	0.42 pmol ends
1 µg of lambda phage DNA	=	0.06 pmol ends

DNA/Protein Conversions

1 kb of DNA = 333 amino acids approx. is 3.7 x 10E4 Da

10 kDa protein approx. is 270 bp DNA

30 kDa protein approx. is 810 bp DNA

50 kDa protein approx. is 1.32 kb DNA

100 kDa protein approx. is 2.7 kb DNA

OD_{260} Units of Nucleic Acid to Concentration

1 OD_{260} Unit = 50μg/ml for dsDNA

1 OD_{260} Unit = 40μg/ml ssRNA

1 OD_{260} Unit = 35μg/ml ssDNA

1 OD_{260} Unit = 20μg/ml for single-stranded oligo

Protein Conversions

Protein Molar Conversion

Micrograms Protein

$$= \text{Protein Size (kDa)} \times \text{Pichomoles protein} \times \frac{10^9\ \mu g}{g} \times \frac{\text{mole}}{10^{12}\ \text{pmol}}$$

$$\text{Picomoles Protein} = \frac{\mu g\ \text{Protein}}{\text{Protein Size (kDa)}} \times \frac{kg}{10^9\ \mu g} \times \frac{10^{12}\ \text{pmol}}{\text{mole}}$$

Formulas for DNA Molar Conversions

For dsDNA

To convert pmol to μg: pmol × N × 660 pg/pmol × 1 μg/10^6pg = μg

To convert μg to pmol μg × 10^6pg/1 μg × pmol/660pg × 1/N = pmol

For ssDNA

To convert pmol to μg pmol × N × 330 pg/pmol × 1μg/10^6pg = μg

To convert μg to pmol μg × 10^6pg/1 μg × pmol/330pg × 1/N = pmol

Protein Molar Conversions

$$\text{Protein Size (kDa)} = \frac{\mu g\ \text{Protein}}{\text{Picomoles Protein}} \times \frac{kg}{10^9\ \mu g} \times \frac{10^{12}\ \text{pmol}}{\text{mole}}$$

100 pmol of 100 kDa protein = 10μg

100 pmol of 50 kDa protein x = 5 μg

100 pmol of 10 kDa protein x = 1 μg

100 pmol of 1 kDa protein = 100ng

Protein/DNA Conversions

1kbDNA	= 333 amino acids = 37 kDa protein
270b DNA	= 10 kDa protein
810 bDNA	= 30 kDa protein
1.35 kbDNA	= 50 kDa protein
2.7 kb DNA	= 100 kDa protein
average MW of an amino acid	= 110 daltons

Dalton (Da) is an alternate name for the atomic mass unit, and kilodalton (kDa) is 1,000 daltons. Thus a protein with a mass of 64 kDa has a molecular weight of 64,000 grams per mole

Agarose Gel (%): Resolution of Linear DNA

Recommended % Agarose	**Optimum Resolution for Linear DNA (Size of fragments in nucleotides;bp)**
0.5	1,000-30,000
0.7	800-12,000
1.0	500-10,000
1.2	400-7,000
1.5	200-3,000
2.0	50-2,000

Polyacrylamide Gel (%): Resolution of Protein

Recommended % Acryl amide	**Protein Size Range**
8	40-200 kDa
10	21-100 kDa
12	10-40 kDa

Length/M.W. of Common Nucleic Acids.

Nucleic Acid	**Number of Nucleotides**	**Molecular Weight**
lambda DNA	48,502(dsDNA)	3.2×10^7
pBR322DNA	4,361(dsDNA)	2.8×10^6
28S rRNA	4,800	1.6×10^6
23S rRNA(*E.coli*)	2,900	1.0×10^6
18S rRNA	1,900	6.5×10^5
16S rRNA(*E.coli*)	1,500	5.1×10^5
5S rRNA(E.coli)	120	4.1×10^4
tRNA(E.coli)	75	2.5×10^4

*Molecular weights based on actual sequence.

Standards

1. Average MW of dsDNA base pair 600.
2. Average MW of ssDNA base 330.
3. Average MW of RNA base 340.

Concentrations of Acids and Bases

Substance	Formula	MW	Moles /liter	Grams /li ter	%by weight	Specific gravity	ml/liter to prepare 1 M solution
Aceticacid, glacial	CH_3COOH	60.05	17.4	1045	99.5	1.05	57.5
Acetic acid	CH_3COOH	60.05	6.27	376	36	1.045	159.5
Formic acid	HCOOH	46.02	23.4	1080	90	1.20	42.7
Hydrochloric acid	HCl	36.5	11.6	424	36	1.18	86.2
Nitric acid	HNO_3	63.02	15.99	1008	71	1.42	62.5
Perchloric acid	$HClO_4$	100.5	11.65	1172	70	1.67	85.8
Phosphoric acid	H_3PO_4	97.9	14.8	1445	85	1.70	67.7
Sulfuric acid	H_2SO_4	98.1	18.0	1766	96	1.84	55.6
Ammonium hydroxide	NH_4OH	35.0	14.8	251	28	0.898	67.6
Potassium hydroxide	KOH	56.1	13.5	757	50	1.52	74.1
Sodium hydroxide	NaOH	40.0	19.1	763	50	1.53	52.4

Common Conversions of Oligonucleotides

Molecular Weight MW = 333 x N

Da Concentration of Oligonucleotides

C (μM or pmol/μl) = A_{260} / (0.01 x N)

C (ng/ml) = (A_{260} x MW) / (0.01 x N)

MW - molecular weight,

A_{260} - absorbance at 260nm

N - number of bases

Melting Temperature of Duplex DNA and Oligonucleotides

For Duplex Oligonucleotide shorter than 25 bp, "The Wallace Rule"

$$T_m \text{ (in °C)} = 2(A+T) + 4(C+G)\text{, where}$$

(A+T) - the sum of the A and T residues in the oligonucleotide,

(C+G) - the sum of G and C residues in the oligonucleotide.

For Duplex DNA, <100 bp long

T_m *(in °C)* = *81.5°C+16.6(log$_{10}$[Na$^+$])*+0.41(%[G+C])-675/n-1.0m

where

n - number of bases in the oligonucleotide

m - the percentage of base-pair mismatches

Conversion Formula

C = A / e x 10E3, *where*

C - mM concentration of compounds

A - observed absorbance at lambda$_{max}$ (nm)

e - molar absorption coefficient (M^{-1} x cm^{-1})

Commonly Used Media, Stock Solutions and Buffers

Growth Media	per liter	final 1X concentration
LB Medium (Adjust pH to 7.0)		
Tryptone	10g	1.0% (w/v)
Yeast extract	5g	0.5% (w/v)
NaCl	10g	1.0% (w/v)
H_2O	to 1 liter	
Low Salt LB Medium		
Tryptone	10g	1.0% (w/v)
Yeast extract	5g	0.5% (w/v)
NaCl	5g	0.5% (w/v)
H_2O	to 1 liter	
Adjust pH to 7.0		
Terrific Broth Medium		
Tryptone	12g	1.2% (w/v)
Yeast extract	24g	2.4% (w/v)
Glycerol	4 ml	0.4% (w/v)
Add H_2O	to 900 ml	

Autoclave, cool to 60°C or less before adding 100 ml of filter sterilized 10X TB phosphate (0.17 M KH_2PO_4, 0.72 M K_2HPO_4).

SOB Medium		
Tryptone	20g	2.0% (w/v)
Yeast extract	5g	0.5% (w/v)
NaCl	0.5g	0.05% (w/v)
250 mM KCl	10 ml	2.5 mM
H_2O	to 900 ml	

Adjust pH to 7.0 and add H_2O to 990 ml.

Autoclave, cool to room temperature and add 10 ml of sterile solution of 1M $MgCl_2$ before use.

M9 Minimal Medium		
5X M9 salts	200 ml	10 mM
Sterile H_2O	to 1 liter	
1M $MgSO_4$	2 ml	1 mM
20% glucose	20 ml	2.0% (w/v)
1M $CaCl_2$	0.1 ml	0.1 mM

5X M9 Salts		
$Na_2HPO_4x7H_2O$	64g	47.8 mM
KH_2PO_4	15g	22mM
NaCl	2.5g	8.6 mM
NH_4Cl	5g	18.7 mM

Additives:	
Antibiotics (if required)	
Ampicillin	to 50 μg/ml
Chloramphenicol	to 20 μg/ml
Kanamycin	to 30 μg/ml
Tetracycline	to 12 μg/ml

Additives:		
Media containing agar or agarose		
Agar (for plates)	15g	1.5% (w/v)
Agar (for top agar)	7g	0.7% (w/v)
Agarose (for plates)	15g	1.5% (w/v)
Agarose (for top agarose)	7g	0.7% (w/v)

Stock Solutions

10 M Ammonium Acetate

Ammonium acetate	385.4g
H_2O	to 500 ml

100X Denhardt Solution

Ficoll 400	10g
Polyvinylpyrrolidone	10g
Bovine serum albumin	10g
H_2O	to 500 ml

Filter sterilize and store at -20°C in 25 ml aliquots

10 mg/ml Ethidium Bromide

Ethidium bromide	0.2g
H_2O	to 20 ml

Mix well and store at 4°C in dark.
CAUTION: Ethidium bromide is a mutagen and must be handled carefully.

1 M KCl

KCl	74.6g
H_2O	to 1 liter

1 M $MgSO_4$

$MgSO_4x7H_2O$	24.6g
H_2O	to 100 ml

10 M NaOH

NaOH	400g
H_2O	to 1 liter

3 M Sodium Acetate (pH 5.2 and 7.0)

Sodium acetate. $3H_2O$	408.1g
H_2O	to 800 ml

Adjust the pH to 5.2 with glacial acetic acid or adjust the pH to 7.0 with dilute acetic acid.

H_2O	to 1 liter

1 M $CaCl_2$

$CaCl_2x2H_2O$	147g
H_2O	to 1 liter

0.5 M EDTA (ethylenediamine tetraacetic acid) (pH 8.0)

$Na_2EDTAx2H_2O$	186.1g
H_2O	to 700 ml

Adjust pH to 8.0 with 10 M NaOH (~50 ml)

H_2O	to 1 liter

1 M Dithiothreitol (DTT)

DTT	15.45g
H_2O	to 100 ml

Store at -20°C

1 M $MgCl_2$

$MgCl_2x6H_2O$	20.3g
H_2O	to 100 ml

5 M NaCl

NaCl	292g
H_2O	to 1 liter

1 M Tris-HCl [tris(hydroxymethyl)aminomethane]

Tris base	121g
H_2O	to 800 ml

Adjust to desired pH with concentrated HCl. Mix and add H_2O to 1 liter

Buffers

Buffers	per liter	final 1X concentration:
10X Stock Phosphate-buffered Saline (PBS)		
NaCl	80g	137 mM
KCl	2g	2.7 mM
Na_2HPO_4	14.4g	100 mM

KH_2PO_4	2.4g	2 mM
H_2O	to 800 ml	
Adjust pH 7.4 with HCl		
H_2O	to 1 liter	

20X SSC

NaCl	175.3g	150 mM
$Na_3citratexH_2O$	88.2g	15 mM
H_2O	to 800 ml	
Adjust pH to 7.0 with 1 M HCl		
H_2O	to 1 liter	

20X SSPE

NaCl	175.3g	150 mM
$NaH_2PO_4xH_2O$	27.6g	10 mM
Na_2EDTA	7.4g	10 mM
H_2O	to 800 ml	
Adjust pH to 7.4 with 10 M NaOH		
H_2O	to 1 liter	

5X Tris-glycine Electrophoresis Buffer

Tris base	15.1g	25 mM
Glycine	72.0g	192 mM
H_2O	to 1 liter	
The pH of diluted solution is 8.3		

5X Tris-glycine-SDS Electrophoresis Buffer

Tris base	15.1g	25 mM
Glycine	72.0g	192 mM
SDS	5.0g	0.1% (w/v)
H_2O	to 1 liter	
The pH of diluted solution is 8.3		

5X Tris-tricine-SDS Electrophoresis Buffer

Tris base	121.1g	0.1M
Tricine	179.2g	0.1M
SDS	10.0g	0.1% (w/v)
H_2O	to 1 liter	
The pH of diluted solution is 8.3.		

50X TAE (Tris-acetate-EDTA) Electrophoresis Buffer		
Tris base	242g	40 mM
Glacial acetic acid	57.1 ml	20 mM
0.5M EDTA (pH 8.0)	100 ml	1 mM
H_2O	to 1 liter	
The pH of diluted solution is ~8.5		

10X TBE (Tris-borate-EDTA) Electrophoresis Buffer		
Tris base	108g	90 mM
Boric acid	55g	90 mM
0.5M EDTA(pH 8.0)	40 ml	1 mM
H_2O	to 1 liter	

10X TPE (Tris-phosphate-EDTA) Electrophoresis Buffer		
Tris base	108g	90 mM
Phosphoric acid (85%)	15.5 ml	23 mM
0.5M EDTA(pH 8.0)	40 ml	1 mM
H_2O	to 1 liter	

TE (Tris-EDTA) Buffer, pH 7.4, 7.6 or 8.0		
1 M Tris, pH 7.4, 7.6, 8.0	10 ml	10 mM
0.5M EDTA(pH 8.0)	2 ml	1 mM
H_2O	to 1 liter	

TE

1x; pH7.4; 7.6; 8.0; (store at 4°C).

	Conc.	Stock	5ml	10ml	50ml	400ml	5ml
Tris Cl	10mM	1M	50µl	100µl	500µl	4ml	0.05ml
EDTA	1mM	0.5M	10µl	20µl	100µl	800µl	0.01ml
H_2O		mQ	4.94ml	9.88ml	49.4ml	395ml	4.94ml

10x

	Conc.	Stock	1ml	5ml	5ml
Tris Cl	0.1M	1M	0.1ml	0.5ml	0.5ml
EDTA	10mM	0.5M	20µl	0.1ml	0.1ml
H_2O	mQ		880µl	4.4ml	4.4ml

Genome Comparisons

Virus	Length, nt	Approx. MW, Da
Bacteriophage PhiX174	5,380	3.5 x 10E6
Bacteriophage Lambda	48,502	3.1 x 10E7
Human Immunodeficiency Virus 1	9,181	3.1 x 10E6
Rous Sarcoma Virus	9,392	3.2 x 10E6
SARS Coronavirus	29,751	1.0 x 10E7
Simian Virus 40 (SV 40)	5,224	3.4 x 10E6
Vaccinia Virus	191,737	1.2 x 10E8
Variolla (Small pox) Virus	185,578	1.2 x 10E8
Archaebacteria	**Length of DNA, bp**	**Approx. MW, Da**
Halobacterium sp. NRC-1	2.01 x 10E6	1.31 x 10E9
Methanosarcina acetivorans C2A	5.75 x 10E6	3.74 x 10E9
Methanococcus jannaschii DSM2661	1.66 x 10E6	1.08 x 10E9
Methanosarcina mazei Go1	4.10 x 10E6	2.66 x 10E9
Methanobacterium thermoautotrophicum delta H	1.75 x 10E6	1.14 x 10E9
Pyrococcus horikoshii OT3	1.74 x 10E6	1.13 x 10E9
Pyrococcus abysii GE5	1.76 x 10E6	1.14 x 10E9
Pyrococcus furiosus DSM3638	1.91 x 10E6	1.24 x 10E9
Sulfolobus solfataricus P2	2.99 x 10E6	1.94 x 10E9
Thermoplasma acidophilum	1.56 x 10E6	1.01 x 10E9
Bacteria	**Length of DNA, bp**	**Approx. MW, Da**
Agrobacterium tumefaciens C58-Cereon	4.91 x 10E6	3.19 x 10E9
Aquifex aeolicus VF5	1.55 x 10E6	1.00 x 10E9
Bacillus subtilis 168	4.20 x 10E6	2.73 x 10E9
Bacillus halodurans	4.20 x 10E6	2.73 x 10E9
Bordetella pertussis Tohama I NCTC-13251	4.07 x 10E6	2.65 x 10E9
Chlamydophila pneumoniae CWL029	1.23 x 10E6	8.00 x 10E8
Deinococcus radiodurans R1	3.28 x 10E6	2.13 x 10E9
Escherichia coli K12	4.64 x 10E6	3.02 x 10E9
Escherichia coli 0157:H7 EDL933	4.10 x 10E6	2.66 x 10E9
Haemophilus influenzae KW20	1.83 x 10E6	1.19 x 10E9
Helicobacter pylori 26695	1.66 x 10E6	1.08 x 10E9
Lactococcus lactis IL1403	2.36 x 10E6	1.53 x 10E9

Mycobacterium leprae TN	3.27 x 10E6	2.12 x 10E9
Mycobacterium tuberculosis H37Rv	4.45 x 10E6	2.89 x 10E9
Mycoplasma genitalium G037	5.80 x 10E5	3.77 x 10E8
Mycoplasma pneumoniae M129	8.16 x 10E5	5.30 x 10E8
Neisseria meningitidis Z2491	2.18 x 10E6	1.42 x 10E9
Salmonella typhimurium LT2 SGSC1412	4.86 x 10E6	3.16 x 10E9
Staphylococcus aureus MW2	2.82 x 10E6	1.83 x 10E9
Streptomyces coelicolor A3(2)	8.67 x 10E6	5.64 x 10E9
Streptococcus pneumoniae R6	2.04 x 10E6	1.33 x 10E9
Streptococcus pyogenes SF370 (M1)	1.85 x 10E6	1.20 x 10E9
Pseudomonas aeruginosa	6.26 x 10E6	4.07 x 10E9
Vibrio cholerae N16961	4.00 x 10E6	2.60 x 10E9

Eukaryotes	**Length of DNA, bp**	**Approx. MW, Da**
Anopheles gambiae PEST (malaria mosquito)	2.78 x 10E8	1.80 x 10E11
Arabidopsis thaliana (flowering plant)	1.15 x 10E8	7.47 x 10E10
Caenorhabditis briggsae (soil-dwelling nematode)	1.04 x 10E8	6.76 x 10E10
Caenorhabditis elegans (round worm)	1.21 x 10E7	7.86 x 10E9
Ciona intestinalis (ascidian tadpole)	1.16 x 10E8	7.54 x 10E10
Drosophila melanogaster (fruit fly)	1.37 x 10E8	8.90 x 10E10
Guillardia theta (chromophyte algae)	5.51 x 10E5	3.58 x 10E8
Homo sapiens (human)	3,15 x 10E9	2.07 x 10E12
Mus musculus (mouse)	~3,00 x 10E9	~1.95 x 10E12
Neurospora crassa OR74A (filamentous fungus)	4.30 x 10E7	2.80 x 10E10
Saccharomyces cerevisiae S288C (budding yeast)	1.21 x 10E7	7.86 x 10E9
Schizosaccharomyces pombe (fission yeast)	1.40 x 10E7	9.10 x 10E9
Plasmodium falciparum 3D7 (human malaria parasite)	2.29 x 10E7	1.49 x 10E10
Oryza sativa japonica (rice)	4.20 x 10E8	2.73 x 10E11

- Frequently used Stock Solutions in molecular biology practical

10 M Ammonium Acetate

To prepare a 10 M solution in 100 ml, dissolve 77 g of ammonium acetate in 70 ml of H_2O at room temperature. To prepare a 5 M solution in 100 ml, dissolve 38.5 g in 70 ml of H_2O. Adjust the volume to 100 ml with H_2O. Sterilize the solution by passing it through a 0.22µm filter. Store the solution in tightly sealed bottles at 4 °C or at room temperature. Ammonium acetate decomposes in hot H_2O and solutions containing it should not be autoclaved.

Ampicillin

Prepare a stock of 100 mg/ml in water. Sterilize by filtration. Store at -20°C but avoid repeated freeze/thaw cycles. Use at a final concentration of 100 µg/ml.

Cresol Red Loading Dye - 2.5X - for PCR Reactions

1M sucrose, 0.02% cresol. Prepare 1% cresol red in water (0.5 g/ 50 ml). To prepare loading dye, dissolve 17 g sucrose in a total volume of 49 ml of water. Add 1 ml 1% cresol red.

EB Buffer (Qiagen recipe)

10 mM Tris-Cl pH 8.3

EDTA Stock

To prepare 1 liter, 0.5M EDTA pH 8.0: Add 186.1 g of disodium EDTA-$2H_2O$ to 800 ml of H_2O. Stir vigorously on a magnetic stirrer. Adjust the pH to 8.0 with NaOH (approx. 20 g of NaOH pellets). Dispense into aliquots and sterilize by autoclaving. The disodium salt of EDTA will not go into solution until the pH of the solution is adjusted to approx. 8.0 by the addition of NaOH. For tetrasodium EDTA, use 226.1 g of EDTA and adjust pH with HCl.

6x Gel Loading Buffer

0.25% Bromophenol blue

0.25%Xylene cyanol FF

15% Ficoll Type 4000

120 mM EDTA

IPTG

IPTG is isopropylthio-b-D-galactoside. Make a 20% (w/v, 0.8 M) solution of IPTG by dissolving 2 g of IPTG in 8 ml of distilled H_2O.

Adjust the volume of the solution to 10 ml with H_2O and sterilize by passing it through a 0.22µm disposable filter. Dispense the solution into 1 ml aliquots and store them at -20 °C.

LB Medium

To make 1 liter, use 10 g tryptone, 5 g yeast extract, 10 g NaCl. Adjust pH to 7.0. Sterilize by autoclaving.

LB Agar

Dispense 15 g per liter of agar directly into final vessel. Prepare LB medium as above and add to agar.

Note *Agar will not go into solution until it is autoclaved (or boiled). If adding antibiotics, autoclave medium first and allow to cool until warm to the touch, then add the antibiotic. Dispense about 30 ml per plate. Allow plates to dry either at 37°C overnight or 20 minutes in a laminar flow hood (lids removed). Store in original Petri plate bags, inverted, at 4°C for up to 2 weeks.*

NaCl

To prepare 1 liter of a 5 M solution: Dissolve 292 g of NaCl in 800 ml of H_2O. Adjust the volume to 1 liter with H_2O. Dispense into aliquots and sterilize by autoclaving. Store the NaCl solution at room temperature.

NaOH

The preparation of 10 N NaOH involves a highly exothermic reaction, which can cause breakage of glass containers. Prepare this solution with extreme care in plastic beakers. To 800 ml of H_2O, slowly add 400g of NaOH pellets, stirring continuously. As an added precaution, place the beaker on ice. When the pellets have dissolved completely, adjust the volume to 1 liter with H_2O. Store the solution in a plastic container at room temperature. Sterilization is not necessary.

20X SB (electrophoresis buffer)

(Buffer diluted to 1X should be 10 mM Sodium hydroxide and pH 8.5)

For 1 liter, weigh out 8 g NaOH and ~40 g boric acid - add water, dissolve and add additional boric acid until pH = 8.0; bring final volume to 1 liter.

SDS Stock

10% or 20% (w/v) SDS. Also called sodium lauryl (or dodecyl) sulfate. To prepare a 20% (w/v) solution, dissolve 200 g of electrophoresis-grade SDS in 900 ml of H_2O. Heat to 68 ° C and stir with a magnetic stirrer to assist dissolution. If necessary, adjust the pH to 7.2 by adding a few drops of concentrated HCl. Adjust the volume to 1 liter with H 2O. Store at room temperature. Sterilization is not necessary. Do not autoclave.

Use a Mask when Weighing this out

20x SSC

0.3M Na(3) citrate

3M NaCl

SOB Medium : per liter

Bacto-tryptone 20 g

Yeast extract 5 g

NaCl 0.584 g

KCl 0.186 g Mix components and adjust pH to 7.0 with NaOH and autoclave.

2 M Mg^{++} Stock

MgCl 2-6H2O 20.33 g

MgSO 4 -7H2O 24.65 g

Distilled water to 100 ml. Autoclave or filter sterilize.

2 M Glucose

Glucose 36.04 g

Distilled water to 100 ml. Filter sterilize.

For SOB Medium + magnesium : Add 1 ml of 2 M Mg ++ stock to 99 ml SOB Medium.

For SOC Medium : Add 1 ml of 2 M Mg ++ stock and 1 ml of 2 M Glucose to 98 ml of SOB Medium.

3M Sodium Acetate - pH 5.2

To prepare a 3 M solution: Dissolve 408.3 g of sodium acetate-$3H_2O$ in 800 ml of H 2O. Adjust the pH to 5.2 with glacial acetic acid. Adjust the volume to 1 liter with H_2O. Dispense into aliquots and sterilize by autoclaving.

Southern Solutions

Depurination solution (for Southern blotting)

250mM HCl

Denaturation solution (for Southern blotting)

1.5M NaCl and 0.5M NaOH

Neutralization solution (for Southern blotting)

1.5M NaCl

0.5M Tris-HCl, pH adjusted to 7.5

50x TAE Prepare a 50x stock solution in 1 liter of H_2O:

242 g of Tris base

57.1 ml of glacial acetic acid

100 ml of 0.5 M EDTA (pH 8.0)

The 1x working solution is 40 mM Tris-acetate/1 mM EDTA.

5X (or 10X) TBE

Prepare a 5x stock solution in 1 liter of H2O:

54 g of Tris base

27.5 g of boric acid

20 ml of 0.5 M EDTA (pH 8.0)

The pH of the concentrated stock buffer should be approx. 8.3.. Some investigators prefer to use more concentrated stock solutions of TBE (10x as opposed to 5x). However, 5x stock solution is more stable because the solutes do not precipitate during storage. Passing the 5x or 10x buffer stocks through a 0.22ìm filter can prevent or delay formation of precipitates.

TE Buffer

10 mM Tris-Cl (pH, usually 7.6 or 8.0)

1 mM EDTA (pH 8.0)

Use concentrated stock solutions to prepare. If sterile water and sterile stocks are used, there is no need to autoclave. Otherwise, sterilize solutions by autoclaving for 20 minutes. Store the buffer at room temperature.

1 M Tris-Cl - used at various pHs

Using Tris base : To make 1 liter, dissolve 121 g Tris Base in 800 ml of water. Adjust pH to the desired value by adding approximately the following:

pH = 7.4 about 70 ml of concentrated HCl
pH = 7.6 about 60 ml of concentrated HCl
pH = 8.0 about 42 ml of concentrated HCl

Make sure solution is at room temperature before making final pH adjustments. Bring final volume to 1 liter. Sterilize by autoclaving.

Using Trizma tables : an alternate procedure for preparing Tris solutions is to combine the proper amount of Tris Base and Tris Hydrochloride to achieve the desired value using Sigma's Tris tables.

WB (10% redistilled glycerol, 90% distilled water, v/v)

In a 1-liter graduated cylinder, add 100 ml of glycerol and 900 ml of distilled water. Cover with parafilm and mix thoroughly. Sterilized by autoclaving, and chill to 4°C.

Western Blotting Solutions

1X Transfer buffer 1: 25 mM Tris, 192 mM Glycine, pH 8.3

Mix 3.03 g Tris, 14.4 g glycine; add dd water to 1 liter - do not adjust pH.

1X Transfer buffer 2: 25 mM Tris, 192 mM Glycine, pH 8.3, 20 % methanol

Mix 3.03 g Tris, 14.4 g glycine; add 200 ml methanol; add dd water to 1 liter - do not adjust pH. (NOTE: methanol is not needed for PVDF membranes)

10X Western Buffer: 200 mM Tris pH = 7.5; 1.5 M NaCl

To prepare 1X Western Buffer, dilute 10X buffer to 1X, adding Tween-20 to 0.1%. Remove 50 ml and set aside for the last two washes. To the remainder, add I-Block to 0.2%, heating gently with constant stirring until dissolved. Bring to room temperature before using.

X-gal 5-bromo-4-chloro-3-indolyl-b-D-galactoside (same recipe for X-phosphate)

Make a 2% (w/v) stock solution by dissolving X-gal in dimethylformamide at a concentration of 20 mg/ml solution. Use a glass or polypropylene tube. Wrap the tube containing the solution in aluminum foil to prevent damage by light and store at -20 ° C. It is not necessary to sterilize X-gal solutions.

For electrophoresis Ethidium bromide, $C_{21}H_{20}N_3Br$; Mw=394.3g/M; (store at NT in the dark);

	Conc.	5ml	10ml	50ml	5ml
EtBr	10mg/ml	50mg	100mg	500mg	0.05ml

CHAPTER 7

Isolation of Bacterial Genomic DNA

1. Bacterial (Prokaryote) Genome

The chromosome of most bacteria is a single super coiled double stranded circular DNA molecule. A complex packaging mechanism is necessary in order to ensure that the entire DNA is folded within the bacterial cell in a manner, which will not inhibit transcription nor allow entanglement of the two daughter strands to occur during replication (Highly condensed >10 mg/ml in region less than l ìm in diameter). This is achieved by two main mechanisms.

- First the DNA is folded into between 40-100 loops
- Second each of the quasicircles is it self super coiled independently of the others each loop are abolished by Ribonuclease.

Deoxyribonucleic remove the super helical nature of the loops, though limited treatment with this enzyme will attack only a small number of the loops and therefore leads to removal of only some of the super coiling. The intact folded chromosome or nucleoid can be isolated in the presence of non ionic detergents and high molarities of salt in two possible forms free and membrane associated. The free form which sediments at 1600-1700s consists of about 60%DNA, 30%RNA and 10%protein the bulk of which is the enzyme RNA polymerase. The membrane bound form contains an additional 20% of membrane-attached protein. Bacterial genome size ranges from 580 to 9900kbp, & the average gene size is constant from 900 to 1100bp of which 87-94% are protein coding (*Rickettisa's* 76% and *Borelia* Plasmid's 41% are exceptions).

In this experiment genomic DNA of *serratia marcescens* is isolated. This bacterium is facultatively anaerobic gram-negative rods about 0.5-0.8µm in diameter and 0.9-2µm in length. It is a chemotrophic bacterium; having both respiratory and fermentative type of metabolism they grow well at 30-37 °C in a rich medium like Luria Bertani broth. The genome of this bacterium is not yet completely sequenced.

2. Difference in Plasmid and Genomic DNA isolation

The difference in size and topology of Plasmid and genomic DNA is an important parameter, which distinguishes their purification procedures. Plasmid size range from 0.1-0.5% of the chromosome. During Plasmid purification, preferential recovery of circular Plasmid DNA over linear chromosomal DNA is done. Treatments with either base or detergent disrupt base pairing and cause the linear chromosomal DNA to denature and separate. In contrast because of its super coiled form, covalently closed circular Plasmid (CCC) DNA is unable to separate and readily reforms a correctly paired super helical structure under renaturation condition. Whereas in case of genomic DNA preparation detergents and other protein denaturants used are in high concentration as compared to that used in Plasmid, hence the lysis solution of genomic DNA can't be used for Plasmid as it may cause severe Denaturation of smaller DNA.

Flow diagram of DNA isolation

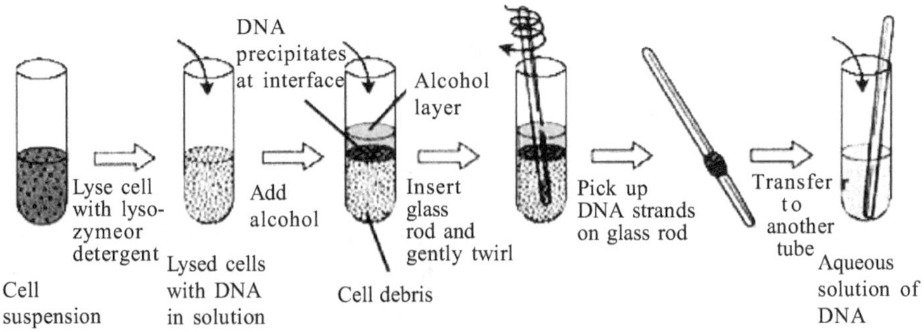

Principle

The common feature in all procedures is that a cell is first broken and then the DNA is separated from other components such as proteins, RNA, lipid and carbohydrates. Purity of the DNA is essential as slight contaminants can inhibit further experiments like Restriction digestion, Polymerase chain reaction, sequencing etc.

The objective of this experiment is to isolate chromosomal DNA with a minimum of breaks. The resuspended cells are first mixed with ethylenediamine tetra acetic acid (EDTA). EDTA forms complexes (chelates) with several kinds of metal ions. Divalent metal contains, such as Mg^{+2} are required cofactors by the majority of DNases. The DNA being extracted is protected from DNases degradation since the complexed Mg^{+2} cannot be utilized by enzyme. The addition of the ionic detergent dissolves the cell membrane & denatures many proteins. RNase is also present to degrade high molecular weight RNA. The proteolytic enzyme (protease) is added to the cell lysate in order to digest proteins that are free in solution or bound to the DNA. RNases & proteases are exceptionally stable enzymes. They remain active in the presence of denaturing detergents such as sarkosyl & at high temperatures. Eventually, the RNase will be degraded by the protease or by phenol: isoamyl: chloroform treatment. However, it is in high enough concentration so that most of the RNA is degraded before significant proteolytic inactivation occurs. In the presence of salts, DNA & RNA precipitate from solutions containing high percentages of isopropanol or ethanol. Smaller molecules such as sugars & amino acids remain in solution.

In this experiment the given cell lysis solution, contains 4M-guanidium thiocyanate salts an effective protein denaturant and a strong inhibitor of ribonuclease and deoxyribonucleases. The procedure described here is a single step extraction protocol which allows isolation of DNA in 2 hours and provides both high yield and purity of DNA thus eliminating the ultra centrifugation step of Cesium Chloride method. Cell lysis occurs due to the action of guanidium thiocyanate and the detergent sarkosyl present in the cell lysis buffer. Upon centrifugation cell debris along with the trapped RNA and proteins are separated. The resulting supernatant mainly consists of genomic DNA and a slight amount of RNA. The nucleic acid is then precipitated using alcohol. This procedure also allows the simultaneous processing of large number of samples too.

A. Materials Required

- Cell lysis buffer,
- DNA rehydrating solution,
- Gel loading dye,
- 50XTAE,
- 100%, 95% and 75% ethanol.

B. Equipment Required

- Electrophoresis unit,
- Transilluminator,
- Microfuge,
- Micropipettes
- Incubator.

C. Procedure

1. Grow cells overnight in 500 mL broth medium.
2. Pellet cells by centrifugation, and resuspend in 5 mL 50 mM Tris (pH 8.0), 50 mM EDTA.
3. Freeze cell suspension at -20°C.
4. Add 0.5 mL 250 mM Tris (pH 8.0), 10 mg/mL lysozyme to frozen suspension, and let it thaw at room temperature. When thawed, place on ice for 45 min.
5. Add 1 mL 0.5% SDS, 50 mM Tris (pH 7.5), 0.4 M EDTA, and 1 mg/mL proteinase K. Place in a 50°C water bath for 60 min.
6. Extract with 6 mL Tris-equilibrated phenol and centrifuge at 10000 xg for 15 minutes. Transfer top layer to new tube (avoid interface). Redo this step if necessary.
7. Add 0.1 vol 3M Na acetate (mix gently), then add 2 vol 95% ethanol (mix by inverting).
8. Spool out DNA and transfer to 5 mL 50 mM Tris (pH 7.5), 1 mM EDTA, 200 g/mL RNAse. Dissolve overnight by rocking at 4°C.
9. Extract with equal volume chloroform (mix by inverting) and centrifuge at 10000 xg for 5 min. Transfer the top layer to a new tube.
10. Add 0.1 vol 3M Na acetate (mix gently), then add 2 vol 95% ethanol (mix by inverting).

11. Spool out DNA and dissolve in 2 mL 50 mM Tris (pH 7.5), 1 mM EDTA.
12. Check the purity of the DNA by electrophoresis and spectrophotometric analysis.

Other Procedure

1. Grow bacterial cell in nutrient broth medium for 24 hrs. or Remove the vial containing the bacterial cell pellet from ice, and thaw at room temperature.
2. Resuspend the cells in 700μl of cell lysis solution at room temperature.
3. Incubate at room temperature for 5 minutes and spin at 10,000 rpm for 10 minutes at room temperature.
4. Collect 500μl of the supernatant in a fresh tube: (a jelly like pellet of cell debris is seen. Avoid decanting this pellet).
5. To the 500μl of the supernatant add 500μl of isopropanol. Mix by inversion till you see white stands of DNA precipitating out.
6. Spool this DNA with the help of a tip and transfer into fresh tube or spin at 12,000rpm for 5minutes and discard the supernatant.
7. Wash the DNA pellet with 95% ethanol by adding ethanol and decanting it. Repeat this step. Give a final wash with 75% ethanol and air dry for 5 minutes.
8. Add 100μl of DNA rehydrating solution and incubate at 55-60 °C degrees for 5 minutes to increase the solubility of genomic DNA.
9. To get rid of insoluble material spin at 12,000 rpm for 10minutes and pipette out the supernatant into a fresh tube.
10. Take 30μl of the freshly isolated DNA along with 5μl of gel loading dye, mix and load into the gel. Take 10μl of control DNA electrophorise along with the isolated samples in 1% agarose gel.

Visualizing DNA

Cut the gel, lift and place on the transilluminator. DNA can be seen as orange band under UV. Wear gloves while handling ethidium bromide stained agarose gels.

CHAPTER 8

Isolation of RNA from Yeast

Introduction

Yeast contain a hard cell wall. Spheroplasts are yeast cells where the cell wall has been enzymatically degraded. These spheroplasts are very fragile so we must be careful with them. Once we have spheroplasts it will be very easy to lyse the yeast cells and quickly release the RNA and proteins in the cell into a controlled environment where we can temporarily keep the cellular RNases from degrading the RNA. We don't want to lyse the cells before they are in this controlled environment.

Growth of Yeast Cultures

Cultures of wild-type and yeast containing various mutations will be grown up the day of the experiment. Each section will be given one wild type yeast culture and a culture containing one of the mutant varieties that we have discussed in class.

Yeast are grown very similarly to bacteria. Like bacteria, they have different stages of growth. Scientists divide these stages of growth into early-log phase, mid log-phase, late log-phase, and the stationary phase. These phases are based on the amount of yeast that are around in the broth and how quickly they are able to grow and divide based on nutrient use. Reading the absorbance of the sample at 600 nm is a way in which you can determine what phase of the yeast growth series you are in and how many cells you have.

Haploid yeast cells contain approximatley 1.2 pg of RNA per cell

Predict approximatly how much RNA you should isolate

Principle

Yeast cells are homogenized with phenol which dissociates proteins from nucleic acids and also denatures the proteins. Upon centrifugation, DNA remains in lower phenol phase, RNA is dispersed in the upper aqueous phase with the denatured proteins being at the inter-phase. RNA from the recovered aqueous phase is precipitated with ethanol and it is free of DNA but may contain polysaccharides as contaminants.

Requirements

1. Cooling centrifuge
2. Dried yeast
3. *Phenol solution* : Dissolve 90% phenol in of water
4. *Potassium acetate* : Dissolve 20% potassium acetate in water and adjust its pH to 5.0 with glacial acetic acid
5. Absolute ethanol
6. Di-ethyl ether

Procedure

1. Suspend 3g of dried yeast in 12 ml of distilled water and incubate at 37°C for 15 min.
2. Add 16 ml of phenol solution. Stir the suspension for 30 min at room temperature and then centrifuge at 3000 x g for 15 min in cold to facilitate the emulsion to separate out into different phases.
3. Carefully remove the 1.8 ml of upper aqueous phase and add 0.2 ml of potassium acetate solution.
4. Add 4ml of cold absolute ethanol to precipitate out RNA and keep it overnight in freeze and then centrifuge at 3000 x g for 15

min. Discard the supernatant and wash the RNA pellet with about 2 ml of cold 70% ethanol, then with 2ml of absolute ethanol and finally with 4 ml of di-ethyl ether. Centrifuge at each washing step at 3000 x g for 15 min and collect the pellet.

5. Air dry the pellet. Dissolve the pellet in 1 ml of 5% chloroform ($HClO_4$). (Pellets can also be dissolved in the sterile distilled water or Tris- EDTA buffer for their further use

6. Measure the concentration of the total RNA in the tube by Orcinol method.

Quantifying RNA

The concentration of RNA is read by measuring the absorbance of a sample at A260 on a spectrophotometer. The RNA sample is placed in a quartz cuvette in order to read the optical density (OD). Common glass and plastic will also absorb at this wavelength so special quartz cuvettes that do not absorb light at this wavelength are used to measure the absorbance of the RNA. Be very careful with these, as they are very expensive.

You will need to use it to determine the concentration of your RNA. Scientists have determined an absorbance coefficient to use to determine the concentration of RNA in your sample preparation. RNA with the concentration of 40 ug/ml has an optical density of 1 at A260. When you get the reading of your sample you can use this information to find out how much RNA you have.

To get your reading put 20 ul of your sample in 1 ml of water and place it in a quartz cuvette. Once you obtain your reading the equation should be used to help you determine your sample concentration

RNA concentration in ug/ul
= O.D. reading x ((40 ug/1000 ul)/1 O.D. unit) x dilution factor

The dilution factor in this case would be 1000/20 = 50 since you dissolved 20 ul in 1000 ul of water.

The absorbance of RNA sample at A280 as well as A260. Both proteins and nucleic acids absorb light at 280 nm. By obtaining the ratio of A260/A280 scientists can thus get an idea of the purity of their RNA sample. For pure RNA without a lot of protein this ratio should be 1.9-2.2.

Checking the RNA for Degradation Using Gel Electrophoresis

We will use 10 ug of RNA to run on a 1.2% agarose gel at 100 V for 1/2 hour. To each sample of RNA add the appropriate amount of loading dye and 1 ul of Ethidium Bromide

CHAPTER 9

Nucleotide Composition of RNA

Materials Required

- RNA sample
- 1 N and 0.1 N HCl
- Boiling water bath
- Whatman #1 filter paper (for chromatography)
- Chromatography tank
- 20 µl micropipette
- Acetic acid: butanol: water (15:60:25) solvent
- UV light source
- UV spectrophotometer

Procedure

1. Place a portion of your RNA sample (approximately 40 mg, hydrated) into a heavy walled Pyrex test tube. Add 1.0 ml of 1 N HCl and seal the tube.
2. Heat the tube in a boiling water bath for 1 hour.

3. Cool the tube, open it and place the contents into a centrifuge tube. Centrifuge the contents at 2,000 RPM in a clinical centrifuge to remove any insoluble residue. The supernatant contains your hydrolyzed RNA.

4. Prepare Whatman filter paper No. 1 for standard one-dimensional chromatography.

5. Using a micropipette, spot 20 µl of your hydrolyzate onto the paper, being careful to keep the spots as small as possible (repeated small drops are better than one large drop). Allow the spots to completely dry before proceeding.

6. Place the paper chromatogram into your chromatography tank and add the solvent (acetic acid:butanol:water). Allow the system to function for an appropriate time (approximately 36 hours for a 20 cm descending strip of Whatman #1). Remove the paper and dry it in a circulating air oven at 40° C for about 2 hours.

7. Locate the spots of nucleotides by their fluorescence under an ultra-violet light source. Expose the paper chromatogram to a UV light source and outline the spots using a light pencil. The order of migration from the point of origin is guanine (light blue fluorescence), adenine, cytilic acid and finally, uridylic acid.

8. After cafefully marking the spots, cut them out with cutters and place the paper cut outs into separately labeled 15 ml conical centrifuge tubes. Add 5.0 ml of 0.1 N HCl to each tube and allow the tubes to sit for several hours to elute the nucleotides from the paper.

9. Pack down the paper with a glass rod (centrifuge in a clinical centrifuge if necessary) and remove an aliquot of the liquid for spectrophotometric assay.

10. Measure the absorbance of each of the four nucleotides at the indicated UV wavelength (having first blanked the instrument with 0.1 N HCl).

Base	Wavelength	Molar Extinction Coefficient
Guanine	250 nm	10.6
Adenine	260 nm	13.0
Cytidylic acid	280 nm	19.95
Uridylic acid	260 nm	9.89

11. Use the molar extinction coefficients to determine the concentration of each base in the sample. Calculate the percent composition of each base, and the purine/pyrimidine ratio.

❑❑❑

CHAPTER 10

DNA Melting Point Determination

Introduction

DNA denaturation, also called DNA melting, is the process by which double-stranded deoxyribonucleic acid unwinds and separates into single-stranded strands during the breaking of hydrogen bonding between the bases. Both terms are used to refer to the process as it occurs when a mixture is heated, although "denaturation" can also refer to the separation of DNA strands induced by chemicals like urea. For multiple copies of DNA molecules, the melting temperature (T_m) is defined as the temperature at which half of the DNA strands are in the double-helical state and half are in the "random-coil" states. The melting temperature depends on both the length of the molecule, and the specific nucleotide sequence composition of that molecule.

Applications of DNA Denaturation

The process of DNA denaturation can be used to analyze some aspects of DNA. Because cytosine / guanine base-pairing is generally stronger than adenosine / thymine base-pairing, the amount of cytosine and guanine in a genome (called the "GC content") can be estimated by measuring the temperature at which the genomic DNA melts. Elevated temperatures are associated with high GC content. GC ratios within a genome are found to be markedly variable. These variations in GC ratio within a genome of higher organisms results in a mosaic like formation with islet regions called isochores. This results in the variations in staining intensity in the chromosomes. The isochores include in them essential protein coding genes, termed housekeeping genes and thus determination of ratio of these specific regions contributes in mapping these essential genesbacterial systematics has recommended use of GC ratios in higher level hierarchical classification. For example, the Actinobacteria are characterised as "high GC-content bacteria". In Streptomyces coelicolor A3(2), GC content is 72%. The GC-content of Yeast (Saccharomyces cerevisiae) is 38%, and that of another common model organism Thale Cress is 36%. Because of the nature of the genetic code, it is virtually impossible for an organism to have a genome with a GC-content approaching either 0% or 100%. A species with an extremely low GC-content is Plasmodium falciparum (GC% = ~20%), and it is usually common to refer to such examples as being AT-rich instead of GC-poor.

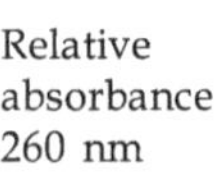

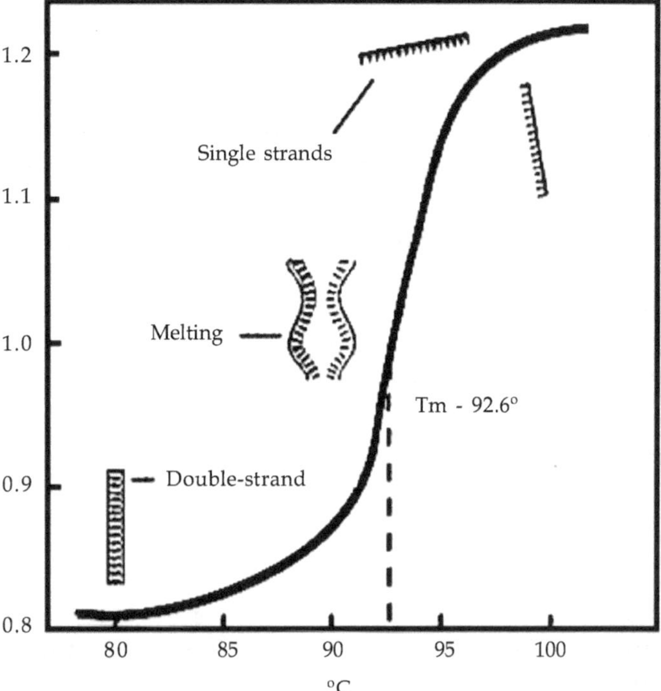

DNA denaturation can also be used to detect sequence differences between two different DNA sequences. DNA is heated and denatured into single-stranded state, and the mixture is cooled to allow strands to rehybridize. Hybrid molecules are formed between similar sequences and any differences between those sequences will result in a disruption of the base-pairing. On a genomic scale, the method has been used by researchers to estimate the genetic distance between two species, a process known as DNA-DNA hybridization. In the context of a single isolated region of DNA, denaturing gradient gels and temperature gradient gels can be used to detect the presence of small mismatches between two sequences, a process known as temperature gradient gel electrophoresis.

Methods of DNA analysis based on melting temperature have the disadvantage of being proxies for studying the underlying sequence; DNA sequencing is generally considered a more accurate method.

The process of DNA melting is also used in molecular biology techniques, notably in the polymerase chain reaction (PCR). Although the temperature of DNA melting is not diagnostic in the technique, methods for estimating T_m are important for determining the appropriate temperatures to use in a protocol. DNA melting temperatures can also be used as a proxy for equalizing the hybridization strengths of a set of molecules, eg. the oligonucleotide probes of DNA microarrays.

Materials

DNA
SSC
UV spectrophotometer (preferably with temperature control)

Procedure

1. Dissolve your DNA preparation in SSC to produce a final concentration of approximately 20 μg DNA/mL.
2. Place the dissolved DNA in an appropriate quartz cuvette along with a second cuvette containing SSC as a blank.
3. Place both cuvettes into a dual beam temperature regulated UV spectrophotometer and measure the absorbance of the sample at 260 nm at temperatures ranging from 25°C to 80°C. Continue to increase the temperature slowly and continue reading the absorbance until a sharp rise in absorbance is noted.

Alternatively

(a) Place the cuvettes into water bath at 25°C and allow to temperature equilibrates. Remove the blank, wipe the outside dry, and rapidly blank the instrument at 260 nm. Transfer the sample to the spectrophotometer (be sure to dry and work rapidly) and read the absorbance.

(b) Raise the temperature of the bath to 50°C and repeat step (a).

(c) Raise the temperature sequentially to 60°C, 65°C, 70°C, 75°C, and 80°C and repeat the absorbance measurements.

(d) Slowly raise the temperature above 80°C and make absorbance measurements every 2° until the absorbance begins to increase. At that point, increase the temperature, but continue to take readings at 1°C intervals.

4. Correct all of the absorbance readings for solvent expansion relative to 25°C.
5. List the corrected values as A_t.
6. Plot the value of A_t / A_{25} versus temperature and calculate the midpoint of any increased absorbance. This midpoint is the melting point (Tm) for your DNA sample.
7. Calculate the GC content of your sample using the formula

$$\text{Percent of G + C} = k\,(Tm - 69.3) \times 2.44$$

Note *Single-strand DNA absorbs more UV light than double strands. Moreover, double strands can be separated by heat (melted) and the temperature at which the strands separate (Tm) is related to the number of guanine-cytosine residues (each having 3 hydrogen bonds, as opposed to the 2 in adenine-thymine). This has led to the development of a rapid test for an approximation of the GC/AT ratio using melting points and the change in* UV_{260} *absorbance (known as "hyperchromicity" or "hyperchromatic shift"). Of course, the separation is also dependent upon environmental influences, particularly the salt concentration of the DNA solution. To standardize this, all Tm measurements are made in SSC buffer. DNA melts between 85°C and 100°C in this buffer (as opposed to 25°Cin distilled water).*

CHAPTER 11

Quantitative Determination of DNA by DPA Method and Spectroscopic Method

To Estimate the DNA Using Diphenylamine

Introduction

DNA estimation is helpful to know about the total amount of DNA of prokaryotic as well as eukaryotic system. One method used for DNA estimation is Diphenylamine assay, is that of *Burton* who modified the original procedure of *Dische*. This procedure is based on treatment of nucleic acid with hot perchloric acid to hydrolyze DNA nucleotides. Diphenylamine reacts specifically with *deoxyribose* in DNA to form a blue-colored complex. Since diphenylamine does not react with ribose, one can assay for DNA in the presence of contaminating RNA.

Principle

The DPA method is based on the colorimetry measurement of products formed by the reaction of ù- hydroxylevulinaldehyde (from the deoxyribose released after depurination of DNA) with diphenylamine in 2.5 N perchloric acid.

When DNA is treated with diphenylamine under acidic condition (PCA) a blue complex is formed with a sharp absorption maximum at

595 nm (600 nm). In acid solution, the straight chain form of the deoxypentose is converted to highly reactive ω – hydroxylevulinaldehyde that reacts with diphenylamine to give blue complex. The reaction yields two mixture of product which together have absorbance maximum at 595 nm (600 nm). Interferences include proteins, lipids, saccharides and RNA. In DNA the only deoxyribose of the purine nucleotides reacts, so that the value obtained represents half of the total deoxyribose present.

Range : 40 to 400μgs.
Stock : 25mg of DNA in 100ml of 5mM NaOH.
Standard : 40 ml of stock + 60 ml of 5mM NaOH (10%)

Reagents

- Diphenylamine reagent (freshly prepared)- Add 10 ml glacial acetic acid to 150 mg diphenylamine in a 50 ml polypropylene tube and mix thoroughly by repeated inversion until complete dissolution. Add 50μl concentrated sulfuric acid and mix thoroughly. Add 50μl acetaldehyde solution and mix thoroughly. Store at room temperature.

Caution

- Prepare fresh and use within 60min. DPA is an irritant: wear appropriate protection.
- Std. DNA conc. 20mg/50ml.
- 2.5 N PCA reagent (Perchloric acid)
- **Acetaldehyde solution:** Add 16 mg acetaldehyde to 10 ml deionized water to have a 16 mg/ml stock solution or 1:40 diluted. Store at 2-8°C for < 1 year.

Procedure

1. Take different aliquots of standard and test samples.
2. Make its volume to 1ml by adding D/W.
3. Then add 0.2ml of 2.5N PCA.
4. Keep at 70^{0}C for 15 minutes in water bath and allow the tubes to cool.
5. Then add 2ml of DPA in each tube.

6. Keep it for 24 hours in dark.
7. Take the O.D. at 590nm.
8. By the standard graph find out the amount of DNA present in the test sample.

Observation Table

No.	Stock DNA (100µg/ml)	0.5N $HClO_4$	Conc. of DNA (µg/ml)		OD at 600nm	OD Test-Blank	Conc. of DA
1	0.0	1.0	0	**Add 2.0 ml of diphenylamine reagent in all the and Keep the tubes in the Dark for 16 to 18 hrs**			
2	0.2	0.8	20				
3	0.4	0.6	40				
4	0.6	0.4	60				
5	0.8	0.2	80				
6	1.0	0.0	100				
7	Unknown 1						
8	Unknown 2						

Spectrophotometric Quntification of DNA

Estimation of DNA Concentration by Spectrophotometric Method

Introduction

It is possible to estimate the concentration of solution of nucleic acids and oligonucleotides by observing absorbance at single wavelength (260nm). This is not good practice. The absorbance of the sample should be measured at several wavelengths. Since ratio of absorbance at 260nm to the absorbance at other wavelength is good indicator of purity of preparation.

Principle

The ration between the reading at 260 nm and 280 nm (OD_{260}: OD_{280}) provides an estimate of the purity of the nucleic acid. Pure preparation of DNA and RNA have (OD_{260}: OD_{280}) value 1.8 and 2.0 respectively.

If there is significant contamination with protein or phenol, OD_{260}: OD_{280} value will be less than the value given above, and accurate quantitation of amount of nucleic acid will not be possible.

Significance absorption at 260nm indicate contamination by phenolate ion, thiocyanate and other organic compounds, where as absorption at higher wavelength (330nm and higher) is usually caused by light scattering and indicate the presence of particulate matter. Absorption at 280nm indicates the presence of protein, because aromatic amino acids absorb strongly at 280nm.

Thus, ratio of absorption at 260nm and 280nm has been used as a measure of purity of isolated nucleic acids. The reverse is not true extinction coefficient of nucleic acids at 260nm and 280nm are so much greater than that of proteins, significant contamination with protein will not greatly change the OD_{260}: OD_{280} of nucleic acids solution.

Nucleic acid absorbs strongly at 260 nm only a significant change in ratio of absorption at two wavelengths. Based on the extinction coefficient, an O.D. of 1.0 at 260 nm corresponds to an approximately 50μg/ml of double stranded DNA.

Requirements

- DNA sample to be estimated
- TE buffer
- Spectrophotometer with all accessories

Procedure

1. To measure the concentration of DNA, make its appropriate dilution with TE buffer (either 1:50 of 1:100).
2. Standardize the spectrophotometer using TE buffer as blank.
3. Measure the absorbance of the sample at 260-280 nm
4. Calculate the DNA concentration of the sample in the following pattern:

   Concentration of DNA (μg/ml) = O.D. 260 X 50X dilution factor

   Or

   Concentration of DNA (μg/μl) = O.D. 260 X 50X dilution factor/1000

Result

1) The OD_{260}: OD_{280} is
2) Conc. of DNA of sample

CHAPTER 12

Estimation of RNA by Orcinol Method

Principle

In this method the estimation of pentose sugar is carried out. Acid hydrolysis of RNA releases ribose (a pentose sugar) which in the presence of strong acid undergoes dehydration to yield furfural. Orcinol, in the presence of ferric chloride as a catalyst, reacts with furfural producing a green colored compound which can be quantitatively estimated at $A_{665\ nm}$. DNA gives a limited positive reaction with Orcinol test.

Materials and Reagents

1. Spectrophotometer
2. Boiling water bath
3. 5% chloroform ($HClO_4$)
4. *Standard RNA solution* : Dissolve yeast RNA (500ìg/ml) in 5% $HClO_4$. make different dilutions to obtain solutions containing 100 – 500ìg RNA/ml with 5% $HClO_4$.

5. *Orcinol reagent* : Dissolve 100mg of ferric chloride ($FeCl_3.6H_2O$) in 100 ml of concentration HCl and then add 3.5 ml of 6% solution of Orcinol prepared in alcohol.

Procedure

1. Take 1.0 ml solution of each of the dilutions of RNA standard solutions, test sample and 1.0 ml of 5% $HClO_4$, as a blank, in different test tubes.
2. Add 1.5 ml Orcinol reagent to all the tubes and mix properly.
3. Keep the test tubes in a boiling water bath for 20 minutes
4. After cooling them, measure the A_{665} against blank.
5. Plot a graph between A_{665} vs. amount of RNA and from this standard curve determine the amount of RNA in the provided sample.

Observation Table

No.	Aliquot (ml)	5% $HClO_4$ (ml)	Conc. of RNA (µg/ml)		OD at $A_{665\,nm}$	OD Test-Blank	Conc. of RNA
B	0.0	1.0	0	Add 1.5 ml of Orcinol reagent and keep the tubes in the boiling water bath for 20 minutes			-----
S_1	0.2	0.8	100				-----
S_2	0.4	0.6	200				-----
S_3	0.6	0.4	300				-----
S_4	0.8	0.2	400				-----
S_5	1.0	0.0	500				-----
U_1	Unknown 1	?	?				?
U_2	Unknown 2	?	?				?
U_3	Unknown 2	?	?				?

CHAPTER 13

Electrophoresis

Electrophoresis is a method of separating out large molecule (DNA or protein). An electric current is passed through a medium containing the molecules, and each molecule travels at a different rate depending on its electrical charge, size and shape. Separation by electrophoresis is based on these differences. In electrophoresis, agarose and acrylamide gels are used for electrophoresis of nucleic acids and proteins.

What is Gel Electrophoresis?

A gel is a solid form. The term electrophoresis describes the migration of charged particle under the influence of an electric field. *Electro* refers to the energy of electricity. *Phoresis*, from the Greek verb *phoros*, means "to carry across." Thus, gel electrophoresis refers to the technique in which molecules are forced across a span of gel, motivated by an electrical current. Activated electrodes at either end of the gel provide the driving force. A molecule's properties determine how rapidly an electric field can move the molecule through a gelatinous medium.

Agarose

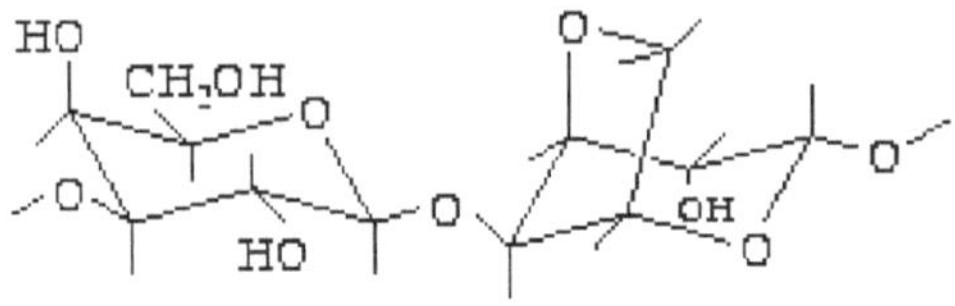

-galactose 3, 6-anhydro L-galactose

Agarose

There are two basic types of materials used to make gels: agarose and polyacrylamide. Agarose is a natural extracted from sea weed. It is very frail and easily destroyed by handling. Agarose gels have very large "pore" size and are used primarily to separate very large molecules with a molecular mass greater than 200 kdal. Agarose gels can be processed faster than polyacrylamide gels, but their resolution is lower. That is, the bands formed in the agarose gels are fuzzy and spread far apart. This is a result of pore size and it cannot be controlled.

Agarose is a linear polysaccharide (average molecular mas about 12,000) made up of the basic repeat unit agarobiose, which comprises alternating units of galactose and 3,6-anhydrogalactose. Agarose is usually used at concentrations between 1% and 3%. Agarose gels are formed by suspending dry agarose in aqueous buffer, then boiling the mixture until a clear solution forms. This is poured and allowed to cool to room temperature to form a rigid gel.

Polyacrylamide

There are two basic types of materials used to make gels: agarose and polyacrylamide. The polyacrylamide gel electrophoresis (PAGE) technique was introduced by Raymond and Weintraub (1959). Polyacrylamide is the same material that is used for skin electrodes and in soft contact lenses. Polyacrylamide gel may be prepared so as to provide a wide variety of electrophoretic conditions. The pore size for the gel may be varied to produce different molecular seiving effects for separating proteins of different sizes. In this way, the percentage of polyacrylmide can be controlled in a given gel. By controlling the percentage (from 3% to 30%), precise pore sizes can be obtained, usually from 5 to 2,000 kdal. This is the ideal range for gene sequencing, protein, polypeptide, and enzyme analysis. Polyacrylamide gels can be cast in a single percentage or with varying gradients. Gradient gels provide continuous decrease in pore size from the top to the bottom of the gel, resulting in thin bands. Because of this banding effect, detailed genetic and molecular analysis can be performed on gradient polyacrylamide gels. Polyacrylamide gels offer greater flexibility and more sharply defined banding than agarose gels.

Polyacrylamide Gel Electrophoresis (PAGE) System

Different samples are loaded in wells or depressions at the top of the polyacrylamide gel. The proteins move into the gel when an electric field is applied. The gel minimizes convection currents caused by small temperature gradients, and it minimizes protein movements other than those induced by the electric field. Proteins can be visualized after electrophoresis by treating the gel with a stain such as Coomassie blue, which binds to the proteins but not to the gel itself. Each band on the gel represents a different protein (or a protein subunit); smaller proteins are found near the bottom of the gel.

Electrophoresis of Proteins

Proteins can be separated and purified. Methods for separating proteins take advantage of properties such as charge, size, and solubility, which vary from one protein to the next. Because many proteins bind to other biomolecules, proteins can also be separated on the basis of their binding properties. The source of a protein is generally tissue or microbial cells. The cell must be broken open and the protein must be released into a solution called a crude extract. If necessary, differential centrifugation can be used to prepare subcellular fractions or to isolate organelles. Once the extract or organelle preparation is ready, a variety are available for separation of proteins. Ion-exchange chromatography can be used to separate proteins with different charges (similar to the way amino acids are separated). Other chromatographic methods take advantage of differences in size, binding affinity, and solubility. Nonchromatographic methods include the selective precipitation of proteins with salt, acid, or high temperatures. In addition to chromatography, another important set of methods is available for the separation of proteins, based on the migration of charged proteins in an electric field, a process called (gel) electrophoresis. Gel electrophoresis is especially useful as an analytical method. Its advantage is that proteins can be visualized as well as separated, permitting a researcher to estimate quickly the number of proteins in a mixture or the degree of purity of a particular protein preparation. Also, gel electrophoresis allows determination of crucial properties of a protein such as its isoelectric point and approximate molecular weight.

Amino acids differ not only in R-group characteristics but also in molecular weight. Different amino acids are linked together in a linear chain by peptide bonds in various combinations and sequences to form specific proteins. A protien may be comprised of amino acids from all of the categories. The net charge of a protein will depend on its amino acid composition. If it has more positively charged amino acids such that the sum of the positive charges exceeds the sum of the negative charges, the protein will have an overall positive charge and migrate to the cathode (negatively charged electrode) in an electrical field. Proteins even with a variation of one amino acids will have a different overall charge, and thus are electrophoretically distinguishable.

If a protein is electrophoresed before the disulfide bonds are broken, it will yield only one band of 125,000 Daltons molecular weight on the gel. That is, the protein will move as one entity and the peptide chains cannot be distinguished. However, after treatment with a reducing agent, the protein will yield four distinct bands with a combined molecular weight of 125,000 Daltons—the same as the original protein.

The gel results will show that some of the high molecular weight bands from samples not treated with the disulfide reducing agent are missing in the samples treated with the disulfide reducing agent. However, two or more bands are imaged on the gel lane containing the treated samples, thereby replacing each of the missing high molecular weight bands from the samples that were not treated. The additional bands that appear in the treated samples represent the individual polypeptides that make up complex proteins. The break up of complex proteins into their respective polypeptides allows us to study the structure of proteins that result from the interaction of several genes. A gene is a discrete unit of hereditary information that usually specifies a protein. A single gene provides the genetic code for only one polypeptide. Thus, a protein consisting of four polypeptides requires the interaction of four genes to synthesize that specific protein.

A molecular weight protein marker is used to prepare a standard separation curve with which various unknown proteins or polypeptide fractions can be identified.

Electrophoresis of Nucleic Acids

Gel electrophoresis is the process by which scientists can sort pieces of DNA cut with restriction enzymes by size. An agarose or polyacrylamide gel is loaded with the DNA fragments and current is passed through the gel. Since DNA is negatively charged, it will migrate towards the positive pole. The DNA will not migrate at the same rate, however, larger pieces of DNA collide with the gel matrix more often and are slowed down, while smaller pieces of DNA move through more quickly.

Since different genes have different nucleotide sequences, restriction enzymes will cut them at different places, generating different size DNA fragments. By using gel electrorphoresis, biologists can tell which gene is which based upon the sizes of the fragments generated when a gene is treated with a restriction enzyme.

Agarose Gel Electrophoresis

Electrophoresis through agarose or polyacrilamide gels is the standard method to separate, identify, quantify and purify DNA fragments. Although agarose gels have a lower resolving power than polyacrilamide gels, they have a greater range of separation and are much simpler and easier to handle than the polyacrilamide gels, and have been more widely used in DNA electrophoresis.

Agarose, a purified form of agar isolated from seaweed, is a linear polymer. DNA can be electrophoresed through gel prepared by melting and re-gelling agarose, a copolymer of D galactose and 3, 6-anhydro L-galactose. Resolution of DNA species on agarose gel made it a widely used reliable method for the identification, separation and subsequent purification of DNA molecule of interest. Special grades of low-gelling-temperature agarose that can be used to analyze very small fragments of DNA (10-50 bp) are also available. The DNA is visualized by adding ethidium bromide (EtBr), a fluorescent molecule which intercalate with the DNA bases, extending the length of linear and nicked circular DNA molecules and making them more rigid. When EtBr is added, UV radiation at 254 nm is absorbed by the DNA and transmitted to the bound dye. The energy re-emitted at 590 nm in the red-orange region of the spectrum. Ethidium bromide is a powerful mutagen and hence the gel should be handled carefully with the gloves. The DNA bands can be visualized under UV and the data can be recorded by gel documentation appliances.

Factors Affecting the Rate of DNA Migration in Agarose Gels

1. Molecular Size of the DNA

Molecules of linear double-stranded DNA, which tend to become oriented in an electric field in an end-on position, migrate through gel matrices at rates that are inversely proportional to the log10 of the number of base pairs. Larger molecules migrate more slowly because of greater frictional drag and because they form their way through the pores of the gel less efficiently than smaller molecules.

2. Agarose Concentration

A linear DNA fragment of a given size migrates at different rates through gels containing different concentrations of agarose.

3. Conformation of DNA

Superhelical circular (form 1), nicked circular (form II), and linear (form III) DNAs of the same molecular weight migrate through agarose gels at different rates. The relative mobilities of the three forms depend primarily on the agarose concentration in the gel, but they ionic strength of the buffer, and the density of superhelical twists in the form I DNA. Under some conditions, the order is reversed.

Amount of agarose in gel (% [w/v])	Efficient range of separation of linear DNA molecules (kb)
0.3	5-60
0.6	1-20
0.7	0.8-20
0.9	0.5-7
1.2	0.4-6
1.5	0.2-3
2.0	0.1-2

Identification of Conformational Forms of DNA

An unambiguous method for identifying the different conformational forms of DNA is to carry out electrophoresis in the presence of increasing quantities of ethidium bromide. As the concentration of ethidium bromide increases, more of the dye becomes bound to the DNA. The negative superhelical turns in form I molecules are progressively removed, the radii of the molecules increase, and their rate of migration decreases. At the critical free-dye concentration, where no superhelical turns remain, the rate of migration of form I

DNA reaches its minimum value. As still more ethidium bromide is added, positive superhelical turns are generated, the DNA molecules become more compact, and their mobility increases rapidly. Simultaneously, the mobilities of form II and form III DNA decrease differentially due to charge neutralization and the greater stiffness imparted to the DNA by the ethidium bromide. For most preparations of form I DNA, the critical concentration of free ethidium bromide is in the range of 0.1mg/ml to 0.5mg/ml.

4. Applied Voltage

At low voltages, the rate of migration of linear DNA fragments is proportional to the voltage applied. However, as the electric field strength is raised, the mobility of high-molecular-weight fragments of DNA increases differentially. *Thus the effective range of separation in agarose gels decreases as the voltage is increased.* To obtain maximum resolution of DNA fragments greater than 2 kb in size, agarose gels should be run at no more than 5 V/cm.. Distance is measured as the shortest path between the electrodes and is not merely the length of the gel itself.

5. Direction of the Electric Field

DNA molecules larger than 50-100 kb in length migrate through agarose gels at the same rate if the direction of the electric field remains constant. However, if the direction of the electric field is altered periodically, the DNA molecules are forced to change course. Because larger molecules of DNA take longer to realign themselves to the new direction of the field, pulsed-field gel electrophoresis can be used to fractionate populations of extremely large molecules of DNA (up to 10,000 kb).

6. Base Composition and Temperature

The electrophoretic behavior of DNA in agarose gels (in contrast to polyacrylamide gels) is not significantly affected by either the base composition of the DNA or the temperature at which the gel is run. Thus, in agarose gels the relative electrophoretic mobilities of DNA fragments of different sizes do not change between 4°C and 30°C. In general, agarose gels are run at room temperature. However, gels containing less than 0.5% agarose and low-melting-temperature agarose gels are rather flimsy, and it is best to run them at 4°C, where they are stronger.

7. Presence of Intercalating Dyes

Ethidium bromide, a fluorescent dye that is used to detect DNA in agarose and polyacrylamide gels, reduces the electrophoretic mobility of linear DNA by about 15%. The dye intercalates between stacked base pairs, extending the length of linear and nicked circular DNA molecules and making them more rigid

Ethidium bromide is a carcinogen and should be handled with care. All solutions containing ethidium bromide should be decontaminated before disposal.

8. Composition of Electrophoresis Buffer

The electrophoretic mobility of DNA is affected by the composition and ionic strength of the electrophoresis buffer. In the absence of ions (e.g., if electrophoresis buffer is omitted from the gel by mistake), electrical conductance is minimal and DNA migrates very slowly, if at all. In buffers of high ionic strength (e.g., if 10x electrophoresis buffer is used by mistake), electrical conductance is very efficient and significant amount of heat is generated. In the worst case, the gel melts and the DNA denatures.

Types of Electrophoresis Buffers

Several different buffers are available for electrophoresis of native double-stranded DNA such as Tris-acetate EDTA (TAE), Tris-borate EDTA (TBE), or Tris-phosphate EDTA (TPE). TAE is the most commonly used buffer. However, its buffering capacity is rather low, and it tends to become exhausted during extended electrophoresis (the anode becomes alkaline, the cathode acidic). Replacement of the buffer or recirculation between the two reservoirs is therefore advisable when carrying out electrophoresis for long periods of time at high current. Both TPE and TBE are slightly more expensive than TAE, but they have significantly higher buffering capacity. Double-stranded linear DNA fragments migrate approximately 10% faster through TAE than through TBE or TPE, but the resolving powers of these systems are almost identical, with the exception that the resolution of supercoiled DNAs is better in TAE than in TBE. The most commonly used buffer for electrophoresis of denatured single-stranded DNA is 50 mM NaOH, 1 mM EDTA (alkaline electrophoresis buffer).

Steps Involved in Agarose Gel Setup (see the below figure)

1) Complete assembly

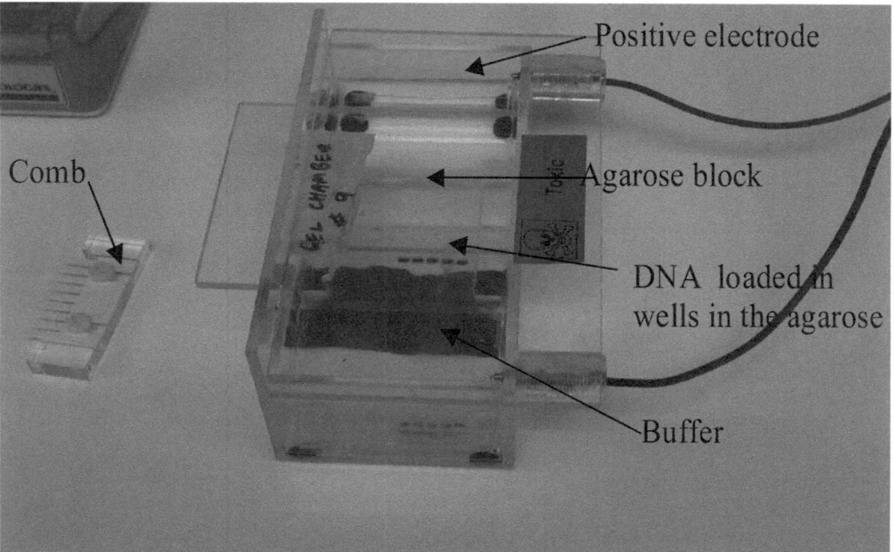

2) Weight the agarose powder as required.

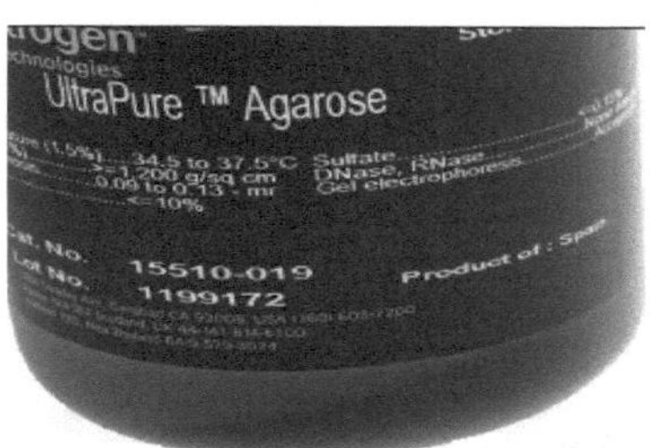

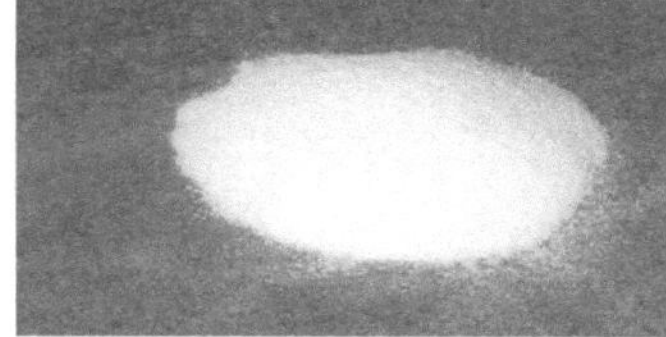

3) Prepare a required volume of electrophoresis buffer) to fill the electrophoresis tank and to prepare the gel.

4) Add the needed amount of powdered agarose (typically 8-15%) to the electrophoresis buffer in an Erlenmeyer flask. The buffer should not occupy more than 50% of the volume of the flask.

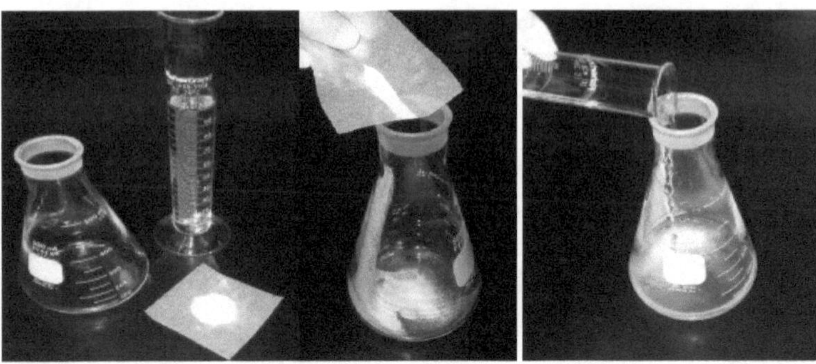

5) Heat the agarose solution in a oven or in water bath to allow all of the grains of agarose to dissolve. If part of the buffer evaporated during the heating, bring the solution back to the original volume through the addition of buffer.

6) Cool down the solution to 60°C. You may add ethidium bromide (Et Br) (10 mg/l in water to a final concentration of 0.5 um/ml) in this step or you can submerge the gel in an Et Br solution once it solidifies.

7) Take the Electrophoresis Equipment

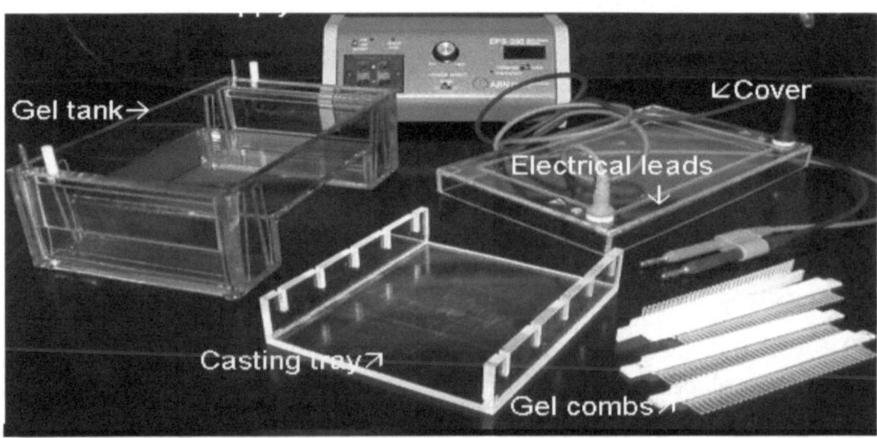

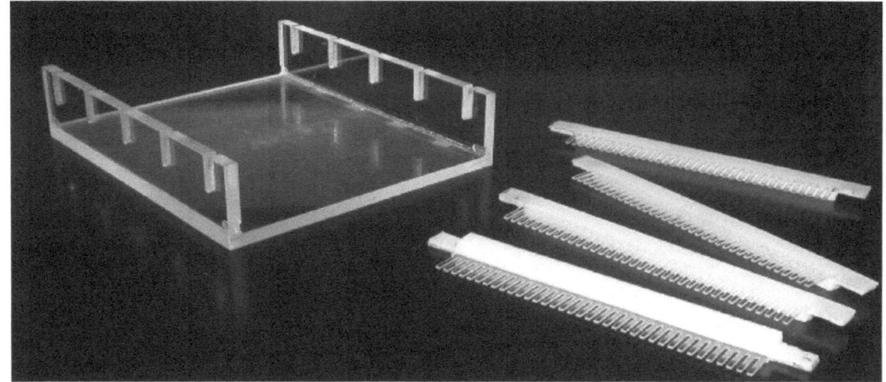

8) Position the comb 0.5-1.0 mm above the plate, thus permitting that a complete well is formed when the agarose solidifies. Air bubbles under or between the teeth of the comb should be avoided. Using a Pasteur pipette seal the glass plate with small amounts of agarose solution. Once the seals are set, pour the

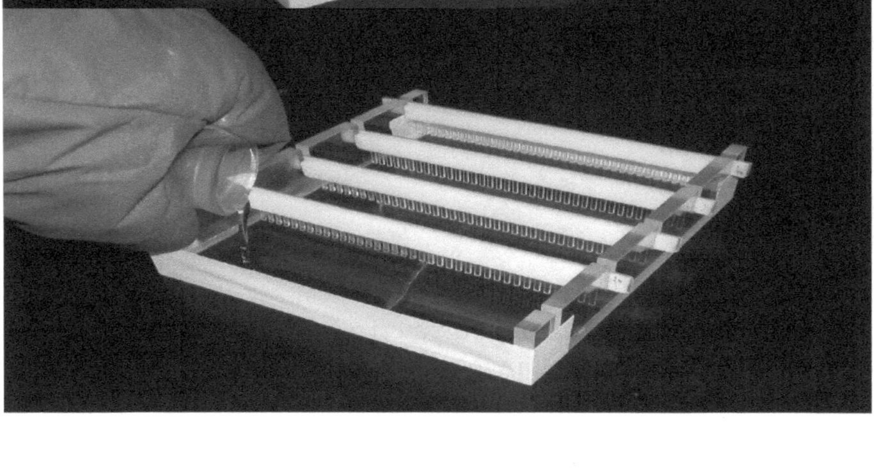

9) Allow the agarose solution to cool slightly (~60°C) and then carefully pour the melted agarose solution into the casting tray. Avoid air

10) Remove the comb and the tape when the gel has completely hardened (30 to 40 minutes at room temperature) and place the gel in the electrophoresis tank. Add enough electrophoresis buffers to the tank to cover the gel (about 1 mm of depth). The top of the wells should be submerged.

When cooled, the agarose polymerizes, forming a flexible gel. It should appear lighter in color when completely cooled (30-45 minutes). Carefully remove the combs and tape.

11) Place the gel in the electrophoresis chamber

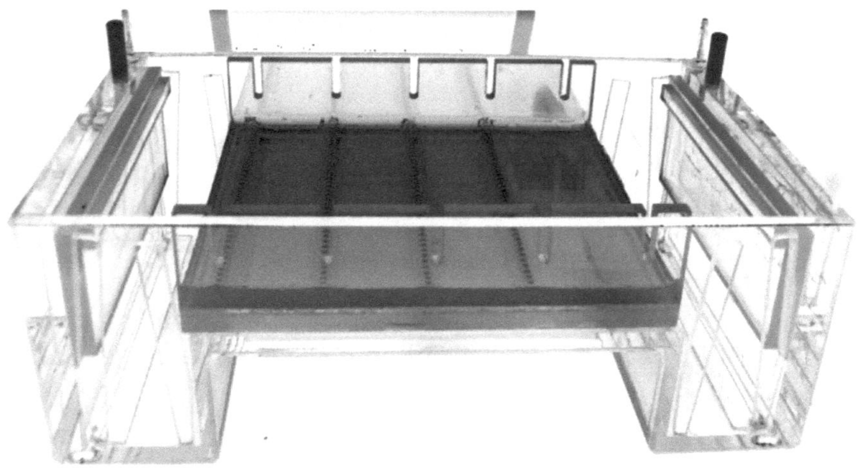

Add enough electrophoresis buffer to cover the gel to a depth of at least 1 mm. Make sure each well is filled with buffer

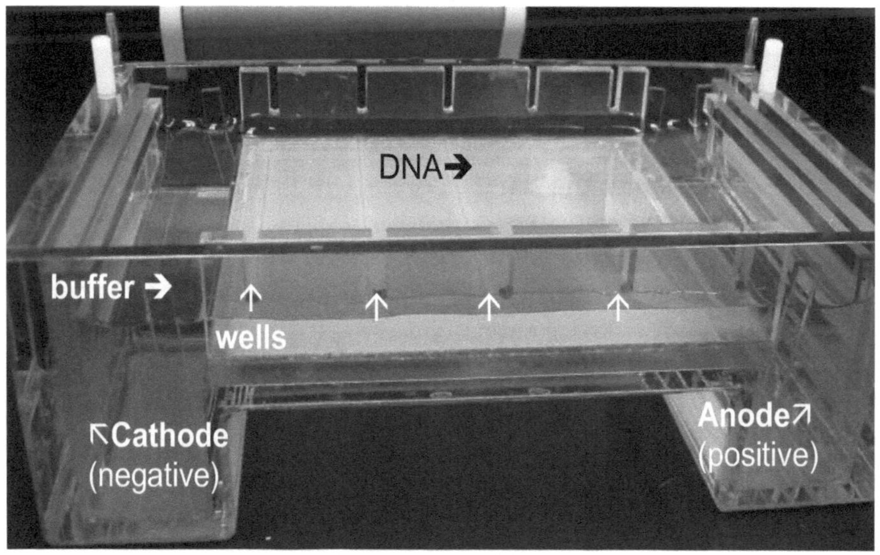

12) Mix the samples of DNA with the 10X sample loading buffer (w/ tracking dye). This allows the samples to be seen when loading onto the gel, and increases the density of the samples, causing them to sink into the gel wells.

10X Loading Buffer: →
- Bromophenol Blue (for colour)
- Glycerol (for weight)

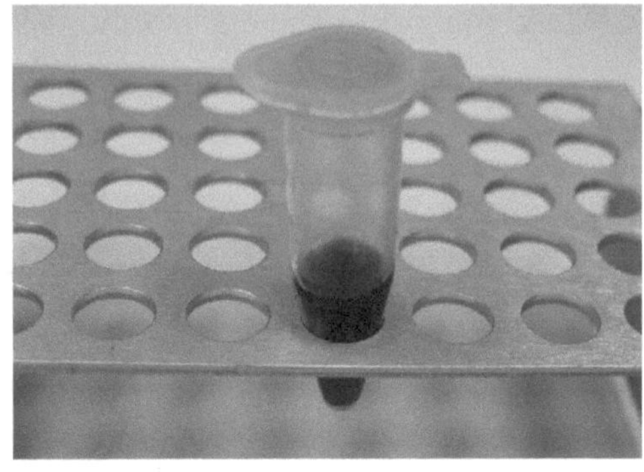

13) Carefully place the pipette tip over a well and gently expel the sample. The sample should sink into the well. Be careful not to puncture the well

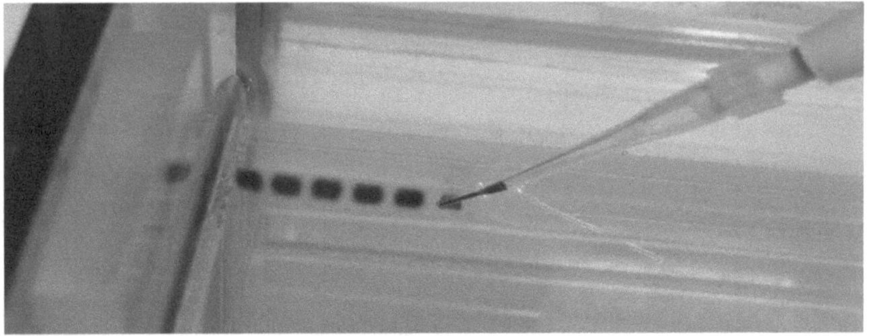

14) Close the lid of the tank. Connect the electrical leads to the power supply. Be sure the leads are attached correctly - DNA migrates toward the anode (red). When the power is turned on, bubbles should form on the electrodes in the electrophoresis chamber.

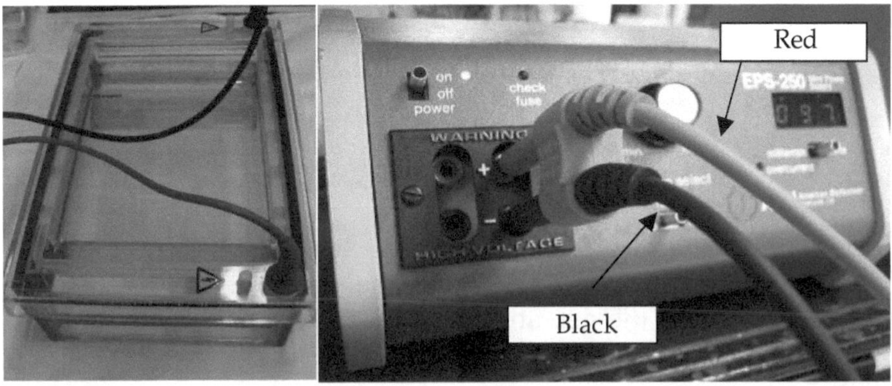

15) If the unit is working you should see bubbles that are formed in the buffer due to the electrical field that has been created; and later you should see the dyes running in the gel. After the current is applied, make sure the Gel is running in the correct direction. Bromophenol blue will run in the same direction as the DNA.

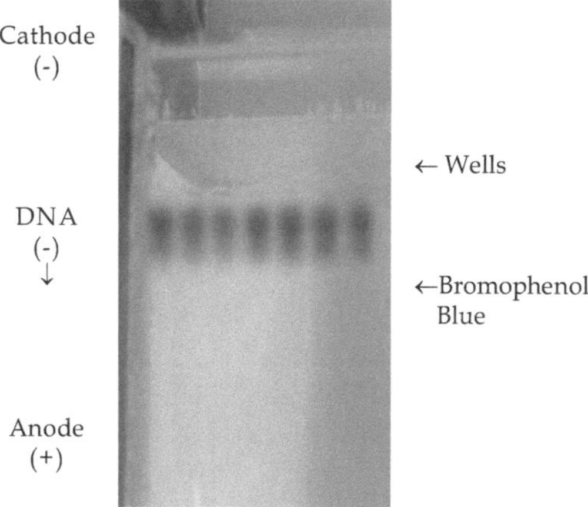

16) If ethidium bromide was present in the agarose gel, DNA may be photographed by with UV light (>500uW/cm2). The location of DNA within the gel may be determined by a film image of light transmitted by a fluorescing DNA. Thus, a wide range of bands or fragments of DNA can be detected on the film (from 200 bp to ~ 50 kb in length). Size and amount of DNA can be calculated by running standards of known size and concentration. DNA bands may be recovered from the gel and utilized for a diversity of cloning designs.

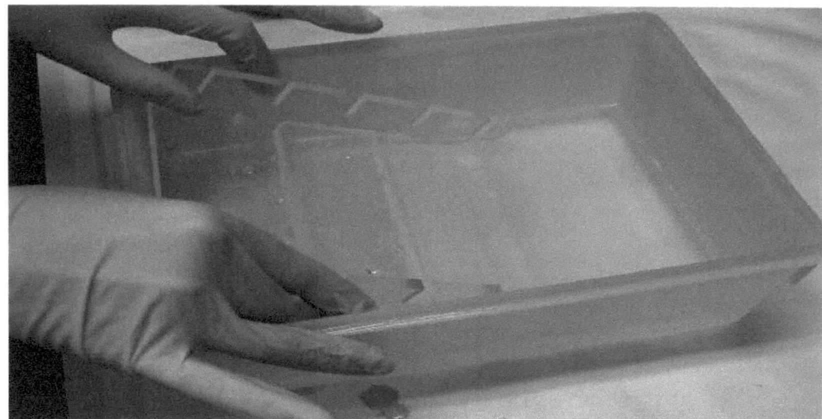

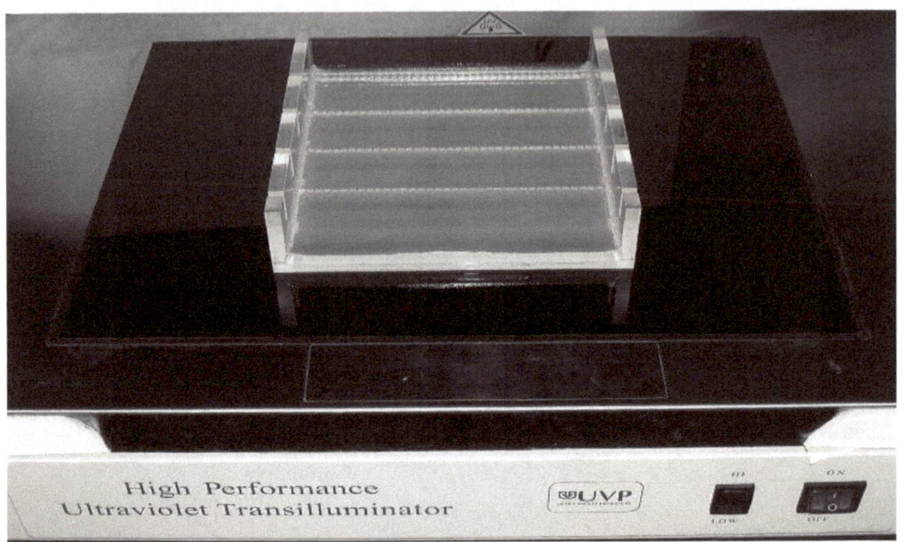

17) Visualization of DNA bands with UV light

DNA Ladder Standard

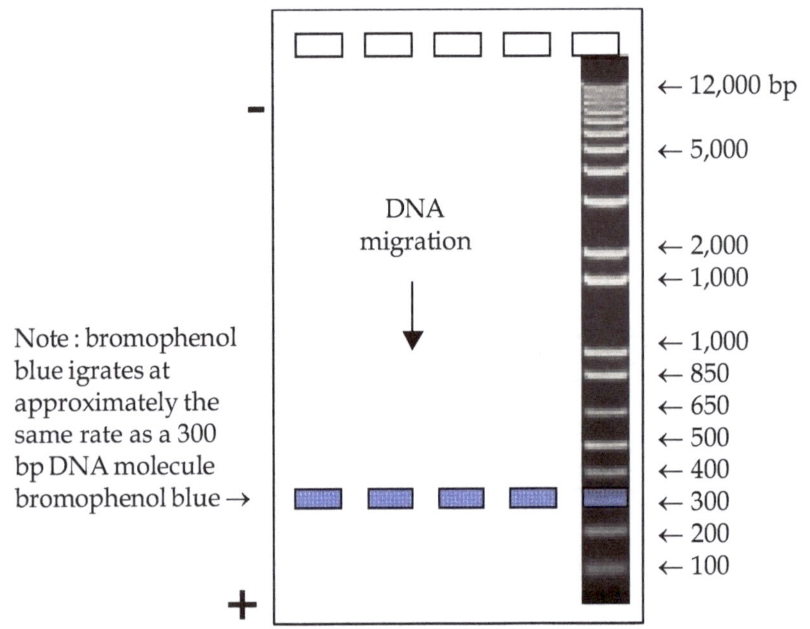

Inclusion of a DNA ladder (DNAs of know sizes) on the gel makes it easy to determine the sizes of unknown DNAs.

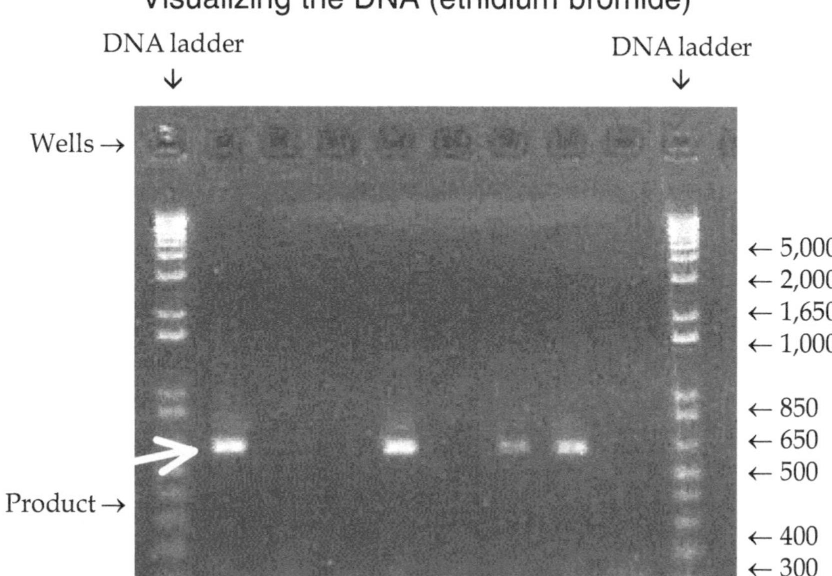

Samples : 1, 4, 6 & 7 were positive for wheat

CHAPTER 14

SDS-PAGE Electrophoresis

Introduction

Electrophoresis is the migration of charged molecules in solution in response to an electric field. Their rate of migration depends on the strength of the field; on the net charge, size and shape of the molecules and also on the ionic strength, viscosity and temperature of the medium in which the molecules are moving. As an analytical tool, electrophoresis is simple, rapid and highly sensitive. It is used analytically to study the properties of a single charged species, and as a separation technique. SDS-PAGE is the most widely used method for qualitatively analyzing any protein mixtures. The method is based on the separation of proteins according to their size and then locating by binding them to a dye. It is particularly useful for monitoring protein purity and determines their relative mass. The SDS-PAGE enables the students to understand the theory behind the techniques in checking the purity of the protein and identifying the protein in a test sample.

Separation of Proteins

Proteins are amphoteric compounds; their net charge therefore is determined by the pH of the medium in which they are suspended. In a solution with a pH above its isoelectric point, a protein has a net negative charge and migrates towards the anode in an electrical field. Below its isoelectric point, the protein is positively charged and migrates towards the cathode. The net charge carried by a protein is in addition independent of its size - i.e.: the charge carried per unit mass (or length, given proteins and nucleic acids are linear macromolecules) of molecule differs from protein to protein. At a given pH therefore, and under non-denaturing conditions, the electrophoretic separation of proteins is determined by both size and charge of the molecules.

SDS-PAGE of Proteins

Separation of Proteins Under Denaturing Conditions

Sodium dodecyl sulphate (SDS) is an anionic detergent which denatures proteins by "wrapping around" the polypeptide backbone - and SDS binds to proteins fairly specifically in a mass ratio of 1.4:1. In so doing, SDS confers a negative charge to the polypeptide in proportion to its length - ie: the denatured polypeptides become "rods" of negative charge cloud with equal charge or charge densities per

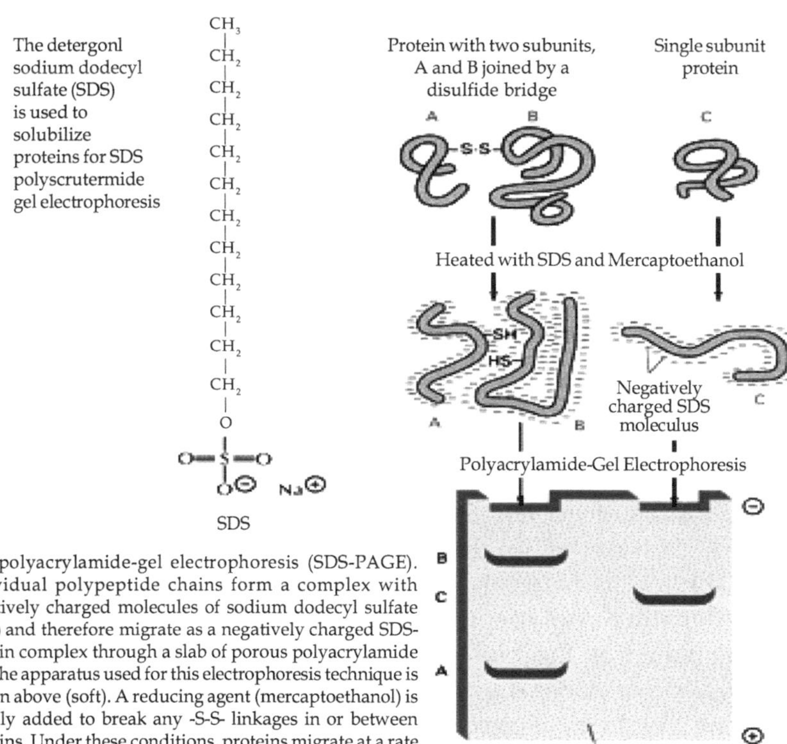

SDS polyacrylamide-gel electrophoresis (SDS-PAGE). Individual polypeptide chains form a complex with negatively charged molecules of sodium dodecyl sulfate (SDS) and therefore migrate as a negatively charged SDS-protein complex through a slab of porous polyacrylamide gel. The apparatus used for this electrophoresis technique is shown above (soft). A reducing agent (mercaptoethanol) is usually added to break any -S-S- linkages in or between proteins. Under these conditions, proteins migrate at a rate that reflects their molecular weight.

unit length. It is usually necessary to reduce disulphide bridges in proteins before they adopt the random-coil configuration necessary for separation by size: this is done with 2- mercaptoethanol or dithiothreitol. In denaturing SDS-PAGE separations therefore, migration is determined not by intrinsic electrical charge of the polypeptide, but by molecular weight.

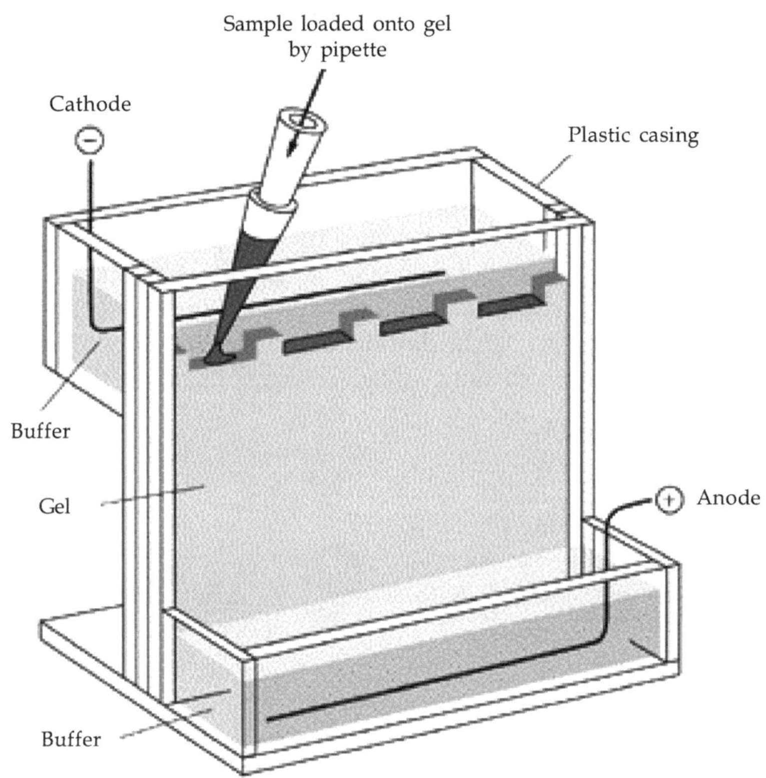

Determination of Molecular Weight

This is done by SDS-PAGE of proteins - or PAGE or agarose gel electrophoresis of nucleic acids - of known molecular weight along with the protein or nucleic acid to be characterised. A linear relationship exists between the logarithm of the molecular weight of an SDS-denatured polypeptide, or native nucleic acid, and its ***Rf***. The Rf is calculated as the *ratio of the distance migrated by the molecule to that migrated by a marker dye-front*. A simple way of determining relative molecular weight by electrophoresis (*M*r) is to plot a standard curve of distance migrated vs. log10MW for known samples, and read off the log*Mr* of the sample after measuring distance migrated on the same gel.

Principle

Crosslinked polyacrylamide gels are formed by the polymerization of acrylamide monomers in the presence of smaller amount of N, N′-methylene bis-acrylamide (generally referred as bis-acrylamide) which serves as a cross linking agent. The polymerization of acrylamide catalyzed by N′, N′, N′, N′ tetramethylenediamine (TEMED) and initiated by ammonium persulfate.

Protocol

Assembling Gel Apparatus

1. Assemble two glass plates (one notched) with two side spacers, clamps, grease, etc. as shown by demonstrators or instructions. Stand assembly upright using clamps as supports, on glass plate. Pour some pre-heated 1% agarose onto glass plate, place assembly in pool of agarose: this seals the bottom of the assembly.
2. Gel concentration of 12.5% in 0.25 M Tris-HCl pH 8.8

Reagent	**Volume (ml: to make 30 ml)**	**Volume (ml: to make 10 ml)**
40% Acrylamide stock	9.4	3.1
water (distilled)	12.3	3.8
1 M Tris-HCl pH 8.8	7.5	2.5
10% SDS	0.3	0.1
Peroxydisulphate 1%	0.5	0.5
TEMED (added last)	20 ul	20 ul
19:1 - 38:1 w:w ratio of acrylamide to N,N′-methylene bis-acrylamide		

Stacking Gels

Gel concentration of 4.5% in 0.125 M Tris-HCl pH 6.8

Reagent:	**Volume (ml to make 15 ml)**	**Volume (ml to make 10 ml)**
40% Acrylamide stock	1.7	1.1
water	10.8	7.1
1 M Tris-HCl pH 6.8	1.9	1.25
10% SDS	0.15	0.1
Peroxydisulphate 1%	0.5	0.5
TEMED (stir quickly)	20 ul	20 ul

Mix as before, then pour onto top of set resolving gel, insert comb, allow to set, remove comb, and fill with electrophoresis buffer. Assemble top tank onto glass plate assembly. Fill with electrophoresis buffer.

3. Electrophoresis buffer : The final TANK buffer composition is 196mM glycine / 0.1% SDS / 50mM Tris-HCl pH 8.3, made by diluting a 10x stock solution. This goes in both top and bottom tanks.

4. Sample Preparation : Grind a little leaf material (e.g. 2 grams) in a mortar. Centrifuge in an Eppendorf tube for 3 min. Take supernatant and mix 100ul 1:1 (v:v) with SDS-PAGE disruption mix: this is 125mM Tris-HCl pH 6.8 / 10% 2-mercaptoethanol / 10% SDS / 10% glycerol, containing a little bromophenol blue. For liquid / purified samples, take e.g. 100 ul After the stacking gel has set, carefully remove the comb wash the wells immediately with distilled water to remove non-polymerized acrylamide. Straighten PAGE apparatus with running buffer in the bottom reservoir. Any bubbles caught between the plates at the bottom of the gel can be removed by squirting running buffer through the syringe fitted with bent needle. Add running buffer to top reservoir.

5. Load sample into the bottom of the wells. The sample can be conveniently loaded using either a micro syringe (washed by pipetting in the bottom reservoir buffer between each sample) or a micropipette fitted with a long narrow tip.

6. Start electrophoresis at 100 volts.

7. When the dye front comes to 0.5 cm above the bottom of the gel, turn off the power pack. Remove the gel plates and gently dry the plates apart. Use a spatula or similar tool to separate the plates. Cut a corner from a bottom of the gel that is closet to the number 1 well.

8. Transfer the gel to a glass or plastic tray containing a minimum of 5 gel volumes of brilliant blue stain at 1X concentration. Shake at room temperature for 30-60 minutes.

9. Remove the stain and add 5 volumes of 1 X Destainer when it gets dark. Continue the process till the background is clear view the gel against a right backgroundand add 50 - 100 ul of disruption mix.

Preparation of Staining of Gels

1. Coomassie Brilliant Blue/Page-Blue 83

Make up stain: 0.2% CBB in 45:45:10 % methanol:water:acetic acid. Cover gel with staining solution, seal in plastic box and leave overnight on shaker (RT) or for 2 to 3 hours at 37 c also with agitation. Destain with 25% 65% 10% methanol water acetic acid mix, with agitation.

2. Copper Chloride (0.3M CuCl2)

Rinse gel in distilled water, immerse in copper chloride solution with agitation for about 20 minutes (RT), rinse with distilled water and immerse in sufficient fresh distilled water to cover the gel (this acts as the destaining step). Seal in a plastic box.

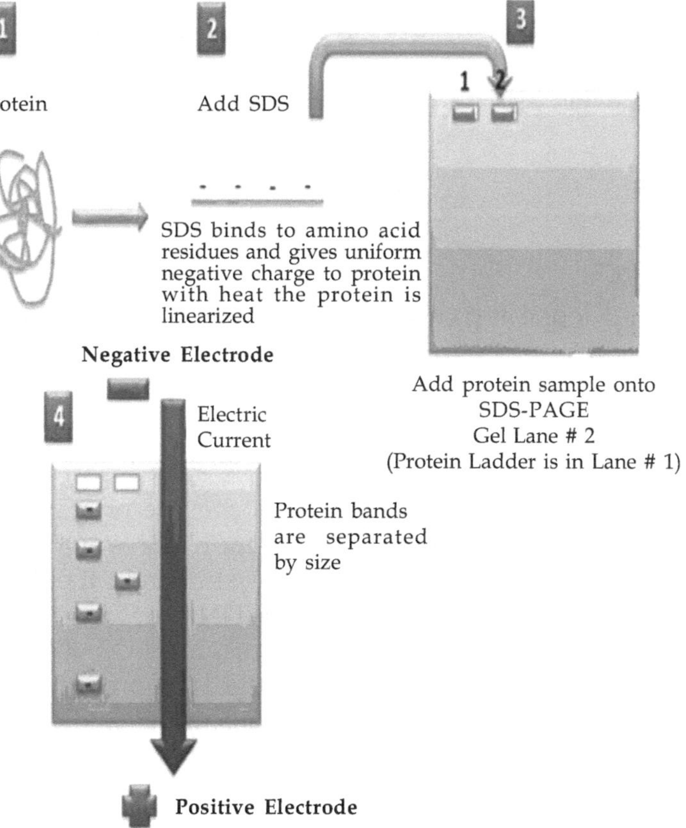

CHAPTER 15

Polymerase Chain Reaction

History of PCR

A 1971 Kleppe and co-workers first described a method using an enzymatic assay to replicate a short DNA template with primers *in vitro*. However, this early manifestation of the basic PCR principle did not receive much attention, and the invention of the polymerase chain reaction in 1983 is generally credited to Kary Mullis.

At the core of the PCR method is the use of a suitable DNA polymerase able to withstand the high temperatures of >90°C (>195°F) required for separation of the two DNA strands in the DNA double helix after each replication cycle. The DNA polymerases initially employed for in vitro experiments presaging PCR were unable to withstand these high temperatures. So the early procedures for DNA replication were very inefficient, time consuming, and required large amounts of DNA polymerase and continual handling throughout the process.

The discovery in 1976 of Taq polymerase (a DNA polymerase purified from the thermophilic bacterium, *Thermus aquaticus*, which naturally occurs in hot (50 to 80 °C (120 to 175 °F) environments) paved the way for dramatic improvements of the PCR method. The DNA polymerase isolated from *T. aquaticus* is stable at high temperatures remaining active even after DNA denaturation, thus obviating the need to add new DNA polymerase after each cycle. This allowed an automated thermocycler-based process for DNA amplification.

Developed the PCR in 1984 by Kary Mullis, PCR is now a common and often indispensable technique used in medical and biological research labs for a variety of applications. These include DNA cloning for sequencing, DNA-based phylogeny, or functional analysis of genes; the diagnosis of hereditary diseases; the identification of genetic fingerprints (used in forensic sciences and paternity testing); and the detection and diagnosis of infectious diseases. In 1993 Mullis was awarded the Nobel Prize in Chemistry for his work on PCR.

In molecular biology, the *polymerase chain reaction (PCR)* is a technique to amplify a single or few copies of a piece of DNA across several orders of magnitude, generating millions or more copies of a particular DNA sequence. The method relies on thermal cycling, consisting of cycles of repeated heating and cooling of the reaction for DNA melting and enzymatic replication of the DNA. Primers (short DNA fragments) containing sequences complementary to the target region along with a DNA polymerase are key components to enable selective and repeated amplification. As PCR progresses, the DNA generated is itself used as a template for replication, setting in motion a chain reaction in which the DNA template is exponentially amplified. PCR can be extensively modified to perform a wide array of genetic manipulations.

Almost all PCR applications emplies a heat-stable DNA polymerase, such as Taq polymerase, an enzyme originally isolated from the bacterium *Thermus aquaticus*. This DNA polymerase enzymatically assembles a new DNA strand from DNA building blocks, the nucleotides, by using single-stranded DNA as a template and DNA oligonucleotides (also called DNA primers), which are required for initiation of DNA synthesis. The vast majority of PCR methods use thermal cycling, i.e., alternately heating and cooling the PCR sample to a defined series of temperature steps. These thermal cycling steps are necessary to physically separate the strands (at high temperatures) in a DNA double helix (DNA melting) used as the template during DNA synthesis (at lower temperatures) by the DNA polymerase to selectively amplify the target DNA. The selectivity of PCR results from the use of primers that are complementary to the DNA region targeted for amplification under specific thermal cycling conditions.

PCR Principles and Procedure

PCR is used to amplify specific regions of a DNA strand (the DNA target). This can be a single gene, a part of a gene, or a non-coding sequence. Most PCR methods typically amplify DNA fragments of up to 10 kilo base pairs (kb), although some techniques allow for amplification of fragments up to 40 kb in size. A basic PCR set up requires several components and reagents. These components include:

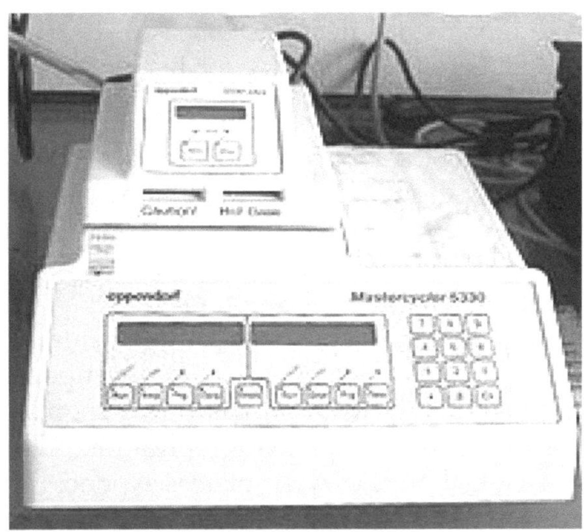

- DNA template that contains the DNA region (target) to be amplified.
- Two primers that are complementary to the 3' (three prime) ends of each of the sense and anti-sense strand of the DNA target.
- Taq polymerase or another DNA polymerase with a temperature optimum at around 70°C.
- Deoxynucleoside triphosphates (dNTPs; also very commonly and erroneously called deoxynucleotide triphosphates), the building blocks from which the DNA polymerases synthesizes a new DNA strand.
- Buffer solution, providing a suitable chemical environment for optimum activity and stability of the DNA polymerase.

- Divalent cations, magnesium or manganese ions; generally Mg^{2+} is used, but Mn^{2+} can be utilized for PCR-mediated DNA mutagenesis, as higher Mn^{2+} concentration increases the error rate during DNA synthesis.
- Monovalent cation potassium ions.

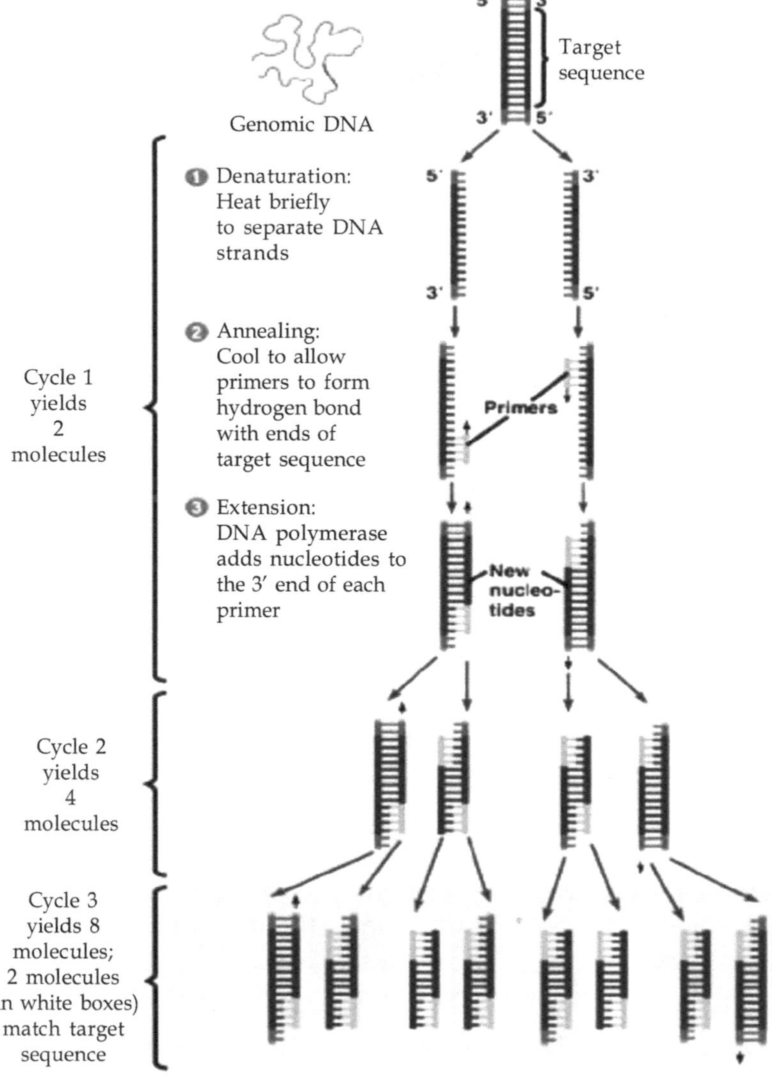

The PCR is commonly carried out in a reaction volume of 10-200 μl in small reaction tubes (0.2-0.5 ml volumes) in a thermal cycler. The thermal cycler heats and cools the reaction tubes to achieve the temperatures required at each step of the reaction (see below). Many modern thermal cyclers make use of the Peltier effect which permits both heating and cooling of the block holding the PCR tubes simply by reversing the electric current. Thin-walled reaction tubes permit favorable thermal conductivity to allow for rapid thermal equilibration. Most thermal cyclers have heated lids to prevent condensation at the top of the reaction tube. Older thermocyclers lacking a heated lid require a layer of oil on top of the reaction mixture or a ball of wax inside the tube.

Procedure

Schematic drawing of the PCR cycle.

1. Denaturing at 94-96°C.
2. Annealing at ~65°C
3. Elongation at 72°C.

Four cycles are shown here. The blue lines represent the DNA template to which primers (red arrows) anneal that are extended by the DNA polymerase (light green circles), to give shorter DNA products (green lines), which themselves are used as templates as PCR progresses.

The PCR usually consists of a series of 20 to 40 repeated temperature changes called cycles; each cycle typically consists of 2-3 discrete temperature steps. Most commonly PCR is carried out with cycles that have three temperature steps The cycling is often preceded by a single temperature step (called *hold*) at a high temperature (>90°C), and followed by one hold at the end for final product extension or brief storage. The temperatures used and the length of time they are applied in each cycle depend on a variety of parameters. These include the enzyme used for DNA synthesis, the concentration of divalent ions and dNTPs in the reaction, and the melting temperature (Tm) of the primers.

Initialization step : This step consists of heating the reaction to a temperature of 94-96°C (or 98°C if extremely thermostable polymerases are used), which is held for 1-9 minutes. It is only required for DNA polymerases that require heat activation by hot-start PCR.

- *Denaturation step* : This step is the first regular cycling event and consists of heating the reaction to 94-98°C for 20-30 seconds. It causes melting of DNA template and primers by disrupting the hydrogen bonds between complementary bases of the DNA strands, yielding single strands of DNA.

- *Annealing step* : The reaction temperature is lowered to 50-65°C for 20-40 seconds allowing annealing of the primers to the single-stranded DNA template. Typically the annealing temperature is about 3-5 degrees Celsius below the Tm of the primers used. Stable DNA-DNA hydrogen bonds are only formed when the primer sequence very closely matches the template sequence. The polymerase binds to the primer-template hybrid and begins DNA synthesis.
- *Extension/elongation step* : The temperature at this step depends on the DNA polymerase used; Taq polymerase has its optimum activity temperature at 75-80°C, and commonly a temperature of 72°C is used with this enzyme. At this step the DNA polymerase synthesizes a new DNA strand complementary to the DNA template strand by adding dNTPs that are complementary to the template in 5' to 3' direction, condensing the 5'-phosphate group of the dNTPs with the 3'-hydroxyl group at the end of the nascent (extending) DNA strand. The extension time depends both on the DNA polymerase used and on the length of the DNA fragment to be amplified. As a rule-of-thumb, at its optimum temperature, the DNA polymerase will polymerize a thousand bases per minute. Under optimum conditions, i.e., if there are no limitations due to limiting substrates or reagents, at each extension step, the amount of DNA target is doubled, leading to exponential (geometric) amplification of the specific DNA fragment.

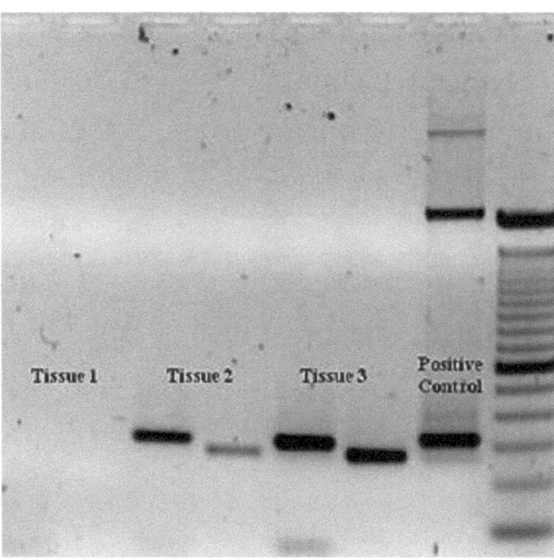

- *Final elongation* : This single step is occasionally performed at a temperature of 70-74°C for 5-15 minutes after the last PCR cycle to ensure that any remaining single-stranded DNA is fully extended.
- *Final hold* : This step at 4-15°C for an indefinite time may be employed for short-term storage of the reaction.

  Ethidium bromide-stained PCR products after gel electrophoresis. Two sets of primers were used to amplify a target sequence from three different tissue samples. No amplification is present in sample #1; DNA bands in sample #2 and #3 indicate successful amplification of the target sequence. The gel also shows a positive control, and a DNA ladder containing DNA fragments of defined length for sizing the bands in the experimental PCRs.

To check whether the PCR generated the anticipated DNA fragment (also sometimes referred to as the amplimer or amplicon), agarose gel electrophoresis is employed for size separation of the PCR products. The size(s) of PCR products is determined by comparison with a DNA ladder (a molecular weight marker), which contains DNA fragments of known size, run on the gel alongside the PCR products.

PCR Stages

10X reaction buffer (with MgCl2)	5 ul
25mM MgCl2(optional)	3 ul
dNTP mix (10mM of each dNTP)	1 ul
Taq DNA Polymerase(5U/ul)	0.25ul
downstream primer (10 um/ul)	1.5ul
upstream primer (10 um/ul)	1.5ul
DMSO	2.5ul
template DNA (10-100ng/ul)	1-2 ul
nuclease-free water (adjust to a final 50 ul)	

Cycle condition:
95 C 5 min denature step

25-35 cycles:
95 C 30 seconds denature step

55 C - 65 C 30 seconds - 60 seconds annealing primer with single strand DNA template,

72 C 30 seconds - 60 seconds extension step (time varies depending on the length of amplified DNA)

72 C 8′ final extension step

Procedure

1. Thaw 10x buffer, dNTPs, and primers. Keep on ice.
   Note: Always Include A Reaction Tube "Blank" Which Will Have All Ingredients Except DNA. This Is An Essential Control For Each PCR Reaction.
2. Create a bulk reagent tube for each set of primers to be used. This will contain the 10x buffer, nucleotides, polymerase, primers, and water that will be required for this set of PCR reactions. For the purposes of determining amounts to use in the bulk reagent tube, consider that you will have 11 reactions for each 10 reactions you will be setting up. Label bulk reaction tubes with primers to be used and number or reactions you will be setting up. Mix well and place on ice.
3. Use 0.2 ml thin walled centrifuge tubes for reaction. Label top of tubes with sample
4. Pipet mix reagent (except template) into each tube.
5. Pipet DNA samples into appropriate tubes, using a new tip for each sample.
6. Mix tube well. Spin for 5 seconds in microfuge to remove liquid from sides of tube.
7. Place tubes in PCR machine, program machine, and begin PCR reaction.

Analysis

1. Analyze the PCR reaction products by agarose gel electrophoresis of a 5 ul aliquot from the total reaction. The products should be readily visible by UV transillumination of the ethidium bromide-stained gel.
2. Store reaction products at -20 C until needed. The reaction products can be further purified using a number of procedures, including the Qiagen PCR gel extract or column purification system

Application of PCR

Isolation of Genomic DNA

PCR allows isolation of DNA fragments from genomic DNA by selective amplification of a specific region of DNA. This use of PCR augments many methods, such as generating hybridization probes for Southern or northern hybridization and DNA cloning, which require larger amounts of DNA, representing a specific DNA region. PCR supplies these techniques with high amounts of pure DNA, enabling analysis of DNA samples even from very small amounts of starting material.

Other applications of PCR include DNA sequencing to determine unknown PCR-amplified sequences in which one of the amplification primers may be used in Sanger sequencing, isolation of a DNA sequence to expedite recombinant DNA technologies involving the insertion of a DNA sequence into a plasmid or the genetic material of another organism. Bacterial colonies *(E.coli)* can be rapidly screened by PCR for correct DNA vector constructs. PCR may also be used for genetic fingerprinting; a forensic technique used to identify a person or organism by comparing experimental DNAs through different PCR-based methods.

Some PCR 'fingerprints' methods have high discriminative power and can be used to identify genetic relationships between individuals, such as parent-child or between siblings, and are used in paternity testing. This technique may also be used to determine evolutionary relationships among organisms.

Amplification and Quantitation of DNA

Because PCR amplifies the regions of DNA that it targets, PCR can be used to analyze extremely small amounts of sample. This is often critical for forensic analysis, when only a trace amount of DNA is available as evidence. PCR may also be used in the analysis of ancient DNA that is tens of thousands of years old. These PCR-based techniques have been successfully used on animals, such as a forty-thousand-year-old mammoth, and also on human DNA, in applications ranging from the analysis of Egyptian mummies to the identification of a Russian Tsar.

Quantitative PCR methods allow the estimation of the amount of a given sequence present in a sample - a technique often applied to quantitatively determine levels of gene expression. Real-time PCR is an established tool for DNA quantification that measures the accumulation of DNA product after each round of PCR amplification.

PCR in Diagnosis of Diseases

PCR allows early diagnosis of malignant diseases such as leukemia and lymphomas, which is currently the highest developed in cancer research and is already being used routinely. PCR assays can be performed directly on genomic DNA samples to detect translocation-specific malignant cells at a sensitivity which is at least 10,000 fold higher than other methods.

PCR also permits identification of non-cultivatable or slow-growing microorganisms such as mycobacteria, anaerobic bacteria, or viruses from tissue culture assays and animal models. The basis for PCR diagnostic applications in microbiology is the detection of infectious agents and the discrimination of non-pathogenic from pathogenic strains by virtue of specific genes.

Viral DNA can likewise be detected by PCR. The primers used need to be specific to the targeted sequences in the DNA of a virus, and the PCR can be used for diagnostic analyses or DNA sequencing of the viral genome. The high sensitivity of PCR permits virus detection soon after infection and even before the onset of disease. Such early detection may give physicians a significant lead in treatment. The amount of virus ("viral load") in a patient can also be quantified by PCR-based DNA quantitation techniques.

Variations on the Basic PCR Technique

- *Allele-specific PCR* : This diagnostic or cloning technique is used to identify or utilize single-nucleotide polymorphisms (SNPs) (single base differences in DNA). It requires prior knowledge of a DNA sequence, including differences between alleles, and uses primers whose 3' ends encompass the SNP. PCR amplification under stringent conditions is much less efficient in the presence of a mismatch between template and primer, so successful amplification with an SNP-specific primer signals presence of the specific SNP in a sequence.
- *Assembly PCR or Polymerase Cycling Assembly (PCA)* : Assembly PCR is the artificial synthesis of long DNA sequences by performing PCR on a pool of long oligonucleotides with short overlapping segments. The oligonucleotides alternate between

sense and antisense directions, and the overlapping segments determine the order of the PCR fragments thereby selectively producing the final long DNA product.

- *Asymmetric PCR* : Asymmetric PCR is used to preferentially amplify one strand of the original DNA more than the other. It finds use in some types of sequencing and hybridization probing where having only one of the two complementary strands is required. PCR is carried out as usual, but with a great excess of the primers for the chosen strand. Due to the slow (arithmetic) amplification later in the reaction after the limiting primer has been used up, extra cycles of PCR are required. A recent modification on this process, known as Linear-After-The-Exponential-PCR (LATE-PCR), uses a limiting primer with a higher melting temperature (Melting temperature | Tm) than the excess primer to maintain reaction efficiency as the limiting primer concentration decreases mid-reaction.
- *Helicase-dependent amplification*: This technique is similar to traditional PCR, but uses a constant temperature rather than cycling through denaturation and annealing/extension cycles. DNA Helicase, an enzyme that unwinds DNA, is used in place of thermal denaturation.
- *Hot-start PCR*: This is a technique that reduces non-specific amplification during the initial set up stages of the PCR. The technique may be performed manually by heating the reaction components to the melting temperature (e.g., 95°C) before adding the polymerase. Specialized enzyme systems have been developed that inhibit the polymerase's activity at ambient temperature, either by the binding of an antibody or by the presence of covalently bound inhibitors that only dissociate after a high-temperature activation step. Hot-start/cold-finish PCR is achieved with new hybrid polymerases that are inactive at ambient temperature and are instantly activated at elongation temperature.
- *Intersequence-specific PCR (ISSR) :* A PCR method for DNA fingerprinting that amplifies regions between some simple sequence repeats to produce a unique fingerprint of amplified fragment lengths.
- *Inverse PCR* : a method used to allow PCR when only one internal sequence is known. This is especially useful in identifying flanking

sequences to various genomic inserts. This involves a series of DNA digestions and self ligation, resulting in known sequences at either end of the unknown sequenc.

- *Ligation-mediated PCR* : This method uses small DNA linkers ligated to the DNA of interest and multiple primers annealing to the DNA linkers; it has been used for DNA sequencing, genome walking, and DNA footprinting.
- *Methylation-specific PCR (MSP)* : The MSP method was developed by Stephen Baylin and Jim Herman at the Johns Hopkins School of Medicine, and is used to detect methylation of CpG islands in genomic DNA. DNA is first treated with sodium bisulfite, which converts unmethylated cytosine bases to uracil, which is recognized by PCR primers as thymine. Two PCRs are then carried out on the modified DNA, using primer sets identical except at any CpG islands within the primer sequences. At these points, one primer set recognizes DNA with cytosines to amplify methylated DNA, and one set recognizes DNA with uracil or thymine to amplify unmethylated DNA. MSP using qPCR can also be performed to obtain quantitative rather than qualitative information about methylation.
- *Miniprimer PCR* : Miniprimer PCR uses a novel thermostable polymerase (S-Tbr) that can extend from short primers ("smalligos") as short as 9 or 10 nucleotides, instead of the approximately 20 nucleotides required by Taq. This method permits PCR targeting smaller primer binding regions, and is particularly useful to amplify unknown, but conserved, DNA sequences, such as the 16S (or eukaryotic 18S) rRNA gene. 16S rRNA miniprimer PCR was used to characterize a microbial mat community growing in an extreme environment, a hypersaline pond in Puerto Rico. In that study, deeply divergent sequences were discovered with high frequency and included representatives that defined two new division-level taxa, suggesting that miniprimer PCR may reveal new dimensions of microbial diversity. By enlarging the "sequence space" that may be queried by PCR primers, this technique may enable novel PCR strategies that are not possible within the limits of primer design imposed by Taq and other commonly used enzymes.
- *Multiplex Ligation-dependent Probe Amplification (MLPA)* : permits multiple targets to be amplified with only a single primer pair, thus avoiding the resolution limitations of multiplex PCR.

- *Multiplex-PCR* : The use of multiple, unique primer sets within a single PCR mixture to produce amplicons of varying sizes specific to different DNA sequences. By targeting multiple genes at once, additional information may be gained from a single test run that otherwise would require several times the reagents and more time to perform. Annealing temperatures for each of the primer sets must be optimized to work correctly within a single reaction, and amplicon sizes, i.e., their base pair length, should be different enough to form distinct bands when visualized by gel electrophoresis.
- *Nested PCR* : increases the specificity of DNA amplification, by reducing background due to non-specific amplification of DNA. Two sets of primers are being used in two successive PCRs. In the first reaction, one pair of primers is used to generate DNA products, which besides the intended target, may still consist of non-specifically amplified DNA fragments. The product(s) are then used in a second PCR with a set of primers whose binding sites are completely or partially different from and located 3' of each of the primers used in the first reaction. Nested PCR is often more successful in specifically amplifying long DNA fragments than conventional PCR, but it requires more detailed knowledge of the target sequences.
- *Overlap-extension PCR* : is a genetic engineering technique allowing the construction of a DNA sequence with an alteration inserted beyond the limit of the longest practical primer length.
- *Quantitative PCR (Q-PCR)* : is used to measure the quantity of a PCR product (preferably real-time). It is the method of choice to quantitatively measure starting amounts of DNA, cDNA or RNA. Q-PCR is commonly used to determine whether a DNA sequence is present in a sample and the number of its copies in the sample. The method with currently the highest level of accuracy is Quantitative real-time PCR. It is often confusingly known as RT-PCR (Real Time PCR) or RQ-PCR. QRT-PCR or RTQ-PCR are more appropriate contractions. RT-PCR commonly refers to reverse transcription PCR (see below), which is often used in conjunction with Q-PCR. QRT-PCR methods use fluorescent dyes, such as Sybr Green, or fluorophore-containing DNA probes, such as TaqMan, to measure the amount of amplified product in real time.

- *RT-PCR* : (Reverse Transcription PCR) is a method used to amplify, isolate or identify a known sequence from a cellular or tissue RNA. The PCR is preceded by a reaction using reverse transcriptase to convert RNA to cDNA. RT-PCR is widely used in expression profiling, to determine the expression of a gene or to identify the sequence of an RNA transcript, including transcription start and termination sites and, if the genomic DNA sequence of a gene is known, to map the location of exons and introns in the gene. The 5' end of a gene (corresponding to the transcription start site) is typically identified by an RT-PCR method, named RACE-PCR, short for *Rapid Amplification of cDNA Ends*.
- *Solid Phase PCR* : encompasses multiple meanings, including Polony Amplification (where PCR colonies are derived in a gel matrix, for example), 'Bridge PCR' (the only primers present are covalently linked to solid support surface), conventional Solid Phase PCR (where Asymmetric PCR is applied in the presence of solid support bearing primer with sequence matching one of the aqueous primers) and Enhanced Solid Phase PCR (where conventional Solid Phase PCR can be improved by employing high Tm solid support primer with application of a thermal 'step' to favour solid support priming).
- *TAIL-PCR* : Thermal asymmetric interlaced PCR is used to isolate unknown sequence flanking a known sequence. Within the known sequence TAIL-PCR uses a nested pair of primers with differing annealing temperatures; a degenerate primer is used to amplify in the other direction from the unknown sequence.
- *Touchdown PCR* : a variant of PCR that aims to reduce nonspecific background by gradually lowering the annealing temperature as PCR cycling progresses. The annealing temperature at the initial cycles is usually a few degrees (3-5°C) above the T_m of the primers used, while at the later cycles, it is a few degrees (3-5°C) below the primer T_m. The higher temperatures give greater specificity for primer binding, and the lower temperatures permit more efficient amplification from the specific products formed during the initial cycles.
- *PAN-AC* : This method uses isothermal conditions for amplification, and may be used in living cells.

- *Universal Fast Walking* : this method allows genome walking and genetic fingerprinting using a more specific 'two-sided' PCR than conventional 'one-sided' approaches (using only one gene-specific primer and one general primer - which can lead to artefactual 'noise') by virtue of a mechanism involving lariat structure formation. Streamlined derivatives of UFW are LaNe RAGE (lariat-dependent nested PCR for rapid amplification of genomic DNA ends), 5'RACE LaNe and 3'RACE LaNe.

CHAPTER 16

Restriction Enzyme Digestion

Introduction

The primary tools used by the molecular biologists in manipulating DNA are restriction enzymes and other DNA/RNA modifying enzymes. Over the years, the number and uses of these enzymes have increased as new molecules have been discovered. Restriction endonucleases are bacterial enzymes that cleave double-stranded DNA. They are present in bacteria presumably to destroy DNA from foreign sources. By cleaving the foreign DNA at specific sites. The host bacterial DNA os protected from cleavage because specific recognition sites are modified, usually by methylation form methylase at one of the bases in the site, making the site no longer a substrate for RE cleavage. Host bacteria used to propagate cloned DNA in the laboratory arte usually mutant in the host restriction genes, thus their intracellular enzyme activities will not destroy the foreign recombinant sequences. The endonuclaase with its accompanying methylase is called a restriction modification (R-M) system.

At least four different kinds of R-M systems exist characterized on the basis of the subunit composition, cofactor requirements and type of DNA cleavage. Most characterized enzymes belong to type II class, together with type IIS class. They comprise the commercially available restriction enzymes used for DNA analysis and manipulation. Type I and Type III enzymes are relatively uncommon and a few additional enzymes fit none of the classes.

Among more than 3000 different enzymes isolated from bacterial strains many share common specificities. Restriction enzymes that recognize identical sequences have been called as isoschizomers.

Principle

All restriction enzymes cleave their DNA substrate to form 5′-phjpsphate and 3′ hydroxyl termini on each strand. The breaks can be staggered, generating either 5′ phosphate extension on each strand or 3′-hydroxyl extension on each strand, or they can be blunt ends.

Type II restriction enzymes have been characterized primarily with respect to their recognition sequences and cleavage specifically rather than their protein properties. The symmetrical recognition sequence of restriction enzymes is termed as palindromes and is of four to eight base pairs. Most. But not all, recognition sequences contain dyad axis symmetry and in most case all the bases within the site are uniquely specified. Those with degenerate or relaxed specificities can recognize multiple bases at some positions.

The enzyme EcoR1 recognizes the sequence

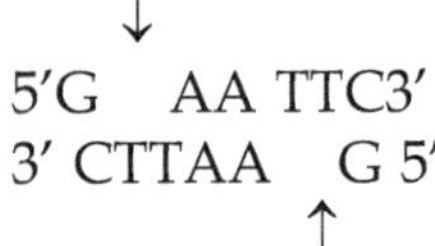

It gives cohesive end or the break is staggered.

Definition of Unit

One unit enzyme is defined as the amount of enzyme required to completely digest one microgram of lambda DNA in a reaction volume of 50ìl in 1 hour under optimal conditions of salt, pH and temperature. All digestions are performed at 37°C, unless noted otherwise.

Factors Affecting Optimum Activity

1. *Temperature* : The optimal digestion of DNA varies over a wide temperature range for different restriction enzymes.

2. *Buffer Systems* : Tris-HCl is the most commonly used buffering agent in incubation mixtures. The buffer system is markedly temperature dependent. The change in pH/10 deg amounts to approximately 0.3.
3. *Ionic Conditions* : Mg^{2+} ions are an absolute requirement for all restriction endonucleases, where as the addition of other salt components depends on different nucleases. If different ionic conditions are required for site-specific cleavage with two or more restriction endonucleases the DNA is to be digested first with the enzyme with an optimum activity at low ionic strength.
4. *Methylation of DNA* : Because restriction endonucleases are a part of prokaryotic restriction/modification system, digestion of DNA can be strongly affected by Methylation of specific adenine or cytidine residues within the recognition sequence of the site specific restriction enzyme of interest.

Principle

EcoR1 has 5 recognition sites on DNA. DNA is the linear double stranded DNA having 48,502 base pairs. In this experiment, the substrate for EcoR1 is DNA. The position of 5 recognition sites on DNA are: 21, 226, 2614, 31747, 39168 and 44972. Upon complete digestion of the substrate under optimal conditions 6 fragments are released whose molecular weights are 21,226bp, 7421bp, 5804bp, 5643bp, 4878bp and 3530bp. Therefore, fragment sizes of the digested sample can be assessed by electrophoresing along with the standard molecular weight marker.

Materials Required

- Substrate DNA such as Lambda DNA or plasmid DNA, DNA samples (1 μg/μl): Dissolve DNA samples in a suitable buffer (10 mM Tris-HCI, pH 8.0) at concentration of 1μg DNA/μl .
- Restriction Enzymes, EcoRI or Hind III
- Assay buffers,
- Nuclease buffers,
- Nuclease free water,
- 50X TAE buffer,
- Agarose,

- Ethidium bromide,
- Gel loading Dye,
- Dry bath, Microfuge,
- Gel tank, combs, cords and power supply, Microwave/heater

Procedure

1. Check the name of the given enzyme and DNA, note the concentrations respectively. Find out the suitable buffer for the enzyme.
2. Thaw the buffer vial completely
3. Add the reaction components in the following order to the vial labeled with enzyme digest.

   Calculate and enter the column to be added below:

Components	Volume to be added	Volume added
Water	Xµl	To make up to 50 µl
Assay buffer	Xµl	1x(1/10th volume)
DNA	Xµl	1µg
Enzyme	Xµl	10-20 units

   Mix the contents gently by finger flicking after each addition

   Note
   - Enzyme is thermolabile.
   - Use fresh tip for each addition

4. Incubate at 37^{0}C (or at optimal temperature of the enzyme activity) for 1 hour.
5. Meanwhile prepare 1.0% agarose gel.
6. Stop the reaction by adding 3µl gel loading dye, mix the contents.
7. Load the samples into the will carefully and run the gel for about 1 hour (till the dye reaches the end of the gel)
8. Visualize the UV light, note down the observation

Precautions

1. Restriction enzymes should be taken out of -20°C freezer only at the time of use and should be immediately kept on ice. Store them back at -20°C immediately after use.
2. Ethidium bromide is a powerful mutagen and carcinogenic. Gloves should be worn while working with solutions containing this dye.
3. Follow all the general precautions for handling nucleic acids.
4. The concentration of restriction enzyme and buffer in the final volume of reaction mixture should be 1x. So, the reaction volumes should be adjusted accordingly.

CHAPTER 17

Extraction of DNA Fragments from Agarose Gel

DNA can be easily isolated and purified after size selection on an agarose gel. The fragment of interest is simply cut out of the gel with a razor blade and purified by a number of different methods.The easiest is to use a method that involves first dissolving the agarose slice in a solution at 50°C, then binding the DNA from the melted agarose to a silica-gel membrane.

Principle

In this method following gel electrophoresis using low gelling/ melting temperature agarose, the band of interest is removed from the gel. The agarose slice is then melted and subjected to phenol extraction.

Materials and Reagents

1. Low gelling agarose
2. 1 x TBE buffer
3. Buffered phenol
4. TE

Procedure

1. Digest DNA sample to completion. Pour, load and electrophorese on 1 % low gelling/melting temperature agarose gel in 1 x TBE buffer.

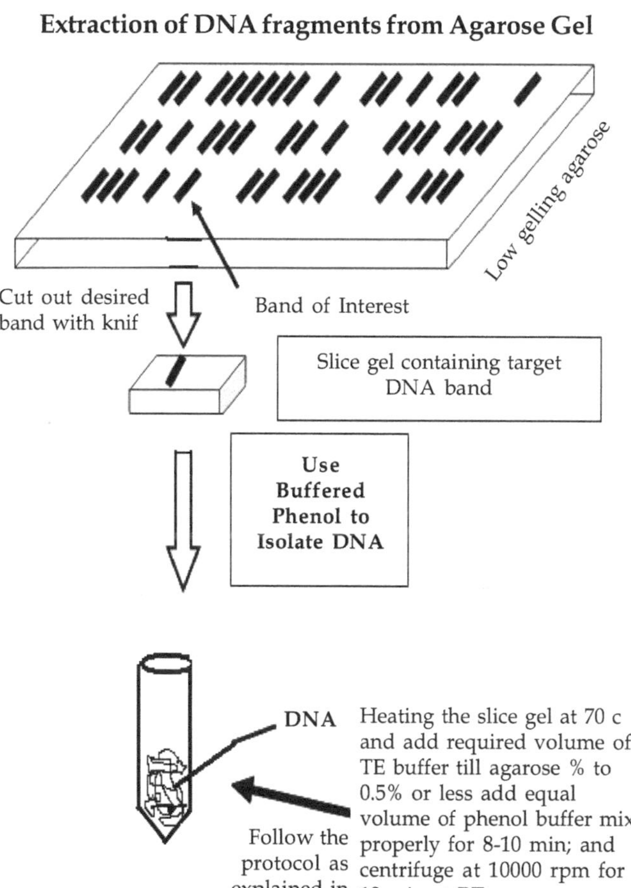

2. Stain the gel using ethidium bromide and cut out the desired band with a scalpel.
3. Melt the gel slice at 70°C and add enough TE buffer to decrease agarose percentage to 0.5% or less.
4. Add an equal volume of buffered phenol; mix vigorously for 5 to 10 min; and centrifuge at 10,000 rpm for 10 min at room temperature.

Note *Use buffered phenol and not the phenol: chloroform : isoamyl alcohol mixture. A thick white interface will be visible after centrifugation.*

5. Collect the aqueous phase and set aside. Re-extract phenol phase and interface with an equal volume of TE buffer. Centrifuge as above.
6. Collect second aqueous phase. If large interface still appears re-extract once more.
7. Combine all the aqueous phases and ethanol precipitate.

Example of DNA Quantitation

Volume of DNA : 100µl

Dilution: 10µl DNA sample + 390µl distilled water (1:40 dilution) A260 of diluted sample (1 cm path length) = 0.60

DNA concentration (µg/ml) = 50 x A_{260} x dilution factor = 50 x 0.60 x 40 = 1200µg/ml

Total amount of DNA = concentration x volume of sample in ml = 1200µg/ml x 0.1 ml = 120µg of DNA.

Determination of Nucleic Acid Purity

- Determine DNA or RNA purity by measuring the A_{260}/A_{280} ratio.
- Measure A_{260}/A_{280} ratios in low-salt buffer. Buffered solutions provide more accurate and reliable values than water since A_{260}/A_{280} ratios are influenced considerably by pH. Lower pH results in lower A_{260}/A_{280} values and reduced sensitivity to protein contamination.
- Pure DNA has an A_{260}/A_{280} ratio of 1.8-2.0 in 10 mM Tris-CI, pH 8.5.
- DNA with protein as impurity would have an A_{260}/A_{280} ratio < 1.8.
- DNA with RNA as impurity would have an A_{260}/A_{280} ratio >2.0.
- Conservative estimates indicate that value between 1.8-.1.9 are appropriate for pure DNA and pure RNA is 1.9-2.0.

Absorbance by Contaminants

- Strong absorbance at 280 nm, resulting in a low A_{260}/A_{280} ratio (1.8), indicates the presence of contaminants such as proteins.
- Strong absorbance at 270 and 275 nm may indicate the presence of contaminating phenol.
- Absorbance of 325 nm suggests contamination by particulates in the solution or dirty cuvettes.

CHAPTER 18

Isolation the Mammalian DNA from Whole Blood

Introduction

This method can be scaled to yield amounts of DNA ranging from less than ten to more than hundreds of micrograms of DNA. However, shearing forces are generated at every step, with the result that the DNA molecules in the final preparation rarely exceed 100-150 kb in length. DNA of this size is adequate for Southern analysis on standard agarose gels, as a template in Polymerase chain reactions (PCRs), and the construction of genomic DNA libraries in bacteriophage Lambda vectors.

Principle

The method described in protocol involves digesting eukaryotic cells or tissues with proteinase K in the presence of EDTA (to sequester divalent cations and thereby inhibit DNases) and solubilizing membranes and denaturing proteins with a detergent such as SDS. The nucleic acids are then purified by phase extractions (isopropanol). Contaminating RNA is eliminated by digestion with an RNase, and low molecular-weight substances are removed by dialysis (or centrifugation).

Requirements

- Blood,
- Cell lysis buffer,
- Potasium acetate solution,
- Proteinase K,
- Ethanol,
- Isopropanol, TE (pH 7.6)

Method

1. Take 40µl of white blood cell pellets in microfuge tubes, from the given sample. Add 600µl of ice-cold cell lysis buffer. Homogenize the suspension quickly by inverting the tubes several times or gently tap the tube.

   The SDS will precipitate from the ice-cold cell lysis buffer producing a cloudy solution. This precipitate will not affect isolation of DNA.

2. (Optional). Add 3µl of proteinase K solution to the lysate to increase the yield of genomic DNA. Incubate the digest for 15-60 minutes for atleast 3 hours but no more than 16 hours at 55 °C.

3. Allow the digest to cool to room temperature ant then add 3µl of 4mg/ml DNase-free RNase. Incubate the digest for 15-60 minutes at 37°C.

4. Allow the sample to cool to room temperature. Add 200µl of potassium acetate solution and mix the contents of the tube by vortexing vigorously for 20 seconds.

5. Pellet the precipitated protein/SDS complex by centrifugation at maximum speed for 3 minutes at 4°C in a microfuge. A pellet of protein should be visible at the bottom of the microfuge tube after centrifugation. If not, incubate the lysate for 5 minutes on ice and repeat the centrifugation step.

6. Transfer the supernatant to a fresh microfuge tube containing 600µl of isopropanol. Mix the solution well and then recover the precipitate of DNA by centrifuging the tube at maximum speed for 1 minute at room temperature in a microfuge.

7. Remove the supernatant by aspiration and add 600µl of 70% ethanol to the DNA pellet. Invert the tube several times and

centrifuge the tube at maximum speed for 1 minute at room temperature in a microfuge.

8. Carefully remove the supernatant by pipettes (or aspiration) and allow the DNA pellet to dry in air for 15 minutes.

9. Redissolve the pellet of DNA in 100 μl of TE (pH 7.6).

   The solubilisation of the genomic DNA pellet can be facilitated by incubation for 16 hours at room temperature or for 1 hour at 65°C.

CHAPTER **19**

Isolation of Plasmid DNA from Bacteria

By Alkaline Lysis Method

Introduction

DNA of prokaryotic cells is relatively simple in comparison to that of eukaryotic cells. In addition to chromosomal DNA, a bacterium may also carry an additional piece of DNA called a plasmid. Bacterial plasmids are double stranded closed circular DNA molecules that range in size from 1 kb to more than 200kb. Plasmids are useful to bacterial cells since they carry genes for antibody resistance, RM system etc that allow bacteria to survive in non-ideal conditions.

Plasmids constructed in the laboratory are used as cloning vehicles. These synthetic plasmids contain only essential features such as origin of replication, antibiotic resistance, multiple cloning sites etc. various copy number of plasmid For e.g. pBR322 20 copy/cell - low copy number, pUC 18, pUC 18- 800 copy/cell - high copy number.

Alkaline lysis, in combination with the detergent sds, has been used for more than 20 years to isolate plasmid DNA from *E.coli* (Birnboim and Doly 1979). This experiment describes how to extract plasmid from bacteria - which involves growth of bacterial culture, harvesting and lysis of bacteria and isolation of plasmid DNA from the lysate using alkaline lysis method. The DNA is analysed on agarose gel.

This method exploits the relatively small size and covalently closed circular nature of plasmid DNA. The bacterial cell pellet obtained after growing and harvesting is treated with different solutions.

Principle

Exposure of bacterial suspensions to the strongly anionic detergent at high pH opens the cell wall, denatures chromosomal DNA and proteins, and releases plasmid DNA into the supernatant. Although the alkaline solution completely disrupts base pairing, the strands of closed circular plasmid DNA are unable to separate from each other because they are topologically intertwined. As long as the intensity and duration of exposure to OH- is not too great, the two strands of plasmid DNA fall once again into register when the pH is returned to neutral.

During lysis, bacterial proteins, broken cell walls, and denatured chromosomal DNA become enmeshed in large complexes that are coated with dodecyl sulfate. These complexes are efficiently precipitated from solution when sodium ions arte replaced by potassium ions (salting out). After the denatured material has been removed by centrifugation, native plasmid DNA can be recovered from the supernatant.

Strands of covalently closed circular (CCC) plasmid DNA unable to separate because of intertwining, renatures immediately and completely when conditions return to normal. The genomic DNA forms an aggregate with SDS and proteins.

Materials Required

LB medium, LB Agar, Bacterial host DH5a plate (Containing plasmid pUC18) Solution I, II and III, Ampicillin, Gel loading dye, RNase, Agarose, 50XTAE, Isoproponal, 1XTE, Autoclaved nuclease free Water, 70% Ethanol, Laminar flow hood, Tabletop centrifuge, Dry bath, Vortex mixer, Microwave/ heater, 37^{0}C Shaker, incubator.

Experimental Protocol

Note : Streaking & inoculation should be done aseptically.

DAY 1 : Streak the given bacterial culture on LB Agar containing Ampicillin (100mg/ml). Incubate at 37^{0}C. (Inverted) overnight.

DAY2 : Inoculate a single colony into 5ml of LB medium containing 100mg/ml Ampicillin. Incubate at 37^{0}C with shaking for 8-16 hrs.

DAY3 : Refer to the flowchart and follow the steps carefully. (For harvesting and alkali lysis)

1. Transfer 1.5ml overnight grown culture into eppendorf tube and centrifuge for 5 minutes at 6000 rpm.
2. Drain the supernatant and gently tap the tube inverted on a paper towel to remove the excess medium.
3. Gently vortex the pellet using vortex mixer so that the suspension disperses uniformly.
4. Add 100ml of solution I to this suspension and mix the contents by finger flicking the tube. No visible clumps of bacteria should remain.

   (Glucose in this solution provides an isoosmotic condition to prevent physical shock. The resuspended solution's pH is raised to basic level with Tris to help denature DNA. EDTA stabilizes the cell membrane by binding the divalent cations. RNase will destroy RNA from the cell contents when the membrane is lysed). Incubate at RT for 5 minutes.
5. Add 200ml of solution II and gently mix the contents by inverting the tube 4-5 times. The cell suspension should look clear at this stage.

   (The sodium dodecyl sulfate (SDS) is an ionic detergent, which dissolves the phospholipids and protein components of the cell membrane. Sodium hydroxide (NaOH) in the solution denatures the Plasmid and the chromosomal DNA into single strands).
6. Add 150ml of solution III and mix the contents by inverting the tube 4-5 times.

   (Sodium acetate from this solution forms an insoluble precipitate of SDS / lipid / protein complex and neutralizes the sodium hydroxide from the previous step. At this neutral pH, the DNA renature. The chromosomal DNA is trapped in the SDS / lipid / protein precipitate. The Plasmid DNA renatures into its double stranded form and escapes being trapped in the precipitate and remains in the supernatant).

7. Centrifuge the tube containing lysate at 10,000 rpm for 15 minutes. (White precipitate is seen on the wall of the tube).
8. Gently transfer the supernatant into a fresh tube without disturbing the pellet. (The pellet is gooey and suddenly can slip into the tube).
9. Add 0.5ml of isopropanol this suspension and mix by inverting the tube. Allow to stand for ~ 10 minutes. (The suspension should turn turbid slightly).
10. Centrifuge at 10,000rpm for 15 minutes.
11. Carefully drain off the supernatant and mark the pellet.
12. Add 200ml of 70% ethanol to the opposite side of the wall from the pellet. Drain off the alcohol and keep the tube inverted on paper towel. (DNA can be dried at 37^{0}C also).
13. When the pellet turns transparent, add 50ml of 1X TE to the pellet and resuspend by finger flicking. (This is your Plasmid DNA preparation, should contain 4-5mg of DNA).
14. Prepare an agarose gel of 1%.
15. Load 10-15ml of your DNA sample after addition of gel loading dye and run the gel at 100V for 30-40 minutes.
16. Visualize with UV doc system and note the observation.

Observation

A good preparation results in only super coiled DNA. Other forms - nicked and linear may be seen depending on the preparation. Some times chromosomal DNA may be seen in the preparation.

Solution I

50 mM glucose

25 mM Tris-CI (pH 8.0)

10 mM EDT A (pH 8.0)

Solution I can be prepared in batches of approximately 100 mL autoclaved for 15 minutes at 10 lb/sq and stored at 4°C.

Solution II

0.2 N NaOH (freshly diluted from 10 N stock)

1% SDS

Close the tube tightly and mix the contents by inverting the tube rapidly for 5 minutes. Make sure that the entire surface of the tube comes in contact with Solution II. Do not vortex. Store the tube on ice.

Solution III

5 M potassium acetate 60 mL

Glacial acetic acid 11.5 mL

Water 28.5 mL

The resulting solution is 3 M with respect to potassium Clod 5 M with respect to acetate.

Close the tube and vortex it gently in an inverted position for 10 seconds to disperse

LB broth with Ampicilin

Tryptone	1 gms
Yeast extract	0.5 gms
NaCl	1 gms
D/W	100 ml
pH	7.5

After autoclave add ampicillin to have final concentration of 100µg/ml.

Culture

pUC 18 containing *E. coli*

TES

10 mM Tris - Cl, pH 8.0

100mM NaCl

1mM EDTA

3 M Sodium Acetate (pH 5.2)

Dissolve 40.81 g of sodium acetate. 3 H_2O in 80 ml water adjust pH to 5.2 with glacial acetic acid. Make up to the volume to 100 ml autoclave and store.

Preparation of Agarose Gel for Electrophoresis

- Prepare 10X TAE buffer, autoclave it and store in a glass bottle and keep at room temperature. (you can use this buffer after dilution up to six month if it is prepared and maintained properly).
- Take 500 ml sterile distilled water and add 10 ml of 10X TAE using sterile pipette (This will give you 1X TAE)
- From this 1X TAE, take 50 ml in EM flask and add 1% (0.5 gm) agarose, boil it in microwave oven for approximately 1 to 2 minutes.
- Allow it to cool down little and then add 2 to 3 ìl ethydium bromide [prepare Etbr stock as : 5 mg Etbr in 1 ml and use it), mix it properly and cast the gel.

Important Notes

- The original protocol requires the addition of lysozyme. Before the addition of solution II, this is not necessary. Do not vortex the tubes after addition of solution II.
- If the plasmid preparation is strictly for analytical purposes and nonenzymatic manipulations are contemplated, then the phenol:chloroform step can be avoided.
- It is important to remove all the supernatant fluid after harvesting the bacterial pellet and all traces of ethanol, etc., after precipitation. The phenol has to be tris-saturated to pH 8.0 and of very good quality.
- Do not disturb the whitish interface while removing the upper aqueous phase after phenol and chloroform treatments.
- While washing with 70% ethanol, do not break the DNA pellet. This step is meant for washing the pellet only to remove traces of ethanol and salts.
- If the pellet is disturbed at this stage, it will be difficult to recover the DNA.
- This preparation will contain a lot of RNA contamination. DNAse-free
- RNAse may be added before the phenol:chloroform step to digest the RNA. Otherwise RNAse may be added, along with the restriction enzyme, during subsequent manipulations.

- The protocol may be upgraded to accommodate up to 10 mL of bacterial culture.
- Sometimes it may become difficult to dissolve the plasmid preparation in IE. Keep it in the freezer overnight. The next day, the DNA will easily dissolve.

CHAPTER 20

Random Amplified Polymorphic DNA (RAPD)

RAPD stands for Random Amplification of Polymorphic DNA. It is a type of PCR reaction, but the segments of DNA that are amplified are random. The scientist performing RAPD creates several arbitrary, short primers (8-12 nucleotides), then proceeds with the PCR using a large template of genomic DNA, hoping that fragments will amplify. By resolving the resulting patterns, a semi-unique profile can be gleaned from a RAPD reaction.

No knowledge of the DNA sequence for the targeted gene is required, as the primers will bind somewhere in the sequence, but it is not certain exactly where. This makes the method popular for comparing the DNA of biological systems that have not had the attention of the scientific community, or in a system in which relatively few DNA sequences are compared (it is not suitable for forming a DNA databank). Because it relies on a large, intact DNA template sequence, it has some limitations in the use of degraded DNA samples. Its resolving power is much lower than targeted, species specific DNA comparison methods, such as short tandem repeats. In recent years, RAPD is used to characterize, and trace, the phylogeny of diverse plant and animal species.

Introduction

Random Amplified Polymorphic DNA (RAPD) markers are decamer (10 nucleotide length) DNA fragments from PCR amplification of random segments of genomic DNA with single primer of arbitrary nucleotide sequence and which are able to differentiate between genetically distinct individuals, although not necessarily in a reproducible way.

How It Works

Unlike traditional PCR analysis, RAPD (pronounced "rapid") does not require any specific knowledge of the DNA sequence of the target organism: the identical 10-mer primers will or will not amplify a segment of DNA, depending on positions that are complementary to the primers' sequence. For example, no fragment is produced if primers annealed too far apart or 3' ends of the primers are not facing each other. Therefore, if a mutation has occurred in the template DNA at the site that was previously complementary to the primer, a PCR product will not be produced, resulting in a different pattern of amplified DNA segments on the gel.

Example

RAPD is an inexpensive yet powerful typing method for many bacterial species.

Selecting the right sequence for the primer is very important because different sequences will produce different band patterns and possibly allow for a more specific recognition of individual strains.

Limitations of RAPD

- Nearly all RAPD markers are dominant, i.e. it is not possible to distinguish whether a DNA segment is amplified from a locus that is heterozygous (1 copy) or homozygous (2 copies). Co-dominant RAPD markers, observed as different-sized DNA segments amplified from the same locus, are detected only rarely.
- PCR is an enzymatic reaction, therefore the quality and concentration of template DNA, concentrations of PCR components, and the PCR cycling conditions may greatly influence the outcome. Thus, the RAPD technique is notoriously laboratory dependent and needs carefully developed laboratory protocols to be reproducible.

- Mismatches between the primer and the template may result in the total absence of PCR product as well as in a merely decreased amount of the product. Thus, the RAPD results can be difficult to interpret.

Developing Locus-specific, Co-Dominant Markers from RAPDs

- The polymorphic RAPD marker band is isolated from the gel.
- It is amplified in the PCR reaction.
- The PCR product is cloned and sequenced.
- New longer and specific primers are designed for the DNA sequence, which is called the Sequenced Characterized Amplified Region Marker (SCAR).

Materials

1. *Equipment*: Thermocycler, Power supply Unit
2. *Water*: Sterile de-ionized or distilled water should be used for preparing all the reagents and pre-mixes
3. *Reaction buffers*: Assay buffer for Taq DNA polymerase (supplied by the manufacturer of Taq DNA polymerase)
4. *Deoxynucleoside triphosphates (dNTP'S):* 2.5 mM each of dCTP, dATP, dTTP, dGTP. Readymade solutions of dNTPS are available from many manufactures. Store at –20 °C
5. *Magnesium chloride*: 25mM stock and store at –20 ° C
6. Taq DNA polymerase Genomic DNA 5-25 ng/ml stocks. DNA of sufficient quality can be obtained in coconut by using SDS protocol.

Methods

Assemble RAPD reactions as follows:

2.5µl DNA stock (25ng/µl)

2.5µl Assay buffer containing 2.5mM MgCl2 (2.5mM)

1µl MgCl2 stock (1.5mM) 1µl primer stock (25pmol)

4µl dNTPS (400µM)

1µl Taq polymerase (1U)

Sterile water to make 25 µl

- Wear gloves throughout RAPD reaction preparation procedure. Assay buffer, dNTPs, MgCl2 and primer solution are thawed from frozen stock. Keep the assembled reaction in themocycler for amplification.
- Amplify DNA in themocycler: Cycling conditions may be modified depending on the thermocycer used.

Temperature Profile

General cycling steps followed are:

Step 1 : Initial denaturation at 94° C for 5.00 min

Step 2 : Denaturation at 94° C for 1.00min

Step 3 : Primer annealing at 55° C for 1.00min

Step 4 : Primer extension at 72° C for 2.00min

Step 5 : Go to 2, 39 times

Step 6 : Final extension at 72° C for 10min

Step 7 : 4° C for ever

After the reaction, DNA is analyzed through gel electrophoresis.

Agarose Gel Electrophoresis

Reagents for Agarose Gel Electrophoresis

1. Agarose, TBE/TAE buffer, ethidium bromide, gel loading dye,
2. To prepare 100ml of a 0.7% agarose solution, measure 0.7g agarose into a glass beaker or flask and add 100ml 1X TBE or TAE.
3. Microwave or stir on a hot plate until agarose is dissolved and solution is clear.
4. Allow solution to cool to about 55° C before pouring. (ethidium bromide can be added at this point to concentration of 0.5μg/ml
5. Place the comb in gel tray.
6. Pour 50° C gel solution into tray to a depth of about 5mm. Allow the gel to solidify for about 20 min at room temperature.
7. To run, gently remove the comb, place the tray in electrophoresis chamber, and cover (just until wells are submerged) with electrophoresis buffer (the same buffer used to prepare the agarose).

8. To the RAPD sample from refrigerator, add 1μl of 6%gel loading dye for every 5 μl of DNA solution. Mix well. Load 20μl of DNA per well. Load also the DNA size standards along side RAPD reactions.
9. Connect the electrodes to the power pack. And electrophoresis at 50-150Volts until the bromophenol blue dye has reached three fourth of the gel length.
10. Stain the gel with ethidium bromide (if not already included in the gel).
11. Examine the gel under UV light (transilluminator).
12. Depending on the objective of the experiment make a note of polymorphism, segregating bands, and appearance of overall pattern with in fingerprint.
13. Bands may be sized by comparison to molecular weight standards. The standards should be used to generate a standard curve for interpolation.
14. After you have run the gel, obtain a photograph, and label and measure the migration of the DNA bands. Make a standard curve plot of the known size markers, and determine the size of the marker bands.

Gel Interpretation

- Bands are sized and matched directly on gels, or photographic films, or photocopies on transparency overlays.
- Note the presence and absence of bands.
- Analyze the data using computer software NTSYS/RAPD Distance

Applications of RAPD

RAPD protocol can be used in genetic mapping and finger printing application in other palms like areca nut and oil palm also.

CHAPTER 21

Amplified Fragment Leanth Polymorphism (AFLP)

Introduction

The AFLP amplified fragment polymorphism also known as selective restriction fragment amplification (SRFA) technique is used to visualize hundreds of amplified DNA restriction fragments simultaneously. The AFLP band patterns, or fingerprints, can be used for many purposes, such as monitoring the identity of an isolate or the degree of similarity among isolates. Polymorphisms in band patterns map to specific loci, allowing the individuals to be genotyped or differentiated based on the alleles they carry. produces highly complex DNA profiles by arbitrary amplification of restriction fragments ligated to double-stranded adaptors with hemi-specific primers harboring adaptor-complementary 5' termini.The technique has been widely used in the construction of genetic maps containing high densities of DNA marker loci.

The AFLP protocol amplifies restriction fragments obtained by endonuclease digestion of target DNA using "universal" AFLP primers complementary to the restriction site and adapter sequence. However, not all restriction fragments are amplified because AFLP primers also contain selective nucleotides at the 3' termini that extend into the amplified restriction fragments. These arbitrary terminal sequences result in the amplification of only a small subset of possible restriction fragments. The number of amplified fragments can therefore be "tailored" by extending the number of arbitrary nucleotides added to the primer termini. Alternatively, the use of endonuclease combinations that vary in their restriction frequency can also be used to tune the number of amplicons. Generally, the abundant restriction fragments produced from complex genomes require of AFLP primers with longer selective regions. Conversely, analysis of small genomes require of only few arbitrary nucleotides added at the primer 3' termini. The resulting AFLP fingerprints are usually a rich source of DNA polymorphisms that can be used in mapping and general fingerprinting endeavors.

The AFLP can be devided in the following steps:

- DNA is digested with two different restriction enzymes
- Oligonucleotide adapters are ligated to the ends of the DNA fragments
- Specific subsets of DNA digestion products are amplified, using combinations of selective primers
- Polymorphism detection is possible withradioisotopes, fluorescent dyes or silver staining

The DNA being examined is digested with two different restriction enzymes, one of which is a frequent cutter (the four-base restriction enzyme) and the other a rare cutter (the six-base restriction enzyme). Various enzyme/primer combinations can be used. MseI and EcoRI are best used in AT-rich genomes as they give fewer fragments in GC-rich genomes. Specific synthetic adapters for each restriction site are then ligated to the digested DNA. Both the restriction and ligation steps can be performed in a single reaction.

Amplification of very small "genomes" (plasmids, cosmids, BACs) requires of primers with no selective nucleotides. AFLP fingerprinting of bacteria and fungi generally requires primers with 2 selective bases. Complex genomes require the use of more than 2 selective bases in one or both primers. In the case of complex genomes it is suggested to carry the amplification in two consecutive steps to increase specificity and the amount of initial template. The AFLP fragments are usually detected by labeling one of the two AFLP primers. For example, radioactively labeled primers can be obtained by phosphorylating the 5' ends with g-33P-ATP and polynucleotide kinase. Do not label the two primers if the generation of doublets resulting from the different mobility of complementary strands in sequencing gels wants to be avoided. Finally, the labelled reaction products are separated by electrophoresis using denaturing polyacrylamide gels and exposed to X-ray films to visualize the AFLP fingerprints.

Materials

1. *Equipment* : Refrigerator and freezer, Laminar flow hood, Centrifuge, Thermocycler, Power supply units, Hotplate or microwave, pH meter, Standard balance, Vertical gel electrophoresis units, UV transilluminator, Automatic sequencer Disposables: PCR tubes, X-ray film
2. Genomic DNA. For genomes smaller or larger than 500 Mb use 50 ng or 100 ng DNA for template preparations, respectively. Determine DNA concentrations by measuring OD260, and confirm the measurement and the integrity of DNA by electrophoresing the sample together with a series of phage l DNA dilutions ranging from 50 ng to 500 ng in agarose gels
3. AFLP primers (50 ng/µl). Primers are named "+0" when having no selective bases, "+1" when having a single selective base, "+2" for having two selective bases, and so on.
   *Eco*RI-primer+0: 5'-GACTGCGTACCAATTC-3'
   *Eco*RI-primer +1: 5'-GACTGCGTACCAATTCA-3'
   *Eco*RI-primers +2: 5'-GACTGCGTACCAATTCAN-3'
   *Eco*RI-primers +3: 5'-GACTGCGTACCAATTCANN-3'
   *Pst*I-primer +0: 5'-GACTGCGTACATGCAG-3'
   *Pst*I-primer +1: 5'-GACTGCGTACATGCAGA-3'
   *Pst*I-primers +3: 5'-GACTGCGTACATCGAGANN-3'
   *Mse*I-primer +1: 5'-GATGAGTCCTGAGTAAC-3'
   *Mse*I-primers +2: 5'-GATGAGTCCTGAGTAACN-3'
   *Mse*I-primers +3: 5'-GATGAGTCCTGAGTAACNN-3'
   *Taq*I-primer +0: 5'-GATGAGTCCTGAGCGAA-3'
   *Taq*I-primers +3: 5'-GATGAGTCCTGAGCGAANN-3'
4. Corresponding adaptors (5 or 50 pmol/µl):
   EcoRI-Adapter: 5' CTCGTAGACTGCGTACC
   CAT CTGACGCATGG-5'
   Pst- Adapter, *Mse*I- Adapter, Taq- Adapter,

5. Double-distilled water.
6. Buffers: 1 M Tris.HAc pH 7.5; 1 M Tris.HCl pH 8.0 and pH 8.3.
7. Magnesium: 0.1 mM MgCl2.
8. TE (10x): 100 mM Tris.HCl, 10 mM EDTA pH 8.0.
9. 100 mM DTT.
10. 5 mM of dNTPs
11. gamma-32P-ATP (~3000 Ci/mmol) or gamma-33P-ATP (~2000 Ci/mmol).
12. 10 mM ATP.
13. T4-buffer (10x): 250 mM Tris.HCl pH 7.5, 100 mM MgCl2, 50 mM DTT, 5 Mm spermidine
14. Restriction-ligation buffer (5x): 50 mM Tris-HAc, 50 mM MgCl2, 250 mM KAc, 25 mM DTT, 250 ng/μL, pH 7.5.
15. Restriction endonucleases: *Eco*RI, *Pst*I, *Mse*I, *Taq*I Enzymes: T4 DNA ligase, T4 polynucleotide kinase, *Taq* DNA polymerase
16. PCR buffer (10x): 100 mM Tris.HCl pH 8.3, 15 mM MgCl2, 500 mM KCl.
17. Molecular weight standards
18. General reagents for polyacrylamide gel electrophoresis.

Methods

Depending on the size of the genome to be analyzed a different set of primers will have to be used. High complexity genomes require of a pre-amplification step. The following table shows primer combinations required for the amplification of different genomes. Note that numbers depict the number of selective nucleotides at the 3' terminus of the individual primers.

Genome	Endonucleases	Pre-amplification	Amplification
Cosmids, BACs, PACs, YACs (0.01-1 Mb)	*Eco*RI-*Mse*I		0-0 or 0-1
Microbes (1-5 Mb)	*Eco*RI-*Mse*I		1-1
Microbes (5-20 Mb)	*Eco*RI-*Mse*I		1-2
Fungi (20-100 Mb)	*Eco*RI-*Mse*I		2-2

Template Preparation

1. *Adapters*: Prepare double-stranded (ds) adapters by mixing individual synthetic oligonucleotides. No denaturing-renaturation step is required. *Eco*RI, *Mse*I, *Pst*I, and *Taq*I adapters have double-stranded regions of 14, 12, 14 and 12 base pairs, respectively. Mix 1500 pmoles of each oligonucleotides to produce 5 pmol/µl solutions. Generally, use 8.5µg, 10.5µg, 8µg or 8µg of the top strand oligonucleotide with 9.0µg, 7µg, 7µg or 7µg of the bottom strand oligonucleotidein 300µl of water for the EcoRI, PstI, MseI and TaqI adapter, respectively.

2. *Digestion of DNA:* Digest genomic DNA (50-100 ng) with restriction endonucleases in a 40µl reaction containing DNA, 8µl 5x RL buffer, 5 units *Eco*RI and 2 units *Mse*I. Mix well and incubate for 2 h at 37°C. For genomes larger than 5000 Mb use 5 units *Pst*I and 2 units *Mse*I. For mammalian and vertebrate genomes use 5 units *Taq*I and 5 units *Eco*RI in subsequent incubation steps at 65°C and 37°C.DNA preparations need to be of sufficient quality to allow complete digestion, since this step is crucial for the production of good quality AFLP fingerprints. Often, contaminating agents can interfere with digestion.

3. *Adaptor ligation:* Ligate adaptors to the digested genomic DNA by adding 1µl *Eco*RI adapter (5 pmol), 1µl *Mse*I adapter (50 pmol), 1µl *Pst*I adapter (5 pmol) or 1µl *Taq*I adapter (50 pmol), 1µl 10 mM ATP, 2µl 5¥ RL buffer, 1 unit T4 DNA ligase and 5µl water to the digestion mix. Incubate another 2 h at 37°C. Overall, DNA is incubated for a total of 4 h with endonucleases, the last 2 h in the presence of T4 DNA ligase and oligonucleotide adapters. Avoid longer incubation periods because of possible "star" activity of *Eco*RI that gives reduced cleavage specificity and aberrant AFLP fingerprints. When using *Taq*I, use high concentration of adapters; if not *Taq*I will not efficiently re-digest aberrant fragment-ligation products at 37°C.

4. *Dilution*: Dilute the ligation reaction mixture 10 times with TE (usually 10µl in 100µl) and use the diluted reaction mixture directly as template DNA for the AFLP reactions. Store diluted DNA at -20°C.

AFLP pre-amplification

1. *Pre-amplification:* Check above for the right combination of primers for your individual application. If genomic complexity is sufficiently low, AFLP preamplification is not required (see Table above); therefore, skip this preamplification section. Assemble the preamplification reaction (50µl total volume) with following components: 5µl ligated DNA, 1.5µl *Eco*RI-primer +0 (75 ng), 1.5µl *Mse*I-primer +C (75 ng), 1.5µl of the *Pst*I one-base extension primer -A (75 ng), 1.5µl *Mse*I-primer +C (75 ng) and/or 1.5µl *Taq*I-primer +A (75 ng), 2µl 5 mM dNTPs, 0.2µl Taq polymerase (1 unit), 5µl 10x PCR-buffer and 34.8µl water. Preamplify the mix for 20 cycles using the following regime: 30 s at 94°C; 60 s at 56°C; 60 s at 72°C. For mammalian genomes use 30 cycles. For genomes larger than 5000 Mb modify the amplification regime as follows: a first cycle of 30 s at 94°C, 60 s at 65°C, 60 s at 72°C, followed by 12 cycles with a stepwise decrease of the annealing temperature in each subsequent cycle by 0.7°C, and 23 cycles of 30 s at 94°C, 30 s at 56°C and 60 s at 72°C. In this step it is advisable to assemble one reaction mix containing primer and dNTPs and another containing the *Taq* polymerase and its buffer.

2. *Dilution:* Preamplification, 10µl of the reaction is diluted with 190µl of TE0.1 to 100µl which is sufficient for 40 AFLP-reactions +2/+3. The diluted reaction mix and the rest of the preamplification reaction is stored at -20°C. If necessary new dilutions of the preamplification reactions may be made to give additional template for the AFLP reactions.

AFLP amplification

1. *Preparation of labelling mix:* Label primers for selective AFLP amplification by phosphorylating the 5' end of the primers with gamma-32P-ATP or gamma-33P-ATP and polynucleotide kinase. Check above for the right primer combination to use in this step. Only one of the two primers of the AFLP reaction should be labelled (e.g., the EcoRI-primer). When possible use the more expensive 33P-labelled primers because they give better product resolution in polyacrylamide gels, and are less prone to degradation due to autoradiolysis. Prepare the following primer labelling mixes (40µl) for 100 AFLP reactions. either 20µl gamma-32P-ATP (~3,000 Ci/mmol) or 10µl gamma- 33P-ATP

(~2,000 Ci/mmol), 5µl 10xT4-buffer, 2µl T4-kinase (10 units/µl) and water to 40µl.

2. *Primer labelling:* Add 10µl of primer (either *Eco*RI- or *Pst*I-primers at 50 ng/µl) to 40µl labelling mix and incubate 60 min at 37°C, followed by incubation at 70°C for 10 min for the inactivation of the kinase. This gives a labelled primer with a concentration of 10 ng/µl.

3. *Prepare AFLP reaction mixes:* Prepare reaction mixes for a minimum of 10 reactions. Working with AFLP reaction mixes is important for the reliability and reproducibility of AFLP reactions and because it facilitates reaction assembly. *Primer* and *dNTPs* mix (50 µl): 5µl labelled primer (10 ng/µl), 6µl unlabelled primer (50 ng/µl), 8µl 5 mM dNTPs and 31µl water. *Taq* polymerase mix (100 µl): 20µl 10x PCR-buffer, 0.8µl *Taq* polymerase (4 units) and 79.2µl water.

4. *AFLP amplification:* Assemble the reaction by adding 5µl of the primers and dNTPs mix and 10µl of the *Taq* polymerase mix to 5µl of pre-amplified ligated DNA. The template DNA should be pipetted first followed by the two mixes. The reagents should be mixed by tapping the base of the tubes on the bench. Pipetting mixes is essential for the rapid start of the AFLP reactions that are assembled at room temperature (to avoid loss of AFLP fingerprint quality). Tubes are amplified in a thermocycler with the following cycle regime a first cycle of 30 s at 94°C, 30 s at 65°C and 60 s at72°C, followed by 12 cycles with a stepwise decrease of the annealing temperature in each subsequent cycle by 0.7°C, and 23 cycles of 30 s at 94°C, 30 s at 56°C and 60 s at 72°C. The reaction is started at a high annealing temperature to obtain optimal primer selectivity. In the following steps the annealing temperature is lowered gradually to a temperature for optimal primer annealing.

Polyacrylamide Gel Electrophoresis of AFLP Products

1. *General:* Amplification products are analyzed on 4.5% denaturing polyacrylamide sequencing gels (see the Separation section for more details). Treat back glass plate of the gels with 2µl of repel silane, and the front plate with 10 ml of bind silane solution (30µl acetic acid and 30µl bind silane in 10 ml ethanol, freshly made immediately before use). The gels should be cast at least

2 h before use and should be prerun for 0.5 h just before loading the samples Perform pre-running and running electrophoretic steps at 110 W. Use TBE (1x) as running buffer.

2. *Sample loading:* Mix AFLP reaction products with an equal volume (20µl) of loading dye. Heat the samples for 3 min at 90°C, and then quickly cool on ice. Rinse the the gel wells with running buffer and push carefully two 24-well sharktooth combs about 0.5 mm into the gel surface to create the gel slots. Rinse the gel slots formed in this way with TBE and load 2µl of each sample per well.
3. *Post-electrophoretic procedures:* Disassemble the gel cassette and remove the front glass plate with the silane-attached gel to the front. Fix the gel by soaking in 10% acetic acid for 30 min and dry it subsequently at room temperature in a fume hood for 10-20 h. Autoradiographic exposure of the 32P-gels to standard X-ray film overnight, without intensifying screens. Exposure of the 33P-gels takes 2 to 3 d in order to generate similar band intensities.

Aplication

- Genetic diversity assessment
- Genetic distance analysis
- Genetic fingerprinting
- Analysis of germplasm collections
- Genome mapping
- Monitoring diagnostic markers

CHAPTER 22

Southern Blotting

Introduction

Southern blotting was named after Edward M. Southern who developed this procedure at Edinburgh University in the 1970s. To oversimplify, DNA molecules are transferred from an agarose gel onto a membrane. Southern blotting is designed to locate a particular sequence of DNA within a complex mixture. For example, Southern Blotting could be used to locate a particular gene within an entire genome. The amount of DNA needed for this technique is dependent on the size and specific activity of the probe. Short probes tend to be more specific. Under optimal conditions, you can expect to detect 0.1 pg of the DNA for which you are probing.

This diagram shows the basic steps involved in a Southern blot.

Let's look at this technique in greater detail.

1. Digest the DNA with an appropriate restriction enzyme.
2. Run the digest on an agarose gel.
3. Denature the DNA (usually while it is still on the gel). For example, soak it in about 0.5M NaOH, which would separate

double-stranded DNA into single-stranded DNA. Only ssDNA can transfer.

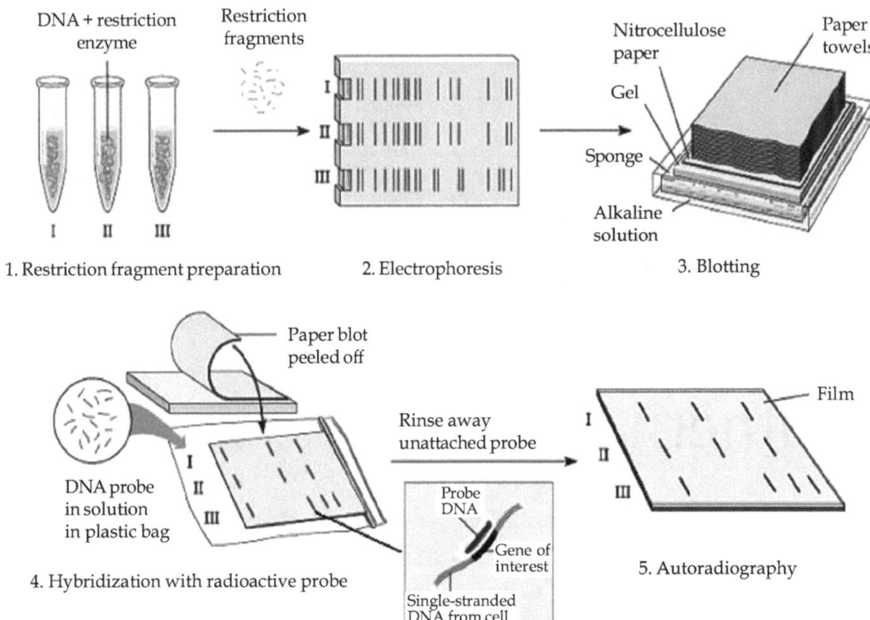

A depurination step is optional. Fragments greater than 15 kb are hard to transfer to the blotting membrane. Depurination with HCl (about 0.2M HCl for 15 minutes) takes the purines out, cutting the DNA into smaller fragments. Be aware, however, that the procedure may also be hampered by fragments that are too small.

Be sure to neutralize the acid after this step, or the base after the prior step if you don't depurinate.

4. Transfer the denatured DNA to the membrane. Traditionally, a nitrocellulose membrane is used, although nylon or a positively charged nylon membrane may be used. Nitrocellulose typically has a binding capacity of about 100μg/cm, while nylon has a binding capacity of about 500 μg/cm. Many scientists feel nylon is better since it binds more and is less fragile. Transfer is usually done by capillary action, which takes several hours. Capillary action transfer draws the buffer up by capillary action through the gel an into the membrane, which will bind ssDNA.

   You may use a vacuum blot apparatus instead of capillary action. In this procedure, a vacuum sucks SSC through the membrane.

This works similarly to capillary action, excepts more SSC goes through the gel and membrane, so it is faster (about an hour). (SSC provides the high salt level that you need to transfer DNA.)

After you transfer your DNA to the membrane, treat it with UV light. This cross links (via covalent bonds) the DNA to the membrane. (You can also bake nitrocellulose at about 80C for a couple of hours, but be aware that it is very combustible.)

5. Probe the membrane with labeled ssDNA. This is also known as hybridization. Whatever you call it, this process relies on the ssDNA hybridizing (annealing) to the DNA on the membrane due to the binding of complementary strands. Probing is often done with ^{32}P labeled ATP, biotin/streptavidin or a bioluminescent probe.

   A prehybridization step is required before hybridization to block non-specific sites, since you don't want your single-stranded probe binding just anywhere on the membrane.

   To hybridize, use the same buffer as for prehybridization, but add your specific probe.

6. Visualize your radioactively labeled target sequence. If you used a radiolabeled ^{32}P probe, then you would visualize by autoradiograph. Biotin/streptavidin detection is done by colorimetric methods, and bioluminescent visualization uses luminesence. ^{32}Plabeled ATP Treat the dsDNA fragment that you are using as a probe with a limiting amount of DNAse, which causes double-stranded nicks in DNA. Add ^{32}P, dATP, and other dNTPs to DNA polymerase I, which has 5' to 3' polymerase activity and 5' to 3' exonuclease activity.

   Nick translation occurs and as the nick is translated down the DNA strand, the polymerase activity continues to nick while the exonuclease activity continues to fill in the nick. As this happens, ^{32}P becomes incorporated into, and thus labels, the DNA. Heat the DNA to make it single stranded, then immediately places it on ice to keep the two strands from reannealing to each other. (If the DNA is on ice, the DNA passes through the annealing temperature too quickly for the DNA to rehybridize into double-stranded DNA).

Prehybridization

To prehybridize, add non-specific ssDNA. Somicated salmon sperm DNA is commonly used. Add 20X SSC, Denhardt's solution (ficol and PVP, which are large molecules to take up space and generate more contact; and BSA, bovine serum albumen, a non-specific protein), SDS (sodium dodecyl sulfate), and formamide.

Altering the concentrations of formamide, SSC, and SDS affects "stringency," or specificity. If you have a higher stringency you should also have a higher degree of similarity between the probe and the target sequence.

Southern (capillary) Blotting of DNA Fragments

Principle: It is a technique for transferring DNA molecules from agarose gel to a solid support such as nitrocellulose filter or nylon membrane. Gel embedded DNA is apurinated in the presence of HCI followed by denaturation in alkali. Denaturation of DNA fragments prior to blotting is essential for dissociating the two polynucleotide chains which can then hybridize with the probe. Denatured DNA molecules move upward by capillary action of buffer and on coming in contact with the nylon membrane they get bound to it. Heating at 800C for 2 h under vacuum facilitates firm binding of DNA to the nylon membrane.

Materials and Reagents

1. Agarose gel containing DNA fragments
2. Nylon membrane (Hybond N or Gene Screen Plus)
3. Whatman 3 MM paper
4. Glass tray (5 cm deep x 20 cm x 30 cm) and glass plate (12.5 x 25 cm).
5. Rough filter papers
6. Parafilm or used X-ray film
7. 0.25 M HCI
8. Denaturating solution: Prepare 1.5 M NaCl solution containing 0.5M NaOH.
9. Neutralizing solution: Make 1.0 M Tris-HCI buffer, pH 8.0 containing 1.5M NaCl

10. SSC solution (20 x): It is made by preparing a solution of 3 M NaCl containing 0.3 M tri-sodium citrate. Adjust the pH to 7.0.
11. SSC solution (2 x): Dilute the above solution 10 times.

Procedure

1. After electrophoresis, transfer the agarose gel to a glass tray containing sufficient HCI (0.25 M) to cover the gel and gently stir for 15 min at room temperature on a platform shaker. Unless otherwise stated, all the subsequent operations are also carried out at room temperature.
2. Pour off HCI and rinse the gel with distilled water.
3. Denature the DNA by adding sufficient volume of denaturating solution (Reagent 8). Gently stir for 30 min.
4. Decant the alkali and rinse with distilled water.
5. Neutralize the gel in neutralizing solution and gently stir for 30 min.
6. Remove the solution and wash the gel with 2 x SSC.
7. Make a bridge on a glass tray with glass plate and place 2 layers of Whatman 3 MM paper cut to proper size on top so that the ends of the paper hang down into the tray. Wet the paper with 20 x SSC and fill the tray with the same solution.
8. Place another piece of Whatman 3 MM paper, which is cut to the size of the gel and pre-wetted in 20 x SSC, on top of the bridge. Carefully transfer the gel on top of the wetted filter paper and remove any entrapped air bubbles by rolling a glass pipette gently over the gel.
9. Cut a piece of nylon membrane (Hybond N or Gene Screen Plus) of exactly the same size as that of the gel. Mark the membrane on the left hand bottom corner with a permanent ink marker. Wet the membrane in distilled water and place it on top of the gel. Cover the membrane with 2 pieces of 3 MM Whatman filter paper cut to the same size and prewetted in 20 x SSC. Avoid entrapment of any air bubbles between the papers.
10. Seal all the sides of the 3 MM Whatman paper along edges of the gel with parafilm or strips of used X-ray. films.

11. Keep a stack of dry rough filter papers cut to the same size on top of the wet filter paper.
12. Place a glass on top of stack of filter papers, keep a weight of about 0.5 kg (500 ml water in a bottle) over it and allow the transfer of DNA from gel to nylon membrane take place overnight.
13. Remove the weight, stack of rough filter paper and the Whatman filter papers from top of the membrane. Carefully remove the membrane and place it in 0.4 M NaOH for 30 sec. Wash it in 2 x SSC.
14. Bake the membrane at 80 °C for 2 h in a vacuum oven.
15. Air dry the membrane briefly, keep between 2 sheets of Whatman 3 mm paper and store at 4 °C in a refrigerator till use.

Precautions

1. Use gloves while handling gel and nylon membrane.
2. Take care that no air bubbles are entrapped between gel, membrane and Whatman papers.
3. Make sure that the buffer moves through the gel and the membrane. The layers of papers, placed on top should not touch the layers of filter paper placed beneath the gel.
4. Adopt all the general precautions used in handling nucleic acids as described in Section.

Dot Blot

A technique of blotting cloned DNA without prior restriction digestion and electrophoresis. Autoradiography reveals dots indicating probe hybridization.

Dot blot is a technique for detecting and identifying proteins, similar to the Western blot technique but differing in that protein samples are not separated electrophoretically but are spotted through circular templates directly onto the membrane or paper substrate. Antigens may be applied directly to nitrocellulose membrane as a discrete spot (dot) to give a simple and reliable assay.

CHAPTER 23

Ligation of Insert DNA to Vector DNA

Introduction

Ligation of DNA fragment is one prime important in genetic engineering because without ligation we could not transfer the interest gene.DNA cloning requires the DNA sequence of interest to be inserted in a vector DNA molecule. For this, both the vector as well as insert DNA is prepare by digestion with compatible restriction enzymes, so that the ends produced during digestion is complementary in both. When setting up ligation, it is important to consider the permutations that can occur and bias the relative concentration of DNA accordingly. Usually a 5- to 10-fold excess of insert over the vector DNA is the norm. This ensures that enough ligated product will be produced in the right orientation.

Materials

Vector digest

Insert DNA

T4 DNA ligase

Ligation buffer: 50 mM Tris-HCI, 10 mM $MgCI_2$, 20 mM DTT, 1 mM ATP, 50/mg/mL Nuclease-free BSA, pH 7.4 (generally supplied with the enzyme)

Protocols

A. Insertion of DNA Into Plasmid Vectors

1. Digest plasmid DNA with an appropriate restriction enzyme and dephosphorylate by treatment with alkaline phosphatase. Combine plasmid DNA and DNA to be inserted in a total volume of 5-10µl. We recommend 100 mM Tris-HCl (pH 7.6), 5 mM $MgCl_2$ for resuspending DNA; however, TE buffer [10 mM Tris-HCI (pH 8.0), 1 mM EDTA] could also be used. Use 0.03 pmol of vector DNA (50 ng pUC18) and 0.1-0.3 pmol of insert.
2. Add 4-8 volumes of Solution A to the DNA solution and mix thoroughly.
3. Add 1 volume of Solution B (5-10 ml) and mix thoroughly.
4. Incubate at 16°C for 30 minutes.
5. The ligation reaction mixture can be used directly for bacterial transformation. Use 10-20µl of the ligation mixture with 100 ml of competent bacterial cells.

Notes

- *Usually, 4 volumes of Solution A are sufficient for DNA suspended in Tris-HCl, $MgCl_2$ buffer. If other buffers are used or the concentration of DNA is high, use 8 volumes of Solution A to obtain optimum efficiency.*
- *The reaction should be carried out at 16°C. Higher temperatures (>26°C) will inhibit the formation of circular DNA. If good results are not obtained, the reaction can be extended overnight. An additional phenol extraction/ethanol precipitation step may be needed if DNA purity is a problem.*
- *A maximum of 20 µl of the ligation mixture may be used to transform 100 µl of competent bacterial cells. Excess ligation mixture may decrease transformation efficiency. For use in electroporation, DNA should be precipitated with ethanol and suspended in a low-salt buffer such as TE.*

Example

In this experiment, 50 ng of EcoR I-digested pUC118 (25 fmol) were mixed with 2.5-250 ng (2.5-250 fmol) of a 1.5 kb EcoR I DNA fragment in total volumes of 5 µl. Solution A (25 µl) and Solution B (5 µl) were added to each DNA mixture and reactions were incubated at 16°C for 30 minutes. A portion of each solution was used directly to transform competent JM109 cells and transformation mixtures were spread onto L-Amp plates containing X-Gal and IPTG. Competent JM109 cells had a transformation efficiency of at least 6.3 x 107 transformants per µg of DNA when assayed with supercoiled pUC118 DNA. The insert:vector molar ratio can be as high as 10.0. Better results were obtained using dephosphorylated vector.

B. Insertion of DNA Into l Phage Vectors

1. Digest l phage vector DNA with an appropriate restriction enzyme and dephosphorylate using alkaline phosphatase if desired. Combine 250 ng (0.01 pmol) of the treated l DNA with the DNA to be inserted (0.03-0.1 pmol) in a total volume of 5-10 µl. Best results are obtained when DNA is resuspended in buffer containing 100 mM Tris-HCl (pH 7.6), 5 mM $MgCl_2$ and 300 mM NaCl. The salt concentration is important for producing concatemeric l DNA. If TE [10 mM Tris-HCl (pH 8.0), 1 mM EDTA] is used, the solution should be supplemented to give a final concentration of 300 mM NaCl.

2. Add 1 volume (5-10 µl) of Solution B to the DNA solution and mix well.

3. Incubate at 26°C[a] for 5-10 minutes.

4. The ligation reaction mixture (up to 5 µl) can be used directly in l in vitro packaging reactions.

C. Self-Circularization of Linear DNA (Intramolecular Ligation)

The procedure for self-circularization of linear DNA is essentially the same as for insertion of DNA fragments into a plasmid vector. However, it is important to use low concentrations of DNA while maintaining a small volume. Low concentrations of DNA in the ligation reaction maximize intramolecular ligation. Small volumes allow for higher bacterial transformation efficiency.

Example

pBR322 DNA was digested with Sca I. Solution A (40µl) and Solution B (5µl) were added to digested pBR322 DNA (350ng/10µl) and the reaction was incubated at 16°C for 30 minutes. The reaction solution was used directly to transform competent *E. coli* HB101 cells. Competent HB101 cells had a transformation efficiency of at least 1 x 108 transformants per µg of DNA when assayed with supercoiled pBR322 DNA.

D. Linker (Adaptor) Ligation

1. Insertion of linker into a plasmid vector.

   Conditions for linker ligation (8 bases or longer) are essentially the same as for insertion of DNA fragments into a plasmid vector (see page 3). However, if the linker is shorter than 8 bases or has a low G+C content, the ligation reaction should be carried out at 4-10°C for 1-2 hours. The recommended molar ratio of phosphorylated linker:dephosphorylated vector is 10-100:1. When using phosphorylated vector, the linker:vector molar ratio should be >100:1.

2. Linker ligation to both termini of a DNA fragment (e.g., ligation of linkers to cDNA).

   a. Combine the DNA fragment (0.01-0.1 pmol) and linker in a volume of 5-10µl. The recommended linker:DNA fragment molar ratio is >100:1.

   b. Add 1 volume (5-10µl) of Solution B and mix well.

   c. Incubate at 16°C for 30 minutes. If the linker is shorter than 8 bases or has a low G+C content the ligation reaction should be carried out at 4-10°C for 1-2 hours.

   Following linker ligation, DNA may be digested directly with restriction endonucleases after inactivating the T4 DNA ligase (heating at 70°C for 10 minutes). If the restriction enzyme being used is known to display star activity, DNA can be precipitated with ethanol prior to digestion.

Ligation of DNA for Cloning

1. Determine concentration of vector and insert. The easiest way to do this is to run a standard agarose gel containing ethidium

bromide, along with standards of known concentration. For the standards, prepare a 10µg/ml stock solution of linear DNA (such as lambda or a digested plasmid). Load 10, 30, and 100 ng as standards on the gel (you may wish to add water or gel buffer to the 10 and 30 ng samples to make loading the gel easier). In nearby lanes, run a minimal amount of your DNA, maybe about 30 ng. If you have purified your DNA using a kit from a prior reaction step (such as using the GeneClean or Wizard kits, you can assuming a 50% yield). Run the gel briefly (let the bromophenol blue dye run about 1 cm). View the gel. By comparing the fluorescence of your samples and the standards, you can judge the DNA concentration with reasonable accuracy. This also confirms that your DNA is not degraded.

2. Calculate the amount of insert and vector needed. Aim for a DNA concentration about 10-20 ng/µl. Typically, try to get to 20 ng/µl in a ligation of 10-20µl. For sticky end ligations, aim for an average insert to vector molar ratio of about 2:1, or 1:1 if the vector DNA has been treated with alkaline phosphatase. A 5:1 ratio is suggested for blunt-end ligations. Please note that this is a molar ratio: you need to consider the size of the DNA, not just its concentration.

   Some theory: Optimal conditions for ligation depend on the length of the molecule, the size of the DNA, whether the ends are cohesive or blunt, and whether both ends of the vector have the same restriction site. This involves calculations related to both the DNA concentration of DNA ends (not the total DNA concentration), the flexibility of DNA, etc.

   A common error is to add too much insert, or have too high an overall DNA concentration. In such cases, there will be a tendency to ligate 2 or more insert fragments into the vector, or to form dimers, trimers, etc. of vector or insert that are non-productive for cloning. In other words, the optimum results occur when the absolute concentration of termini is high enough to favor intermolecular ligation but not so high as to cause the formation of extensive oligomers. Therefore, insert:vector molar ratios between 2 and 6 are best. Ratios below 2:1 result in lower ligation efficiency. Ratios above 6:1 promote multiple inserts or the formation of long chains.

For critical cloning experiments, such as making a library, it is generally useful to test several ratios of insert to vector. Nevertheless, here are some guidelines:

In general, for ligations of fragments with cohesive ends, one can achieve the greatest number of recombinants when the molar concentration of insert is 2x that of the vector. For a 3 kb vector, the optimum concentration of vector (not total DNA) is 10-20 ng/μl; for a 10 kb vector, the optimum is 5-10 ng/μl (a lower concentration of DNA is required for larger vectors as the likelihood that the ends of the same molecule will interact is less than for smaller vectors).

3. Set up the ligation. Normally aim for a minimum volume of 10 ìl. Anything greater than 25 μl is usually just a waste of reagents. Also, Step 2 will help indicate the best volume.

   The following is a typical recipe:

   10 X ligase reaction buffer (containing ATP, BSA)

   1 μl vector, insert, and water 8.5μl

   T4 DNA ligase (NEB, low conc.) 0.5μl total volume 10 μl

   Mix gently by flicking with your finger, and spin for 2 seconds in a microcentrifuge to get the liquid down to the bottom of the tube.

   Incubate at 4 °C to 20 °C (room temperature) for 4 to 24 hours.

   Considerations in picking the best temperature: The optimal incubation temperature for T4 DNA ligase is 16 °C; when very high efficiency ligation is desired (e.g. making libraries) this is recommended. However, ligase is active over a broad range. For routine purposes such as subcloning, convenience often dictates incubation time and temperature; ligations performed at 4 °C overnight, or at room temperature for 30 min to a couple of hours also can work well. Nevertheless, for sticky end ligations, between 4 °C to 12 °C often works best since this helps the DNA termini to anneal. It follows that for blunt ends, there is no harm doing ligations at room temperature.

   After ligation has proceeded (don't go longer than 24 hour, even at 4 °C, freeze the ligation just in case DNAses are present.

4. *Other tidbits*: Not all manufacturer's sell T4 ligase at the same concentration, or use the same unit definition. Be careful. T4 DNA ligase is impaired at high salt concentrations. That is one reason why it is generally good to purify DNA from restriction digests, which are often in 50-100 mM salt. Ligations can be performed in low-salt restriction buffers (such as NEB 1 and 4), or other low-salt buffers such as most Klenow buffers, as long as ATP is added to 1 mM. Ligations work moderately in NEB buffer 2.

   ATP is provided with most ligation buffers (for example, ligase from NEB). Note, however, that ATP breaks down over time so buffers should be discarded after ~2 years. For the same reason, thaw the buffers in your hand or at room temperature, not at 37C. ATP concentrations above 0.5 mM inhibit ligation of blunt ends.

   Rapid ligation can be achieved by adding PEG6000 to the ligation (5% w/v). This is normally what is added to many of the "fast ligation" kits on the market. However, at high DNA concentrations one may get too many oligomers, rather than simple circles.

   The DNA concentrations described above are for vector-insert ligations, not other applications such as linker addition.

   Make sure that at least one of your molecules contains a 5' phosphate. Oligos, for example, lack phosphate at their 5' ends. There are ligases other than T4 DNA ligase. Make sure that you use the correct one.

CHAPTER 24

Bacterial Transformation

To Prepare the Competent Cell and Carry Out Transformation

Introduction

Transformation results from the uptake of purified DNA by bacterial cells and is the most frequently used procedure for introducing recombinant DNA molecules into bacteria. It is now widely used to transfer small plasmids from one bacterial strain to another. DNA is extracted from donor cell and used to treat the recipient cells that have been rendered more susceptible to DNA uptake. These competent cells allow DNA to enter through pores or channels in the cell membrane and in the case of plasmids, permit subsequent plasmid replication.

Principle

1. Some species of bacteria naturally take up DNA at a certain stage of growth called competence. Some species however are not naturally competent at any stage of growth. Competence can be artificially induced in these cells by treating them with $CaCl_2$ prior to adding DNA. The Ca^{2+} destabilizes the cell membrane

and a Ca –Phosphate - DNA complex is formed which adheres to the cell surface and is resistant to DNases. The DNA is taken up during a heat shock step when the cells are exposed briefly to a temperature of 42 °C. Immediately chilling on ice ensures closure of pores. Selection for cells containing transformed DNA is greatly enhanced by the selection marker carried by the DNA. pUC series and pBR 322 have ampicillin resistance factor which enables only the transformed cells to grow on LB Ampicillin plates.

2. The pUC plasmids also have the gene for β-galactosidase (lac Z) from E.coli. The lac Z gene has a series of unique restriction sites engineered into it such that the plasmid can be cut within the lac Z gene. If these plasmids are transformed into a lac Z strain of E-coli. They will make them Lac+. Although neither the host nor the plasmid encoded fragments are themselves active, they can associate to form an enzymatically active protein. This type of complementation is known as alpha complementation.

3. Lac+ bacteria that result from alpha complementation can be recognized as they form blue colonies in presence of X-gal (as β-galactosidase cleaves this chromogenic substrate) and IPTG (that acts as an inducer for the expression of the enzyme).

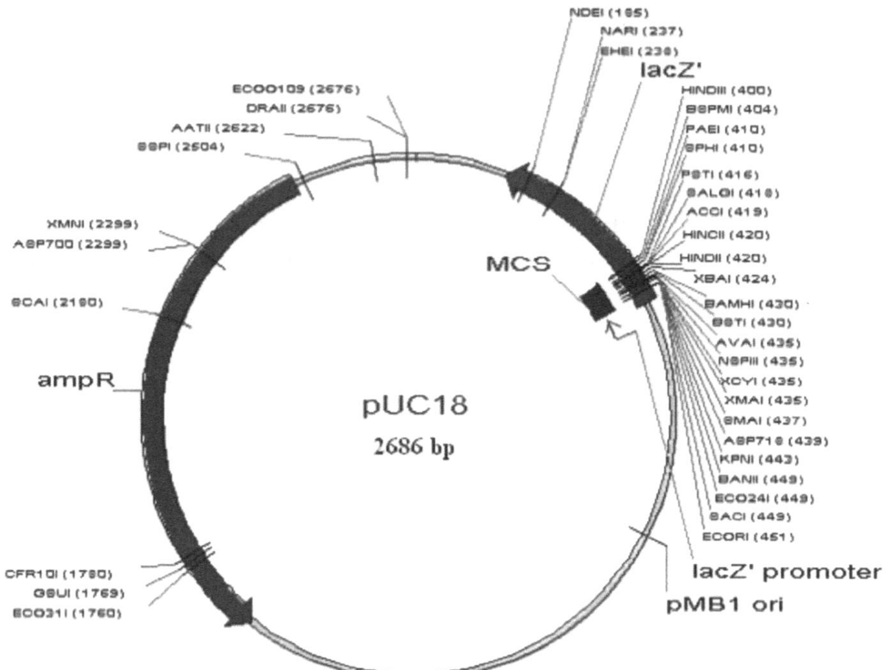

4. Any plasmid which has a DNA fragment cloned into it (lac Z genes site) will not have a functional lac Z gene and thus will produce white colonies which are unable to cleave X-gal.

Protocol

The experiment should be done strictly under aseptic conditions.

Preparation of Competent Cells

This procedure allows the preparation of cells that can be used immediately

Revival of Host Strain

1. Streak DH 5α strain provided as lyophilized vial onto a fresh LB plate to get a single colony. Incubate the plate overnight at 37^{0}C.
2. Pick a single colony the following day and incubate in the evening into 5ml LB medium. Incubate the tube O/N at 37^{0}C for about 16 to 18 hrs.
3. On the third day, begin the preparation of competent cells as outlined.

A. Transfer the 5ml O/N saturated culture into 100ml LB medium and incubate at 37^{0}C Grow until the OD (A_{600}) reaches 0.23-0.26 (it takes around 2-3 hrs to reach the required OD).

B. Once the OD is reached quickly chill the culture flask on ice and leave it on ice in refrigerator for 10-20 mins (from this point all work done at 4^{0}C)

C. Transfer the culture aseptically into sterile centrifuge tubes and spin down at 6000 rpm for 8 min (if you have a refrigerated centrifuge spin at 4 ^{0}C)

D. Discard the supernatant and to the cell pellet add 0.1 M $CaCl_2$ (~ 6% of the volume you have spun) suspend the cell pellet gently in $CaCl_2$ and this should be done by keeping the tubes in ice bucket. Keep on ice for 30 mins. Centrifuge at 6000 rpm for 8 mins.

E. Discard the supernatant and resuspend the pellet gently in $CaCl_2$. (0.8% of volume you have spun)

(For a 100ml culture id you spin in a single tube add 6ml of 0.1M $CaCl_2$ during the first resuspension and for the final resuspension 0.5ml is sufficient)

F. Aseptically transfer 100ml of the above competent cells made into 5 eppendorff vials. All transferring should be done on ice.

G. $CaCl_2$ Solution: Make required quantity of 0.1M $CaCl_2$ from the stock solution provided to you. Use autoclaved water only.

Note *While resuspending keep your centrifuge tube on ice. O.1M $CaCl_2$ should be ice cold.*

Transformation Procedure

1. To the 100μl of competent cells prepared add 5μl of the DNA provided. Gently tap and keep the vial on ice for 20 minutes.
2. Prepare a water bath at 42 °C. Keep the vial in the water bath such that the competent cells are immersed for 2 months. The temp should be maintained accurately during this period.
3. Quickly remove the vial after heat shock and chill the vial on ice for 20 minutes.
4. Add 1ml of LB aseptically to the vial and incubate the culture for 1 hour at 37 °C to allow bacteria to recover and express the antibiotic resistance.
5. Preparation of LB Amp. Plates: when the agar is around 44 °C, add 100μg per ml of antibiotic, 40μl of X-Gal & IPTG each for every 20 ml of agar, mix well and pour into petri plates.

Plating

Transfer 10μl, 25μl, 50μl, and 100μl of step 4 to each plate. Add 100μl of LB broth on top of the inoculums, mix well and spread thoroughly using a pipette or spreader. A control plate with competent cells that receive no plasmid DNA should also be plated to rule out contamination of cells.

Antibiotic Preparation

Dissolve required quantity of the antibiotic in sterile water and store at 4 °C Make antibiotic fresh, just before use, Add only after the Medium is Cool Enough (<45°C for Agar).

CHAPTER 25

Detection of Plasmid by Transformation

Introduction

The most common method of bacterial transformation utilizes high levels of calcium chloride to facilitate the entry of plasmid DNA vectors into *E.coli* in conjunction with structural alteration of the bacterial cell wall. In general, there are three basic steps in the introduction of plasmid DNA unto cells.

(1) Preparation of "*competent*" cells to enable cells to accept DNA.
(2) Transformation of these competent cells, and
(3) Selection of transformants.

Most *E.coli* strains exhibit transformation efficiency (measured as transformants per microgram plasmid DNA) in the range of 10^5 to 10^8.

The factors influencing this efficiency depends upon the condition that render cells competent.

1) These are harvesting cells during the logarithmic phase of growth.
2) Maintaining cells at a temperature of 4°C during treatment
3) Prolonged exposure to ice-cold $CaCl_2$

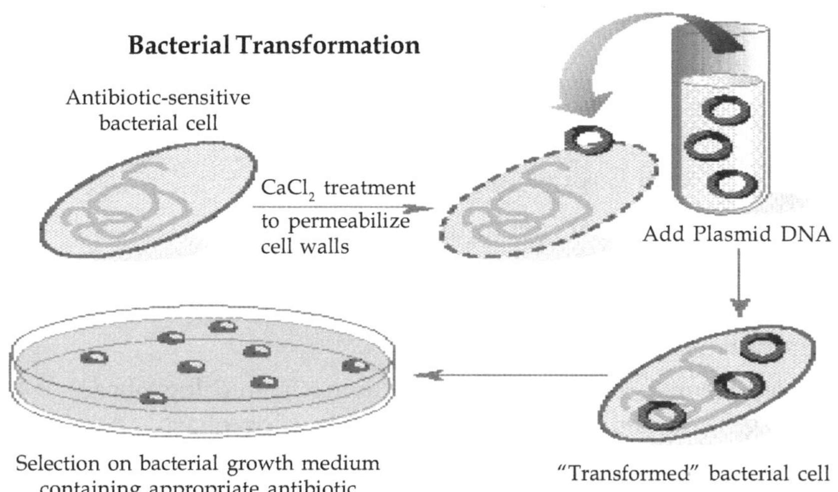

Calcium chlorides provudes the cells with an overall positive charge that serve to attract the negatively charged DNA. Transformation is induced by destabilization of the lipids in the cell enveloped through heat shock treatment.

A period of recovery time is given to enable cells to express the selectable marker. Then the cells are placed onto a selective medium that will only allow survival of successful transformants. Transformation efficiencies vary between strains of *E.coli* and the vectors. Factors such as plasmid size, quantity, purity, DNA configuration and selectable markers also influence the outcome of a transformation experiment.

LB Broth

Tryptone	1 gms
Yeast extract	0.5 gms
NaCl	1 gms
D/W	100 ml
pH	7.5

LB Agar with Ampicillin

Add 2% agar in the above composition and after autoclave add ampicillin to have final concentration of 100µg/ml.

Culture

DH5α *E. coli*

Plasmid

pUC 18 plasmid.

Solution

Sterile 50 mM $CaCl_2$ solution

Sterile 100 mM $CaCl_2$ solution

Procedure

Preparation of Competent Cells

1. Take 10ml *E. coli* DH5α cells growing in mid log phase. 0.4 to 0.5 O.D. at 600 nm. represents the mid log phase.
2. Centrifuge the 5 ml of cells for 5 min at 3000 rpm, discard the supernatant and take the pellet.
3. To the pellet, add 5 ml pre-cooled 50 mM $CaCl_2$ solution (equal culture volume) and suspend the cells very gently.
4. Centrifuge at 3000 rpm, 4° C for 5 min.
5. Resuspend the cells with pre-cooled 0.5 ml of 50 mM $CaCl_2$ and leave it in ice. The cells are now competent for taking up exogenous DNA.

Transformation

- Take 20µl ligated recombinant DNA.
- 200µl of 100 mM $CaCl_2$ in an eppendorf tube.
- Add 200µl of competent cells.
- Leave in ice for 30 minutes.
- Give a heat shock at 42° C for 1 minute.
- Add to 2 ml of plain LB in 30 ml screw cap tube.
- Keep inside the incubator for incubation for one hour at 37° C.
- Plate 100µl of the expressed culture in LB containing 100µg/ml ampicillin. Pate the host cell without plasmid also as the control

Result

Growth of the pUC 18 transformed DH 5 α *E. coli* on plate containing ampicillin indicates the presence of the plasmid in the host

DH5 α *E. coli*. As the DH 5 α *E. coli* without the plasmid pUC18 can not grow on the LB medium containing the ampicillin.

Observation

No.	DH 5 α *E. coli* Plate	pUC 18 transformed DH 5 α *E. coli* Plate
Growth (+) or (-)		

Transformation-Electrophoresis

High-Efficiency Transformation by Electroporation

Electroporation with high voltage is currently the most efficient method for transforming *E. coli* with plasmid DNA. The procedure described may be used to transform freshly prepared cells or to transform cells that have been previously grown and frozen. With freshly grown cells, it routinely gives more than 109 bacterial transformants per microgram of input plasmid DNA.

Materials

- Single colony of *E. coli* cells
- LB medium
- H_2O, ice cold 10% glycerol, ice cold SOC medium
- LB plates containing antibiotics
- 1-liter centrifuge bottle,
- 50-ml narrow-bottom polypropylene tube
- Microcentrifuge tubes, chilled ice cold
- Centrifuge (or equivalent)
- Electroporation apparatus with a pulse controller or 200- or 400-ohm resistor
- Chilled electroporation cuvettes, 0.2-cm electrode gap
- Additional reagents and equipment for growth of bacteria in liquid media.

Prepare the Cells

1. Inoculate a single colony of *E. coli* cells into 5 ml LB medium. Grow 5 hr to overnight at 37 °C with moderate shaking

2. Inoculate 2.5 ml of the culture into 500 ml LB medium in a sterile 2-liter flask. Grow at 37 °C, shaking at 300 rpm, to an OD600 of 0.5 to 0.7.
3. Chill cells in an ice-water bath 10 to 15 min and transfer to a prechilled 1-liter centrifuge bottle. Cells should be kept at 4 °C for all subsequent steps.
4. Centrifuge cells 20 min at 4200 rpm, 4 °C
5. Pour off supernatant and resuspend the pellet in 5 ml ice-cold water. Add 500 ml ice-cold water and mix well. Centrifuge cells as in step 4.
6. Pour off supernatant immediately and resuspend the pellet by swirling in remaining liquid. Because the pellet is very loose, the supernatant must be poured off immediately.
7. Add another 500 ml ice-cold water, mix well, and centrifuge again as in step 4.
8. Pour off supernatant immediately and resuspend the pellet by swirling in remaining liquid.
9. Introduction of Plasmid DNA into Cells
   a. If fresh cells are to be used for electroporation, place suspension in a prechilled, narrow-bottom, 50 ml polypropylene tube, and centrifuge 10 min at 4200 rpm 2 °C. Fresh cells work better than frozen cells. Estimate the pellet volume and add an equal volume of ice-cold water to resuspend cells (on ice). Aliquot 50- to 300-ml cells intoprechilled micro centrifuge tubes.
   b. If frozen cells are to be used for electroporation, add 40 ml ice-cold 10% glycerol to the cells and mix well. Centrifuge cells as described in step a. Estimate the pellet volume and add an equal volume of ice-cold 10% glycerol to resuspend cells (on ice). Place 50- to 300-l aliquots of cells into pre chilled micro centrifuge tubes and freeze on dry ice (not in liquid nitrogen). Store at -80 °C.
10. Transform the Cells : Set the electroporation apparatus to 2.5 kV, 25 F. Set the pulse controller to 200 or 400 ohms. The pulse controller is necessary when high-voltage pulses are applied over short gaps in high-resistance samples (see Background Information).

11. Add 5 pg to 0.5 microgram plasmid DNA in 1 ml to tubes containing fresh or thawed cells (on ice). Mix by tapping the tube or by swirling the cells with the pipettor.

12. Transfer the DNA and cells into a cuvette that has been chilled 5 min on ice, shake slightly to settle the cells to the bottom, and wipe the ice and water from the cuvette. The volume of DNA added to the cells should be kept small. Adding DNA up to one-tenth of the cell volume will decrease the transformation efficiency 2- to 3-fold. Also, since the resistance of the sample should be high, make sure that addition of the DNA to the cells does not increase the total salt concentration in the cuvette by >1 mM.

13. Place the cuvette into the sample chamber. If using a homemade apparatus, connect the electrodes to the cuvette.

14. Apply the pulse by pushing the button or flipping the switch.

15. Remove the cuvette. Immediately add 1 ml SOC medium and transfer to a sterile culture tube with a Pasteur pipet. Incubate 30 to 60 min with moderate shaking at 37C. If the actual voltage and time constant of the pulse are displayed on the electroporation apparatus, check this information. Verify that the set voltage was actually delivered, and record the time constant of the pulse so that you may vary it later if necessary (see Critical Parameters).

16. Plate aliquots of the transformation culture on LB plates containing antibiotics.

CHAPTER **26**

Conjugation in *E.coli* by Plate Method

Introduction

Genetic variability is essential for the evolutionary success of all organisms. In diploid eukaryotic the processes of crossing over exchange of genetic material between homologous chromosomes and meiosis contributes to his variability. In haploid, asexually reproducing prokaryotic organisms, genetic recombination may occur by conjugation transduction, and transformation. In this experiment only the process of conjugation is considered.

Conjugation is a matting process during which a one direction transfer of genetic material occurs at physical contact between two sexually differentiated cell types. This differentiation or existence of different mating strains in some bacteria, is determined by the presence of a fertility factor of F factor, within the cell. Cells that lack the F factor are recipients (females) of the genetic material during conjugation and are designated as F^- . cells possessing the F factor have ability to act as genetic donors (males)during the mating. If this F factor is extra chromosomal (a Plasmid or episome), the cells are designated as F^+; most commonly only the F factor is transferred during conjugation. If this factor becomes incorporated into the bacterial chromosome there is a transfer of chromosomal genes, although generally not involving the entire chromosome or the F factor. The resulting cells are designated as Hfr, for high frequency recombinants.

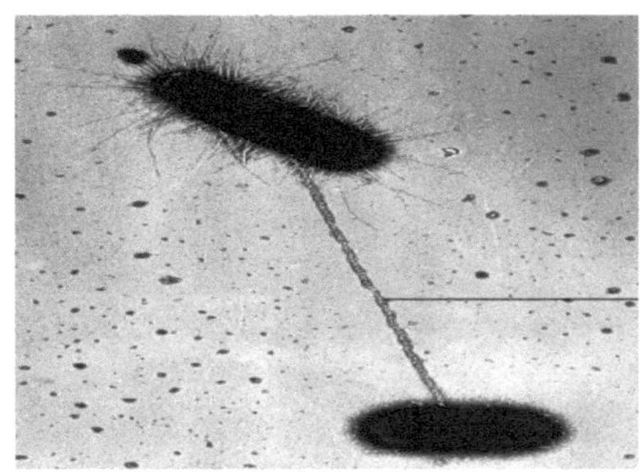

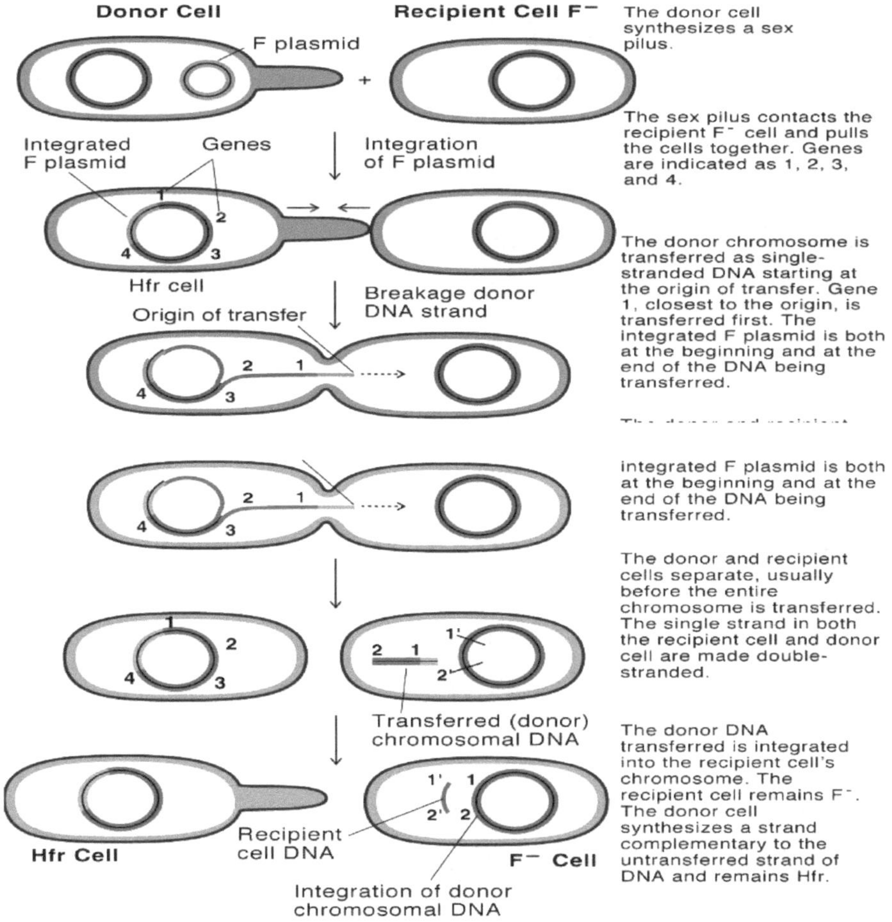

Donor Cell
Recipient Cell F⁻
F plasmid

The donor cell synthesizes a sex pilus.

Integrated F plasmid
Genes
Integration of F plasmid
1 2 3 4
Hfr cell

The sex pilus contacts the recipient F⁻ cell and pulls the cells together. Genes are indicated as 1, 2, 3, and 4.

Breakage donor DNA strand
Origin of transfer
2 1 3 4

The donor chromosome is transferred as single-stranded DNA starting at the origin of transfer. Gene 1, closest to the origin, is transferred first. The integrated F plasmid is both at the beginning and at the end of the DNA being transferred.

2 1 3 4

integrated F plasmid is both at the beginning and at the end of the DNA being transferred.

The donor and recipient cells separate, usually before the entire chromosome is transferred. The single strand in both the recipient cell and donor cell are made double-stranded.

1 2 3 4
2 1 1' 2'
Transferred (donor) chromosomal DNA

The donor DNA transferred is integrated into the recipient cell's chromosome. The recipient cell remains F⁻. The donor cell synthesizes a strand complementary to the untransferred strand of DNA and remains Hfr.

1' 1 2' 2
Hfr Cell
Recipient cell DNA
F⁻ Cell
Integration of donor chromosomal DNA

In this experiments mix culture is prepared representing a cross between an Hfr prototrophic (wild type) strain of *E.coil* that is streptomycin sensitive and F^- auxotrophic streptomycin resistant (mutant) *E.coil* strain and. Following a incubation period isolation of only the prototrophic and streptomycin resistant recombinants will be performed by plating the mixed culture on a minimal medium containing streptomycin. The streptomycin is incorporated into the medium to inhibit the growth of the wild type streptomycin-sensitive parental Hfr cells.

Requirements

Cultures

Young culture of *E.Coli*

F^- *E.Coli*	:	Recipient strain , F^-, *thi, his, trp, pure, leu, metA,ilv, argG,rspl, str*^{r} (MTCC 1588)
Hfr $^+$*E.Coil*	:	Prototrophic , *Str* ^{S}, *rif* ^{r}, Donor srain (MTCC 1650)

Media

Minimal Agar

Part I	**pH 7.0**
KH_2PO_4	3.0 gm
Na_2HPO_4	6.0 gm
NH_4Cl	2.0 gm
NaCl	5.0 gm
D/W	800.00 ml

Part II	**pH 7.0**
Glucose	8.0 gm
$MgSO_4.7H_2O$	0.1 gm
Agar	15.0 gm
D/W	200.00 ml

Autoclave solution A and B separately and mix before plating.

To the combine sterile molten media add 50 mg (1ml of 50 mg / ml of) sterile streptomycin solution before pouring agar.

Glassware

Sterile pipette 10.0 ml and 0.1 ml
Sterile Petri dish
Sterile flask
Spreader
Sterile D/W for dilution

Procedure

1. Prepare the young culture of the both of organisms. (Take care that Hfr strain should not be kept on shaker)
2. Mix 1 ml of F^- strain and 0.3 ml of Hfr in sterile test tube /sterile flask.
3. Incubate the mixture at 37 °C for 100 minutes
4. With the help of spreader spread it on plate containing Minimal agar + glucose + streptomycin. Streak the plate with pure Hfr and F plate also as control
5. Incubate the plate for 37 °C for 24 hrs.
6. Check the presence of growth on each plate after 24 hrs.

Result

No growth on plate with Hfr and F^- *E. coli* and growth with mixed culture on minimal + Streptomycin indicates the conjugation between Hfr and F^-. Because pure Hfr strain can not grow on the minimal + streptomycin as they are streptomycin sensitive and pure F $^-$ strain can not grow on the minimal + streptomycin due to their auxotrophic nature.

Observation

Observe all the plates for the presence (+) or absence (-) of colonies record your result in table

	Hfr *E. coli* plate	F^- *E. coli* Plate	Mixed culture plate
Growth (+) or (-)			

CHAPTER 27

Isolation of Lactose non Fermentor Mutants of *E.coli* by Physical Mutagenesis

Introduction

Mutation refers to any heritable change in the base sequence of a DNA molecule. A mutagen is a physical of chemical agent that causes mutations to occur (or increase the frequency of their occurrence). Mutagenesis is the process of producing a mutation. If the mutation occurs in nature without the addition of a mutagen, it is referred as a spontaneous mutation, and the resulting mutants are spontaneous mutants. If it caused by a mutagen, the process is induced mutagenesis.

Ultraviolet light is a fairly potent mutagen. The main DNA lesions are two chemically different types of covalent dimers: cyclobutane-pyrimidine dimers and pyrimidine dimers. The pyrimidine dimers only account for 20% of the total dimers but are correspondingly more mutagenic. In *E. coli* the number of mutants induced by ultraviolet (UV) light can be reduced by exposure to visible light (photoreactivation), which implicates cyclobutane pyrimidine dimers since dimers can not be photoreactivated.

Aromatic amino acid of proteins, and purine as well as and pyrimidine bases of nucleic acid absorb ultra violet radiation. Pyridine bases, particularly thymine of DNA is the main target compound involved in bactericidal action of ultra violet radiation. The energy absorbed by thymine causes it to under go a photochemical reaction with adjacent (adjoining) thymine molecule in same DNA strand. DNA replication is blocked because the formation of thymine: thymine (T:T) dimmers. As the dosage increase, cytosine-thymine (C:T) to (C:C) cytosine-cytosine dimmers can be found to add the lethality of ultra violate radiation.

Induction of *lac*$^-$ Mutant in *E.coli* by Ultra Violate Radiation

E.coli can ferment lactose (glucose+ galactose) with production of acid and gas. The enzymes needed for uptake and catabolism of disaccharide lactose are inducible enzyme performing following function.

lactose Operon Systems

- *ß*-galactoside → Cleave lactose into glucose and galactose
- Permease → Transport lactose into cells through the cell membrane
- Transacetylase → Function is unknown
- If mutation occurs in operon, *E.coli* (*lac*$^+$) become *E.coli* (*lac*$^-$) mutant. *Lac*- mutant can be easily identified by visual examination of colonies. Both fermenter and non-fermenter colonies are able to grow on media because a non-fermented carbon source (peptone) is also induced but only fermenter produce acid. To differentiate lactose fermenter and non fermenter colonies media in common use like Mac-Conkey and EMB agar can be use. The *lac*$^-$ mutant of *E.coli* gives yellow colored lactose non-fermenter colonies.

Requirements

- UV lamp
- Sterile Petri plates (empty)
- 10 ml actively growing culture of *E.coli* (Lac$^+$)
- N-agar plates
- Dilution tubes
- Mac-Conkey agar plate

- Lactose broth
- Alcohol, glass spreader, sterile pipettes hand gloves and other lab requirements.

Procedure

- Take 10 ml of actively growing culture of *E.coli* (having OD: 0.5-0.7 at A_{600nm}) into a sterile Petri plates. Put one clean, rust free iron pin before sterilization of the plates.
- Prior to exposure of culture to UV, withdraw 0.5 ml of the culture under aseptic condition and label it as **O second** reading.
- Put culture containing Petri plate in UV chamber on a stage (magnetic stirrer) previously set at the distance of 30 cm. (Note: Before exposure of culture, switch on the UV light to ensure the sterilization of the chamber).
- Now, expose the culture to the UV light (254nm) and withdraw 0.5 ml of the culture after the exposure of 30 sec., 60 sec., and 90 sec. into previously sterilized and carbon paper wrapped test tubes.
- Immediately keep all the tubes in the dark for 30 min to prevent photoreactivation.
- Prepare serial dilution in to the sterile tubes containing 4.5 ml distilled water.
- From the dilutions take 0.1 ml of suspension and spread it on to the N-agar plates and incubate at 37°C for 24 hours.
- Next day select the plate having well isolated colonies and label it as "**Master Plate**". Make replica of this plate on the Mac Conkey agar plate and incubate at 37°C for 24 hours.
- Next day examine growth of *lac*$^-$ mutant (yellow, lactose non-fermenter colonies)
- Pick up *lac*$^-$ mutant colony and inoculate into sterile lactose broth tubes and incubate it for 24 hours at 37°c for the verification of mutants.

Replica-Plating Technique

This technique is often used to isolate nutritional mutants (auxotrophs) as well as various other type of mutants, e.g., antibiotic resistant mutants. For convenience, if one wants to isolate nutritional mutants employing replica-plating techniques, he is required to follow the following steps:

1. Bacterial cultures are diluted, and the cells are spread on the surface of semisolid nutrient agar medium in a Petri dish (called master plate). The medium in the master plate is a complete medium i.e. containing all the nutritional components required by the bacterial population. After a sufficient incubation period, each bacterium produces a visible colony on the surface of the agar in the master plate
2. A piece of sterile velvet cloth is stretched over a cylindrical block of wood or metal that is slightly smaller in diameter than the Petri dishes used in the process.
3. The master plate is now inverted and gently pressed onto sterile velvet. Since the fibres of the velvet act as fine inoculating needle, some cells from each colony of the master plate stick to the velvet.
4. Other Petri dish (called replica plate) is taken containing a minimal medium i.e., a medium deficient with specific nutritional component.
5. The replica-plate is now inverted and gently pressed onto the velvet thus stamping the bacterial cells onto the surface of its minimal medium. The replica plate is identically oriented at the application on the velvet with respect to mark placed on its rim so that the colonies that appear on the replica plate after incubation occupy positions congruent with those of their siblings on the master plate.
6. After sufficient incubation, it is observed that a colony that develops on the complete medium of the master plate fails to develop on the minimal medium of the master plate that lacks a specific nutritional component. Such colony is marked on the master plate and is isolated; it represents mutant for that specific nutritional component not used in the minimal medium of the replica plate.

CHAPTER 28

Isolation of Auxotrophic Mutant by Chemical (5 Bu) Mutagenesis

Introduction

The isolation of mutants is often critical to the study of cellular process, especially when such studies are performed at the molecular level. The phenotype of a particular mutant is attributable to the loss of a particular cellular function; the loss of that function is actually attributable to a mutation in a gene necessary for that function. In this experiment we will isolate auxotrophic mutant which can grow only if provided with specific amino acid. For example *arg*⁻ mutant have lost the catalytic function of an enzyme involved in arginine biosynthesis due to mutation in the structural gene that encodes the enzyme.

Mutagenic Agent

Mutation is an alteration in the coding sequence of a gene. Mutation occur spontaneously at a frequency of about 10^{-8} per bacterial cell per generation. The rate of mutation can be increased with mutagenic agents. Mutants are strains with any structural alteration of their DNA that gives rise to a mutant phenotype. Mutagens are chemical agents (e.g. nitorsoguanidine, 5 bromouracil), physical (e.g.UV) and biological agents (e.g. trnasposons) that cause mutation to occur.

A chemical substance resembling a base is called a base analogue. A base analogue may be incorporated in to newly synthesized DNA instead of a normal base. The pyrimidine analogoue 5 bromourcil (5-BU) is structurally very similar to thymine. If bacteriophage are grown in the presence of 5-BU they incorporate the substance as if it were thymine. 5 BU does not have a lethal action because it is incorporated in the place of T and functions almost normally. 5- BU can however undergo internal rearengment (tautomerization) from the usual keto state to the rare enol enolstate. 5 –BU now paires with guanine instead of adenine, the natural partner of thymine. Thus there is 5-BU-G pairing instead of T-A pairing.

Ampicillin Enrichment for Auxotroph

Ampicillin is an antibiotic which is analog of penicillin and affect cell wall biosynthesis and kill only cells that are growing; cells that are not growing are unaffected by this drug. We can use this property to enrich for auxotrophic mutants in a population of mutagenized bacteria, and then isolate mutant defective in arginine or any other amino acid biosynthesis. Following mutagenesis, the cell will be allowed to devide several times so that the mutant phenotype will be expressed. The rare mutants in the population which require an amino acid (like arginine) will not grown in a minimal + glucose medium that lackes that amino acid. Thus transfer of a mutagenized culture to a minimal medium without arginine and subsequent treatment with penicillin results in death of prototrophic cells (which are growing) but not auxotrophic (e.g. *arg*⁻ and *his*⁻). To enrich for the desired e.g. *arg*⁻ auxotrophs, the treated culture will be grown up in the minimal medium with arginine; this will select against the other types of auxotrophs. In a single cycle of ampicillin treatment , the desired auxotrophic mutant can be enriched by a factor of 10^3 to 10^4.

Requirement

Growth Medium

Minimal Medium

Part I	**pH 7.0**
KH_2PO_4	3.0 gm
Na_2HPO_4	6.0 gm

NH_4Cl	2.0 gm
NaCl	5.0 gm
D/W	800.00 ml
Part II	**pH 7.0**
Glucose	8.0 gm
$MgSO_4.7H_2O$	0.1 gm
Agar	15.0 gm
D/W	200.00 ml

Autoclave solution A and B separately and mix before plating.

Minimal + Arginine Containing Medium

Prepare 10% stock solution of Arginine autoclave it at 10 lbs for 10 minutes. Add 1 ml from the stock solution in the 1000 ml of minimal medium prepared according to the above composition before plating

LB Broth

Tryptone	1 gms
Yeast extract	0.5 gms
NaCl	1 gms
D/W	100 ml
pH	7.5

LB Agar

Add 2% agar to above composition

Other Requirement

1. 5 BU
2. *Ecoil* strain (DH 5 α)
3. N saline
4. LB agar plate
5. Minimal agar plate
6. Minimal + Arginine plate
7. Ampicillin
8. Sterile centrifuge tubes and other glassware

Procedure

Ist Day

1. Prepare young culture from overnight grown *E. Coli* culture till OD comes to 0.5- 0.7 in LB broth
2. Place 100µl of 5BU (2.5 mg/ml of DMSO) in eppendorf tube containing 3 ml of young culture.
3. Incubate for 20 min and centrifuge after completion of incubation time centrifuge at 3000 rpm for 10 min discard the supernatant wash the cell with N-saline and resuspend the cell in 2ml of N-saline.
4. Inoculate 2 ml mutagenised culture in 10 ml of L broth (in 250 ml flask) incubate it 2 hrs in shaking condition with 100rpm at 37 °C.
5. After incubation centrifuge the LB broth containing the 5BU treated cell, Discard the supernatant wash and resuspend the cell with 2ml of N saline.
6. Inoculate 10ml of minimal medium with 2 ml of 5BU treated washed cell. Incubate it for 4hrs. in shaking condition with 100rpm and 37° C temperature.
7. After 4 hrs add 0.2 ml of Ampicillin (2mg/ml) and continue the incubation overnight.

2nd Day

8. Take overnight grown 5BU and Ampicillin treated culture and centrifuge it at 3000rpm for 10 min wash with N Saline and resuspend it in 2ml of N-saline.
9. Incubate 10 ml minimal +Arginine broth with 2 ml of ampicillin treated washed cell. Incubate it for 4hrs. in shaking condition with 100rpm and 37°C temperature.
10. After incubation Prepare different dilutions from above suspension and plate them on LB-agar and incubate it for 24 hrs.

3rd Day

11. After 24 hrs. select the well isolated colony give them number and transfer the colony on Minimal agar and Minimal + Arginine Agar

12. Incubate the plates for 24 - 48 h at 37° C.
13. Note the growth of strains patched on different plates and record the results.

Result

Frequency of arg $^{-}$ Mutant in the given culture is

Observation Table

Colony No. on LB plate	Growth on Minimal agar plate (+) or (-)	Growth on minimal + arginine agar plate (+) or (-)

Calculation of Frequency of Mutation

Frequency of mutation for Argi $^{-}$Auxotrophic Mutation = No. of Arg $^{-}$ Auxotophs/ Total no. of colony transferred from LB agar plate

Note *Colonies which are present on LB agar and minimal + arginine but absent in minimal agar should counted as the Auxotrophic mutant*

Microorganisms Used

F^- *E.Coli*	:	Recipient strain , F^-, *thi, his, trp, pure, leu, metA,ilv, argG,rspl, str*^{r} (MTCC 1588) (LB agar)
Hfr $^+$*E.Coil*	:	Prototrophic , *Str* ^{S}, *rif* ^{r}, Donor srain (MTCC 1650) (LB Agar)
pUC 18	:	Contains a plasmid pUC 18, used as a vector for cloning and the recombinants could be selected by blue/ white colour screening (MTCC 1601) (LB Agar containing 100µg/ml ampicillin)
DH5α	:	rec^- strain, F^- , nal^r , F^r, recA, gyrA (MTCC 1652) (LB Agar)

CHAPTER 29

Mutagenesis (Direct Selection)

To isolate a streptomycin resistant mutant of *E.coli* by Direct selection using Gradient Plate Method.

Introduction

Zylbalskii devised this technique.

A mutation is any change in the nucleotide sequence of DNA. These changes are heritable. Although most mutations are harmful to the organism, a small percentage is beneficial under certain environmental conditions. Mutation can occur during chromosome duplication and they may be spontaneous of they may be induced by mutagenic agent. In this experiment, we select for a mutation causing streptomycin resistance in *E. coli*. Resistant mutants will be detected directly using thy gradient plate method.

Principle

Microbiologist developed a highly efficient method for isolation of antibiotic resistant mutant. The gradient is said as gradient of antibiotic formed on an agar plate. A gene imparting resistant toward an antibiotic is present in plasmid or chromosome. The occurrence of mutant with different level of streptomycin resistant involves mutation of gene with unequal potential. This means that under certain cells degree of mutation is more while few shows less mutation. Some cells show intermediate resistant while others show completes resistant. The resistant in cell against streptomycin of this type, while in case of penicillin resistant increases step arise & so resistant mutant of high concentration of penicillin cannot be isolated/ obtained. While it is possible with streptomycin obtaining high resistance to penicillin stepwise & so this stepwise increase in resistance involves mutation of gene with equal potential. This theory of mutation through multi step procedure. The gene direct the organism to produce an enzyme, which inactivates an antibiotic & thus resistance to antibiotic, is observed in case of cells. Resistance of antibiotic is a plasmid-mediated resistance.

Materials

I) Materials Required for Gradient Plate Preparation

- Empty sterile Petri plate
- Nutrient agar deeps (10 ml per tube) 1% streptomycin solution (in water)
- Sterile 1ml pipette
- Marking pen

II) Materials Required for Mutant Isolation

- 24 hour, nutrient broth culture of *E.coli*
- Sterile 1 ml pipette
- Bent glass rod or spreader
- Beaker containing 95% isopropyl alcohol (or ethanol)
- Bunsen burner & striker (or matches)
- Marking pen
- Inoculating loop

III) General Lab Equipment

- Incubator set at 37^{0}C
- Autoclave
- Water bath set at 80^0 C for maintaining deeps if they are premelted.

Method

I) Procedure for Gradient Plate Preparation

1. Melt two nutrient agar deep and maintain them in liquid state.
2. Place the wood spacer under one edge of the empty, sterile Petri plate.
3. Cool one of the melted deeps to approximately 45⁰C.
4. Pour the contents of the cooled, nutrient agar deep into the plate. Keep the plate in slanted position. Be sure that the medium covers the entire bottom surface of the plate.
5. Move on to step 6 while you are waiting for the agar to solidify.
6. Cool the second melted deep to approximately 45⁰C.
7. Add streptomycin solution to the final conc. Of 100µg/ml (0.1ml) into the second agar deep. Cap the deep & discard the pipette properly. Rotate the deep gently between the palms of your hands to mix the agar & the streptomycin solution.
8. When the agar in the Petri dish has solidified, pour the contents of the second deep carefully over the top.
9. When the contents of the second deep have solidified, label the high & low streptomycin conc. Areas of the plate. Use a marking pen & write on the bottom of the plate.

II) Procedure for Mutant Isolation

1. Place the glass spreader in the beaker containing 95% isopropyl alcohol.
2. Using the aseptic technique, pipette 0.2 ml of the *E. coli* broth onto the surface of the gradient plate. Discard the pipette.
3. Remove the glass spreader from the beaker & gently shake off the excess alcohol. While holding the spreader down & away from your body, quickly pass it through the flame of the Bunsen burner. The alcohol residue will burn off quickly.
4. Use the glass spreader to spread the *E.coli* broth over the surface of the gradient plate. Spread the organism over the entire surface of the plate. Return of spreader to the beaker containing alcohol when you are finished.

5. Invert and incubate the plate at 37 °C for 48 hours.
6. Observe the plates in the area of high streptomycin conc. For any streptomycin resistant mutants.

CHAPTER 30

Mutagenesis (Indirect Selection)

To isolate a streptomycin resistant mutant of *E.coli* by Indirect selection using Replica Plate Method.

Introduction

J. Lederberg & his wife E. M. Lederberg devised this technique in 1952.

U.V. light coming from the sun emits strong radiation to extent the antibacterial effect in the microbial population. However it has a very short penetration power, UV light is used as a mutagenic agent & was used by Beedle & Tatum to obtain biochemical mutation. The mutation rate is only once in 10^{-9} cells per population. Then hence using mutagens like UV light by 10 to 1000 folds will increase this mutation rate.

In isolation of the mutant using UV light the dose of UV light is measured by its units/nm^2 organisms. The time selected for the mutation is calculated on the basis of 0.1 to 1% survival & a suitable biochemical character is used to evaluate the mutation.

Principle

Principle involves the use of multiple inoculating "wire" instead of a single one to transfer a sample cells from each colony to one or many subsequent dishes in such a way that the position of cells deposited on each new plate corresponds accurately to that of the colonies on the original plate. Thus replica plating of the master plate is obtained.

In this technique, velvet is used as inoculator; the individual fiber is functioning as inoculation wire. The velvet should have dense & short pile. Disc of velvet larger than petridish (about 14 cm diameter for standard 9 cm petridishes) is sterilized by autoclaving. Then the velvet dish is laid over a wooden, metal or plastic block about 8 cm in diameter for 9 cm petridish & fastened in place with a rubber band.

Materials

- N-agar plates, UV treated test culture, Spreader, alcohol, n-agar plate with antibiotic, Velvet cloth, replica block, Pipette.

Procedure

(A) First Day

1. Transfer the given test *E.coli* culture from the test tube to sterile Petri plate.
2. Expose the petriplate (by opening the lid) with constantly striving the plate for 80 seconds at a distance of 30 cm from UV source.
3. Cover the test tube with black paper to avoid photoreaction
4. Prepare the dilution10^{-4} to 10^{-10} & spread 0.1ml of the above dilution on N-agar plate with the help of spreader.
5. Incubate at 37^0C for 24 hrs in dark.

(B) Second Day

6. Select the master plate (with well isolated 30-60 colonies) which is to be replicated is placed on replica block in particular orientation determined by reference mark on the block and enough pressure is applied to force the fibers in to the colony.
7. Plate is then remove and is replaced by another N-agar containing antibiotic plate.

8. Similarly replica of original plate is obtained on simple N-agar plate by using same velvet block (or master can be place in the refrigerator instead of making new N-agar plate).
9. Incubate the plate at 37^{0}C for 24 hours. Next day colonies in antibiotic a containing medium are the indications of resistance mutants.
10. Look for the similar colony growing exactly at the same place on N-agar plate or master plate.

CHAPTER 31

Agrobacterium Mediated Gene Transfer from Bacteria to Plant

Limitations of Bacteria

- Normally genetically engineered bacteria are very useful for the large scale production of valuable eukaryotic proteins.
- Bacteria grow rapidly, They grow on relatively simple media, they are easy to handle, they are easily transformed, they can grow on wide range of substrates, they multiply, fast, they contain simple and well characterized genome, they can be easily modified but even though, they have their own problems...

(1) Bacteria do not have intrudes and so they do not have any system or machinery to remove introns.

(2) Many eukaryotic proteins undergo different types of modifications after they are produced. These modifications are known as "post translation al modifications".

(3) Each bacterial cell usually contains many copies of the plasmid but there is no mechanism to ensure the equal distribution of plasmids in daughter cells.

Genetic Manipulation of Plant Cells

Ti Plasmid (Tumor Inducing Plasmid)

- Ti plasmid is considered as the most efficient Plant vector. It was first time discovered in 1983.
- This plasmid is found in *Agrobacteriaum tumefaciens*.
- This organism is a typical plant pathogen that enters the dicotyledonous plants from their root system and damages them.
- Agrobacterium enter through the fresh wound and attaches itself to the wall of an intact cell. Later on it transfer a short piece of Ti plasmid into the nucleus of plant cell. This DNA piece is known as T-DNA.
- T-DNA carries few genes which are expressed within the plant Construction of Ti plasmid :

(1) One gene codes for an enzyme which is responsible for the synthesis of opine from amino acids.

- Opines are not found normally in plants and not metabolized by plants, but they can be used as a nutrient by Agrobacteriaum.
- Total 24 types of opines are detected in nature the particular opine produced by plant depends on the strain of bacterium.
- Some strains produce a gene responsible for the synthesis of "Nopaline" and others have the gene for the synthesis of "Octopine". These genes are respectively known as Nos & OCS.

(2) The other gene found in T-DNA is responsible for the induction of disorganized growth of infected plant cells that forms callus, Gail or tumor. It is called one gene.

- It is also observed that certain Agrobacteriaum strains contain genes for plant growth hormones Auxin & cytokinine instead of onc gene, such hormones also contribute to accelerate plant growth.

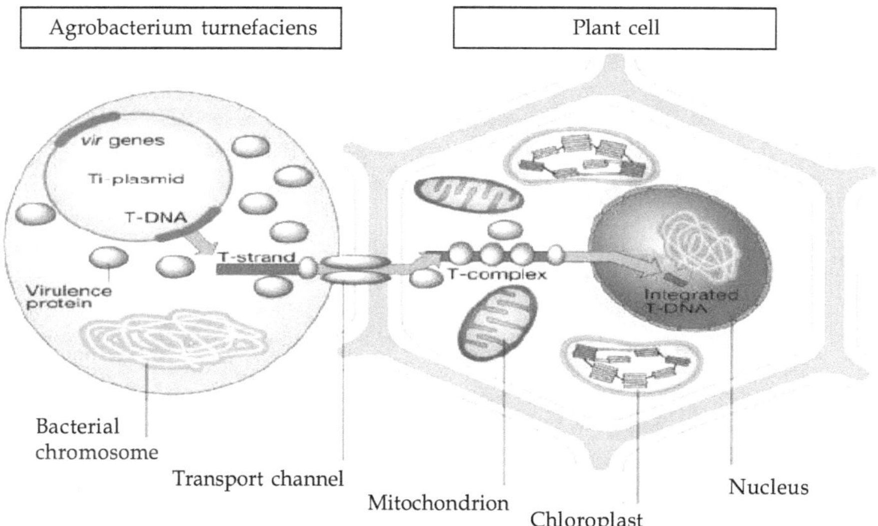

Transformation of Ti Plasmid

- Inside the plant cell; the T-DNA does not remain independent but it becomes integrated into plant chromosome.
- The integration occurs due to the presence of two repeated sequences of 25 base pairs which are located dear the ends of T-DNA.
- Once Ti plasmid is nicked, T-DNA is released from and it travels into plant cell.
- T-DNA becomes circular in its intermediate form, and ultimately it is inserted into plant chromosome DNA.
- Genes that remain on Ti plasmid include those for attachment of bacterium to plant cell wall, those for transfer of T-DNA and those for uptake and catabolism of opine.
- The gene which is responsible for the transfer of T-DNA lies on Ti plasmid and it is called vir-Region.

Schematic represent of transfer of gene using Ti- plasmid

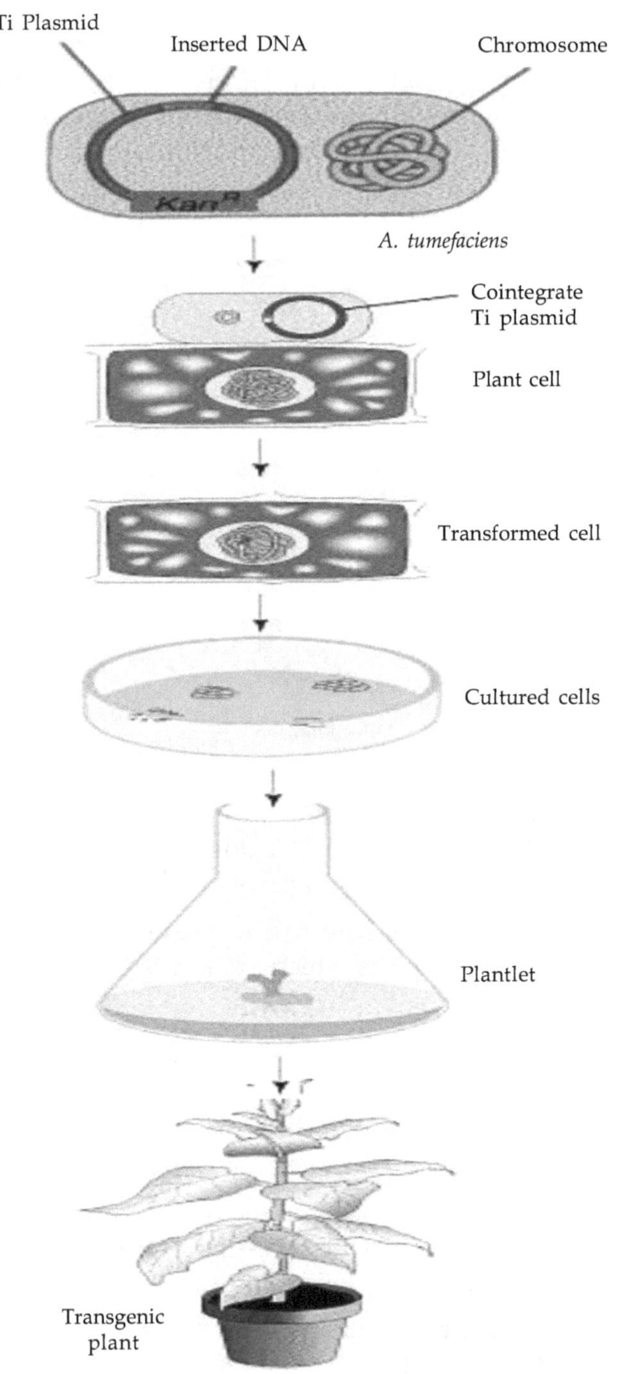

- This transformation is irreversible and callus cells can continuously produce opine once they are infected (One type of opine) is also present on Ti plasmid and it is called noc gene.

Onc : Gene for tumor induction
Vir : Gene for the transfer of (Oncogene) T-DNA (Virulence)
nos : Nopaline synthesis
noc : Nopaline catabolism

Ti Plasmid as a Vector

- Usefulness of Ti plasmid as a vector during plant genetic manipulation is due to many features of Ti-plasmid these features include.

(1) The genes in T-DNA are eukaryotic in nature. Even though these genes are of bacterial origin, they are transcribed by plants RNA polymerase. These genes contain introns and they are removed correctly during m-RNA splicing.

(2) Only some part of T-DNA is required if one wishes to use it as a vector.

- Opine production genes and tumor formation genes can be replaced by foreign DNA.
- Thus nos/ocs & onc genes can be deleted.
- If nos or ocs genes are removed, their promoters are retained to ensure the gene expression of foreign DNA.
- Such experiments have been successfully carried out for the expression of kanamycin & methotrexate resistance in plants.
- Phaseolin is synthesized only in developing seeds. Therefore it seems that their promoter is regulated one.

Direct Transformation

- In this technique, plant protoplasts are prepared by treating plant cells with poly ethylene Glycol (PEG).
- Protoplasts can take up DNA from their surrounding medium.
- This uptakes DNA can be stably integrated into plant chromosomal DNA.
- This technique has been used to transform plant cells with a gene for kanamycin resistance, linked to a strong promoter.

- Plants regenerated from this plant were resistant to kanamycin, and this resistance was inheritable.
- The limitation of direct transformation is that the uptake DNA is randomly integrated in plant genome.
- As along as one can select the desired plant among all undesirables, this method should not present a great problem.

Preparation of Shoot Apex Explant

1. Cotton variety Coker 312 was used in the transformation experiments.
2. Seeds of Coker 312 were treated with 20% hydrogen peroxide for 2 hours and rinsed three times with double-distilled water. The seeds were then placed on a rotor shaker at 100 rpm overnight. After removing the seed coat, seeds were germinated in MS basal medium
3. Shoot apices were cultured on MS medium with 0.1mg/L Kinetin for 3 days.

Agrobacterium Strain and Plasmid

Agrobacterium strain LBA4404 harboring a 'super-binary' vector pTOK233 was used to develop the optimized transformation protocol. This strain has been successfully used in transformation of rice The T-DNA of pTOK233 contains a hygromycin-resistance gene (*HPT*), a kanamycin-resistance gene (*NPTII*), and a *GUS* gene which has an intron in the N- terminal region of the coding sequence and which is fused to the CaMV35S promoter

The intron –gus gene expresses GUS activity in plant cells, but not in cells of *A. tumefaciens Agrobacterium* strain EHA 105 harboring both NPT II and *bar* genes was used to transfer a herbicide resistance trait into cotton. The *bar* gene was originally cloned from the bacterium *Streptomycin hygroscopius*. It encodes for phosphinothricin acetyltransferase (*PAT*) (Thompson *et al.* 1987) that detoxifies phosphinothricin or glufosinate, the active ingredient of the herbicides Liberty and Basta (DeBlock *et al.* 1987). Therefore, plants expressing the *bar* gene are tolerant to herbicides Liberty and Basta.

Pretreatment of Shoot Apex

The shoot apical meristem (SAM) is a population of cells located at the tip of the shoot axis. The shoot apex is divided into three layers, To obtain germline transformation, the shoot apices were wounded in the middle tip by using a scalpel to expose layer III cell before co-culturing with *Agrobacterium.*

Agrobacterium Co-cultivation and Transgenic Plants Regeneration

1. The *Agrobacterium* strains were cultured in LB medium (contains 10g/L Bacto Tryptone, Bacto, 5g/L Yeast extract and 10g/L NaCl).
2. Twenty ml of LB medium plus antibiotics (50mg/L kanamycin and 50 mg/L hygromycin for strain LBA 4404 or kanamycin 50mg/L for strain EHA 105) was inoculated with *Agrobacterium* and incubated in a 100ml Erlenmeyer flask overnight (about 17 hours) on a shaker set for 180 to 220 rpm at 28°C.
3. Then 2ml of the overnight culture was withdrawn and used to inoculate 50ml of LB medium without antibiotics.
4. Acetosyringone was added to the culture at a final concentration of 100 μM. After incubation for 3 to 4 hours at 28°C with shaking, those cultures were diluted with additional LB medium (containing 100 μM acetosyringone) to a concentration (OD_{600} 0.6) for transformation. Equal numbers of shoot apices were randomly distributed to two independent treatments, one with *Agrobacterium* co-cultivation and one without *Agrobacterium* co-cultivation
5. Shoot apices were inoculated by placing one drop of *Agrobacterium* solution onto each shoot apex in co-culture medium (MS + 100 μM acetosyringone) and incubating at 28 °C under dark conditions for approximately 1 to 4 days.
6. After co-cultivation, explants were washed three times with sterile distilled water. Cleaned apices were blotted dry using a sterile paper towel and cultured on the selection medium consisting of MS with 400 mg/L timentin and 50 ml/L kanamycin.
7. Shoot apices not inoculated with *Agrobacterium* were plated on the selection medium as a negative control. Timentin was included in the selection medium to suppress the *Agrobacterium* growth.

The Petri dishes were incubated at a temperature of 28 °C under an 18 hours photoperiod and sub-cultured every 3 weeks. The process was repeated until controls, not co-cultivated with *Agrobacterium*, were totally dead.

8. After this period the surviving shoot apices were transferred to an MS medium without kanamycin to root the plants. Rooted plants were then transferred to soiland grown to maturity in a greenhouse.

β-Glucuronidase (GUS) Histochemical Analysis

The histochemical assay for GUS gene expression was performed by established methods (Jefferson, 1987; Kosugi *et al.*, 1990). Following co-cultivation, apices were harvested for GUS staining.

1. The apices were incubated overnight in a solution containing 25 mg/l X gluc, 10 mM EDTA, 100 mM NaH_2PO_4, 0.1% Triton X-100 and 50% methanol, pH 8.0) at 37 °C.
2. The number of apices that stained with blue spots was recorded. Young leaves of putative transgenic plants were also collected for GUS staining to confirm the transformation event.

Kanamycin and Glufosinate Leaf Test

In the putative transgenic plants, expression of the transgene (*NPT II*) or *bar* gene was analyzed by first establishing the lowest concentration of Kanamycin or glufosinate that would kill untransformed plants. Leaves of control plants were painted with a cotton swab when they had two totally opened true leaves using 0, 0.1, 1, 2, or 3% (W/V) of kanamycin or 0, 0.1, 0.2, 0.3, 0.4, and 0.5 ml/L Liberty. The lowest level (2%) of kanamycin that caused damage to the controls was used to evaluate for resistance to kanamycin in the greenhouse. The lowest level (0.3 ml/L) of Liberty was used to evaluate for resistance to glufosinate. Plants were evaluated for resistance 7 days after leaf application of kanamycin or Liberty.

Polymerase Chain Reaction Analysis

1. DNA was isolated from young leaves of putative transgenic plants using the DNAeasy Plant Mini Kit (Qiagen, Santa Clarita, CA).

2. The DNA samples were tested for the presence of the T-DNA region using a pair of *nptII* specific primers (upstream 5′-AGAACTCGTCAAGAAGGCGA-3′anddownstream 5′CTGAATGAACTGCAGGACGA-3′) to amplify the 700 bp *nptII* fragments.
3. Regenerated plants transformed by EHA101 were screened for the presence of the *bar* gene by PCR using the *bar* gene specific primers (upstream 5′- CATCGTCAACCACTACATCGAG-3′ and downstream 5′- CAGCTGCCAGAAACCCACGTCA-3′).
4. The 25µl amplification mixture contained 2.5 ul 10X PCR II buffer (50 mM Tris (PH 8.3); 500 mM KCl);1.5 mM MgCl2; 1.0 mM dNTP mix (Pharmacia Biotech); 0.2 uM primer; 0.5 unit of AmpliTaq DNA polymerase (Promega); and 20 ng of genomic DNA as template.
5. DNA was amplified in a Perkin Elmer Geneamp PCR System 9600, programmed for a first denaturation step of 2 minutes at 94 °C followed by 45 cycles of 94 °C for 1 minute, 35 °C for 1 minute, and 72 °C for 2 minutes.
6. After the completion of 45 cycles, a final extension at 72 °C was carried out for 5 minutes. The completed reactions were then held at 4 °C until electrophoresis was done.
7. PCR products were separated by loading 12 uL of each sample and 2 uL of loading buffer type II on a 1.2 % agarose gel prepared with 1.0X TBE buffer. The sample were subject to electrophoresis at 90-100V for 4 hours in 1.0X TBE buffer. The gel was stained with ethidium bromide and visualized under UV light.

Southern Blot Analysis

DNA was isolated from young leaves of putative transgenic plants using the DNAeasy Plant Mini Kit (Qiagen, Santa Clarita, CA) and completely digested with *HindIII*. Based on the construct of the plasmid, Hind III digested genomic DNA will result in a 3.1 Kb fragment in LBA 4404 transformed plants and a 1.8 Kb fragment in EHA 105 transformed plants. Twenty µg of genomic DNA was digested with *Hind* III overnight in a 37 °C water bath. The digested DNA fragments were electrophoresed on an 0.8% agarose gel in 0.5x Tris-Borate-EDTA (TBE) buffer, and transferred to a nylon membrane by the alkaline transfer method (Reed and Mann, 1985). The [^{32}P]- labeled probes for

LBA 4404 transformed plants were made from a 0.5-kb polymerase chain reaction (PCR) product (Primer: 5'-CTG TAG AAA CCC CAA CCC GTG-3' and 5'-CAT TAC GCT GCG ATG GAT CCC-3') containing the *GUS* coding region. The probes for EHA 105 transformed plants were made from a 430 bp PCR product (Primer: 5′- CAT CGTCAACCACTACATCGAG-3′ and 5′- CAGCTGCCAGAAA-CCCACGTCA-3′). The band was excised from agarose gel and purified using a Pre A gene Kit (Bio-Rad, Hercules, CA). The probe was then labeled with ^{32}P-dCTP.

CHAPTER 32

Western Blotting

Introduction

Western blot analysis can detect one protein in a mixture of any number of proteins while giving you information about the size of the protein. It does not matter whether the protein has been synthesized in vivo or in vitro. This method is, however, dependent on the use of a high-quality antibody directed against a desired protein. So you must be able to produce at least a small portion of the protein from a cloned DNA fragment. You will use this antibody as a probe to detect the protein of interest.

Western blotting tells you how much protein has accumulated in cells. If you are interested in the rate of synthesis of a protein, Radio-Immune Precipitation (RIP) may be the best assay for you. Also, if a protein is degraded quickly, Western blotting won't detect it well; you'll need to use (RIP). See the section on RIP for more information, as well as a helpful comparative chart that illustrates the differences between these two techniques.

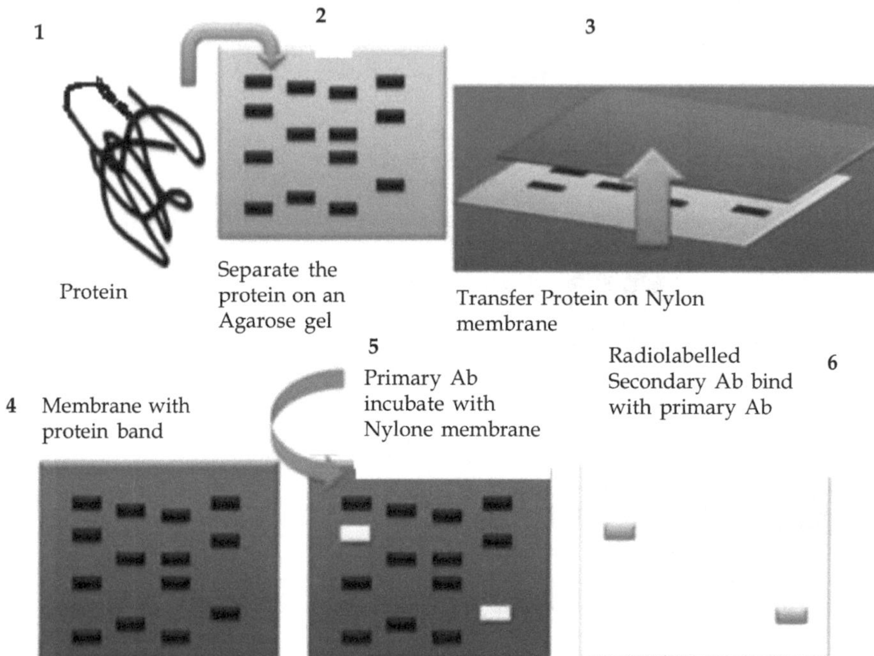

1. Separate the proteins using SDS-polyacrylamide gel electrophoresis (also known as SDS-PAGE). This separates the proteins by size.

2. Place a nitrocellulose membrane on the gel and, using electrophoresis, drive the protein (polypeptide) bands onto the nitrocellulose membrane. You want the negative charge to be on the side of the gel and the positive charge to be on the side of the nitrocellulose membrane to drive the negatively charged proteins over to the positively charged nitrocellulose membrane. This gives you a nitrocellulose membrane that is imprinted with the same protein bands as the gel.

   One thing to be aware of is that proteins bind better to nitrocellulose at a low pH. You may need to go through some trial-and-error to find the optimal pH. You also need to be sure there are no air bubbles between the nitrocellulose and the gel or your proteins will not transfer.

3. Incubate the nitrocellulose membrane with a primary antibody. sticks to your protein and forms an antibody-protein complex with the protein of interest.

4. Incubate the nitrocellulose membrane with a secondary antibody. This antibody should be an antibody-enzyme conjugate. The secondary antibody should be an antibody against the primary antibody. This means the secondary antibody will "stick" to the primary antibody, just like the primary antibody "stuck" to the protein. The conjugated enzyme is there to allow you to visualize all of this. It's kind of like a molecular flare stuck on the antibodies so you can visualize what s going on.
5. To actually see your enzyme in action, you'll need to incubate it in a reaction mix that is specific for your enzyme. If everything worked properly, you will see bands wherever there is a protein-primary antibody-secondary antibody-enzyme complex, or, in other words, wherever your protein is.
6. Put x-ray film on your gel to detect a flash of light, which is given off by the enzyme. The reaction usually runs out in about an hour.

Making a Primary Antibody

This description assumes you have available purified protein. Run the protein on an SDS-PAGE gel. Stain the gel with KCl. The KCl forms a precipitate with the SDS. Since the area with the protein has a low concentration of SDS, the area with the protein will not show a precipitate. This will allow you to see the protein band as a clear band against a milky white precipitate on the rest of the gel.

Carefully cut out the band and soak it in 1 mL PBS buffer. Crush it and make an emulsion with 1 mL Freund's Complete Adjuvant (which is an oily substance). The complete adjuvant contains microbacteria (an immune stimulant) to increase the immune response. Inject this subscapularly into a rabbit. This is your first inoculation. Only use the complete adjuvant for the first inoculation. *NEVER* inject a rabbit with complete adjuvant more than one time.

Rest the rabbit for one month, then repeat the process using an *incomplete* adjuvant. You can expect to see good antibody titers about 10 days after the second booster.

Bleed the rabbit. You can expect about 30 to 40 ml per bleeding, and about 50 percent of the volume is serum. Now you have rabbit antisera.

To get your primary antibody, dilute the rabbit antisera in blotto (aka Carnation Nonfat Dry Instant Milk) and apply it to your nitrocellulose blot. Make sure you dilute 1:500 to 1:100 in blotto; less dilution will give you background binding and really muddy up your results.

Secondary Antibody

This is much easier than the procedure for the primary antibody. Grab a catalogue and look for a goat-anti-rabbit antibody conjugated to horseradish peroxidase (HRP). The goat-anti-rabbit is your secondary antibody (the one that "sticks" to the primary antibody) and the HRP is the conjugated enzyme that will allow you to visualize your protein.

To separate the proteins through SDS-PAGE and detection followed by characterization of proteins through Western blotting.

Principle

Identification of protein separated by gel electrophoresis is limited by the small pore size of the gel, as the macromolecule probe for protein analysis cannot permeate the gel. This limitation is overcome by blotting the protein into an adsorbent porous membrane. The apparatus consists of a tank containing buffer, in which is located a cassette. Clamping the gel and the membrane tightly together, a current is applied from electrodes, and repeated on either side of the cassette to avoid heating effects. The proteins are separated according to their electrophoresis mobility and blotted onto the membrane identified, using suitable immunochemicals to locate the protein of interest. The individual techniques are explained below.

SDS-PAGE

Stage 1. Prepare a PAGE gel slab and fix to a vertical electrophoretic apparatus. Treat the sample with suitable buffer and load onto the gel slots.

Stage 2. Apply electric current. After a few minutes, proteins in the sample migrate according to their electrophoretic mobility in the stacking gel. The stacking gel has a polyacrylamide concentration resulting in higher pore size and a lower pH of 7. This enables the protein to concentrate into sharp bands due to isotachopharesis, or band-sharpening effect. At the end of the stacking gel, it meets the separating gel, which has a higher polyacrylamide concentration and higher pH. In the separating gel, the proteins travel according to their size.

Stage 3. When the dye front reaches the bottom of the separating gel, the proteins in the sample are resolved depending on their size. However the protein cannot be visualized directly. The gel needs to be stained with suitable stainer to visualize all the proteins. The identification of protein of interest can be done using a suitable probe and a developing system.

Western Blotting

Blotting is the transfer of resolved proteins from the gel to surface of suitable membrane. This is done commonly by electrophoresis (known as electroblotting). In this method, the transfer buffer has a low ionic strength which allows electrotransfer of proteins. Methane in the buffer increases binding of proteins to niotrocellulose and reduces gel smelling during transfer. The use of the membrane as a support for protein enables the case of manipulation efficient washing and faster reactions during the immunodetection, as proteins are more accessible for reaction.

(a) The membrane is in close contact with PAGE gel containing proteins. The proteins are electrotransferred to nitrocellulose membrane.

(b) At the end of electrotransfer, all proteins would have migrated to the NC membrane. The protein was transferred to the corresponding position on the membrane as on the gel. A mirror image of the gel was formed. However, the protein location and detection can only be assessed after immunodetection.

Immunodetection

The transferred proteins are bound to the surface of NC membrane and are accessible for reaction with immunochemical reagents. All the unoccupied sites on the membrane are blocked with inert proteins, detergents, or other suitable blocking agents. The membrane is then probed with a primary antibody and a suitable substrate so the enzyme identifies the ag-ab complex form on the membrane.

Applications

To characterize proteins and to identify specific antigens for antibodies.

Preparation of Reagents

1. *Blotting Buffer.* Add 25 ml of blotting buffer component A and 25 mL of component B to 150 ml of distilled water.
2. *Other Buffers.* Dilute the required amount of buffer concentrate to 1X concentration with water.
3. *Antibody.* Dilute primary Ab and label secondary HRP conjugate in an assay buffer.
4. *Substrate.* Dilute TMB/H_2O_2, 10X concentration 10 times with heat just before use.

Procedure

1. Run SDS: Polyacrylamide gel electrophoresis.
2. Electro blots.
3. Assemble the blotting sandwich within the blotting cassette. Care should be taken to avoid air bubbles between the gel and NC membrane.
4. Insert the cassette into the apparatus filled with blotting buffer so that the gel faces the cathode.
5. Connect the power supply and use a voltage of 50 V for 5 hours for blotting.

Immunodetection

1. Remove the NC membrane gently from the cassette and place it in the blocking buffer for 2 hours at room temperature, or overnight in the cold.
2. Suspend the primary antibody in 10 ml with the assay buffer, using a suitable tube.
3. Immerse the blot in the 10 Ab solute and gently agitate for 30 minutes.
4. Wash the blot by immersing it in wash buffer for 3-5 minutes. Repeat 2 more times.
5. Prepare 1:1000 dilutions of labeled 20 Ab in the assay buffer. Prepare sufficient (10 ml) volume of diluted Ab to cover the blot.

6. Immerse the blot in 20 Ab solute and agitate gently for 30 minutes.
7. Wash the blot in wash buffer for 3–5 minutes and repeat 4 times.
8. Immerse the washed blot in 10 mL of substrate solution with gentle shaking. Bands will develop sufficient color within 5–10 min.
9. Remove the blot and wash with distilled water. Dry.
10. Although the colored hands fade with time, the rate of color loss can be retarded if the blots are kept in the dark.

Interpretation and Result

The proteins separated through the SDS-PAGE have been successfully transferred onto the nitrocellulose membrane and the transferred proteins detected by immunodetection, which was confirmed by the development of color bands on the nitrocellulose membrane.

- *1X Transfer buffer*: 25 mM Tris, 192 mM Glycine, pH 8.3. Mix 3.03 g Tris and 14.4 g glycine; add water to 1 liter - do not add acid or base to pH - it should be >8.0. Use 0.5X for transfer in 20% methanol.
- *10X Western Buffer:* 200 mM Tris pH = 7.5; 1.5 M NaCl (containing 0.1% Tween-20 and 0.2% I-Block). To prepare 1X Western Buffer, dilute 10X buffer to 1X, adding Tween-20 to 0.1%. Remove 50 ml and set aside for the last two washes. To the remainder, add I-Block to 0.2% (Cat #T2015, Applied Biosystems - formerly Tropix). To dissolve I-Block, heat solution in a beaker briefly in a microwave to about 60°C, then stir until dissolved (solution will be cloudy). Bring to room temperature before using.
- *Primary antibody*: For his tagged proteins - Anti-His monoclonal antibody - BD Bioscience #631212
- *Secondary antibody*: Goat anti-mouse alkaline phosphatase conjugated - Biorad #170-6520.
- *Substrate*: ECF chemifluorescent substrate - Amersham #RPN5785. Mix substrate with accompanying buffer as per manufacturer's recommended instructions, prepare 1 ml aliquots and store at -20°C.

CHAPTER 33

Northern Blotting

Introduction

The northern blot is a technique used in molecular biology research to study gene expression by detection of RNA (or isolated mRNA) in a sample. With northern blotting it is possible to observe cellular control over structure and function by determining the particular gene expression levels during differentiation, morphogenesis, as well as abnormal or diseased conditions. Northern blotting involves the use of electrophoresis to separate RNA samples by size and detection with a hybridization probe complementary to part of or the entire target sequence. The term 'northern blot' actually refers specifically to the capillary transfer of RNA from the electrophoresis gel to the blotting membrane; however the entire process is commonly referred to as northern blotting. The northern blot technique was developed in 1977 by James Alwine, David Kemp, and George Stark at Stanford University. Northern blotting takes its name from its similarity to the first blotting technique, the Southern blot, named for biologist Edwin Southern. The major difference is that RNA, rather than DNA, is analyzed in the northern blot.

Flow Chart for Northern Blotting

1. RNA is isolated from several biological samples (e.g. various tissues, various developmental stages of same tissue etc.) RNA is more susceptible to degradation than DNA. The decision whether to isolate total RNA or mRNA depends on the scientific question at hand. mRNA comprises 2-5% of total RNA, and its isolation requires a further purification step using a poly A tail hybridization.
2. The RNA samples are separated according to their size on an agarose gel
3. The gel is then blotted on a nylon membrane or a nitrocellulose filter.

4. The membrane is placed in a dish containing hybridization buffer with a labeled probe (radioactively or chemically labeled). The probe is specifically designed for the sequence of interest. Thus, it will hybridize to the RNA on the blot that corresponds to the sequence of interest.

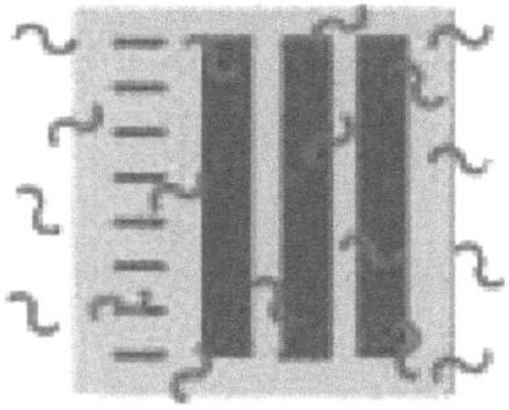

5. The membrane is washed to remove unbound probe. Access probe is removed
6. The labeled probe is detected via autoradiography (if a radioactive probe is used) or via a chemiluminescence's reaction (if a chemically labeled probe is used). In both cases this results

in the formation of a dark band on an X-ray film. We can now compare expression patterns of the sequence of interest in the different samples.

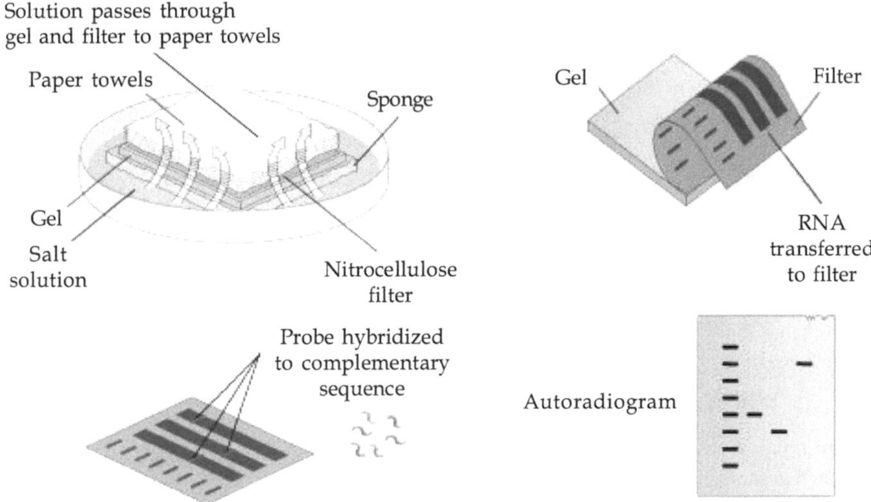

Procedure

- General blotting procedure starts with extraction of total RNA from a homogenized tissue sample.
- The mRNA can then be isolated through the use of oligo (dT) cellulose chromatography to maintain only those RNAs with a poly (A) tail.
- RNA samples are then separated by gel electrophoresis. Since the gels are fragile and the probes are unable to enter the matrix, the RNA samples, now separated by size, are transferred to a nylon membrane through a capillary or vacuum blotting system. Capillary blotting system setup for the transfer of RNA from an electrophoresis gel to a blotting membrane.
- A nylon membrane with a positive charge is the most effective for use in northern blotting since the negatively charged nucleic acids have a high affinity for them.
- The transfer buffer used for the blotting usually contains formamide because it lowers the annealing temperature of the probe-RNA interaction preventing RNA degradation by high temperatures.

- Once the RNA has been transferred to the membrane it is immobilized through covalent linkage to the membrane by UV light or heat.
- After a probe has been labeled, it is hybridized to the RNA on the membrane.
- Experimental conditions that can affect the efficiency and specificity of hybridization include ionic strength, viscosity, duplex length, mismatched base pairs, and base composition. The membrane is washed to ensure that the probe has bound specifically and to avoid background signals from arising. The hybrid signals are then detected by X-ray film and can be quantified by densitometry. To create controls for comparison in a northern blot, samples not displaying the gene product of interest can be used after determination by micro arrays or RT-PCR.

Preparation of Equipment and Reagents

Immerse the gel tanks, combs etc. in 0.2 M NaOH for 15 minutes to destroy any contaminating RNases before rinsing in distilled water. It is not necessary to use DEPC-treated water.

Make up 10 X Northern Running Buffer (containing MOPS)

Final concentration

0.2M MOPS

10 mM EDTA

50 mM Na Acetate

For 500 ml

20.93 g

10 ml of 0.5 M

8.33 ml of 3M

The buffer brought to a pH of 7.0. For 500 mL of 10 X MOPS, this means add approximately 2 g of NaOH. Make up to 500 mL in DEPC-treated water to DEPC-treat solutions; add 0.1% DEPC to the solution in a bottle which can be autoclaved. Mix the solution well and allow it to stand, with the cap tightly closed, for at least 30 minutes. Then loosen the cap and autoclave. This must be done in a fume hood, as DEPC is very toxic. Allow the DEPC stock solution bottle to warm to RT before opening to avoid condensation. Autoclaving of the MOPS buffer is not necessary and turns the solution yellow. Store at 4°C

Cast the Gel

Cast a 12-cm, 0.7–1.0% agarose gel, which requires at least 100 mL of agarose solution. Ethidium bromide is not added to the gel, but rather to the sample buffer. A thin, low-percentage agarose (0.6%–0.7%) gel is critical for good transfer of large MW RNAs (>4–5kb).

For 100 ml

1.00 g agarose (1% gel)

10 ml 10 X running buffer

100 ml DDW

6 mL 35% formaldehyde (check that it is a fresh batch that doesn't contain precipitates)

Dissolve the agarose in the microwave, let the solution cool to less than 60.25°C, then add the formaldehyde and cast the gel Prepare RNA Samples for Loading RNA is stored at -80°C as an ethanol precipitate. Determine the amount of RNA to be loaded in each well (e.g., 2 mg of poly A(+) RNA). Based on your RNA yield estimations, precipitate the appropriate amount of RNA ethanol solution for at least 15 minutes at 13000 g at 4°C. Remove all supernatant and resuspend each sample in 12 mL of sample buffer. RNA sample buffer.

For 500 ml

50 ml 10 X running buffer

250 ml deionized formamide

90 ml formaldehyde

108 ml water (DEPC-treated)

2 ml ethidium bromide (stock concentration, 10 mg/ml)

Heat the samples at 65°C for 5 minutes. Then, to each, add 3 ml of RNA loading buffer. This consists of 0.25% bromophenol blue, 0.25% xylene cyanol in 20% Ficoll in DEPC-treated water.

Running the Gel

Fill the tank with 1X MOPS (prepare 1X with DDW). above protocols added formaldehyde to the running buffer, but this is unnecessary. Run the gel at between 100 to 200 V for several hours, until the xylene cyanol dye front has migrated 2 to 4 cm into the gel and the bromophenol blue is about two-thirds down the gel. Circulate the buffer from end to end every half an hour, especially if running the gel at 200 V. When the RNA has run an appropriate distance, photograph the gel, including a ruler aligned with the wells. Presoak the gel in 20 X SSC while setting up the transfer (about 15 minutes).

Northern Transfer

Setting up Transfer

The physical set-up for Northern transfer is like to that for Southern transfer, except that 20X SSC is used as the transfer buffer,

Note *RNA is hydrolyzed in strongly alkaline solutions within seconds) and the membrane we use is Gene Screen Plus.*

Post-transfer Handling of the Membrane

After overnight transfer, the position and orientation of the wells is marked on the membrane and it is rinsed gently in 2X SSC for 5 minutes before being air dried in the fume hood. The membrane is arid at 80°C for 2 hours and is then ready for pre hybridization and hybridization.

1. Run Northern in the usual way, transfers to Gene screen plus in 20 X SSC, and bake in the oven at 80°C for 2 hours.
2. Hybridization buffer is as follows: 50% Formamide, 3 X SSC, 10X Denhardt's, 10 mM phosphate buffer, pH 8.0, 2 mM EDTA, 0.1% SDS, 200mg/ml herring sperm DNA. 800 U/ml preservative-free sodium heparin. Prehybridize membrane for 4-6 hours at 60°C. Hybridize in fresh buffer for 18-24 hours at 65°C-20 ng of riboprobe to the bag in a small volume of buffer (e.g., 3-5 ml). This can include hydrolysis, but this may or may not make a difference to the degree of background. Wash at high stringency, i.e., 0.1 X SSC, 0.1% SDS at 65°C. Be careful to wash all formamide-containing hybridization buffer off at low stringency first, i.e., use a 2 X SSC wash initially, and then increase stringency progressively. A good signal can be obtained with little background using this washing regimen, but is lost when washed at 75°C. Expose against x-ray film using an intensifying screen at -70°C for 24 hours or longer. A good signal can be obtained after 36 hours, equal to that obtained with Northern using cDNA probes after 96 hours of exposure.

Applications

- Northern blotting allows one to observe a particular gene's expression pattern between tissues, organs, developmental stages, environmental stress levels, pathogen infection, and over the course of treatment.

- The technique has been used to show overexpression of oncogenes and downregulation of tumor-suppressor genes in cancerous cells when compared to 'normal' tissue, as well as the gene expression in the rejection of transplanted organs. If an upregulated gene is observed by an abundance of mRNA on the northern blot the sample can then be sequenced to determine if the gene is known to researchers or if it is a novel finding.
- The expression patterns obtained under given conditions can provide insight into the function of that gene. Since the RNA is first separated by size, if only one probe type is used variance in the level of each band on the membrane can provide insight into the size of the product, suggesting alternative splice products of the same gene or repetitive sequence motifs.
- The variance in size of a gene product can also indicate deletions or errors in transcript processing, by altering the probe target used along the known sequence it is possible to determine which region of the RNA is missing.
- BlotBase is an online database publishing northern blots. BlotBase has over 700 published northern blots of human and mouse samples, in over 650 genes across more than 25 different tissue types. Northern blots can be searched by a blot ID, paper reference, gene identifier, or by tissue. The results of a search provide the blot ID, species, tissue, gene, expression level, blot image (if available), and links to the publication that the work originated from. This new database provides sharing of information between members of the science community that was not previously seen in northern blotting as it was in sequence analysis, genome determination, protein structure, etc.

Solutions

10 x MOPS Buffer

0.2 M MOPS (Na salt)

0.05 M Na Acetate

0.01 M Na_2EDTA

adjust to pH 7.0

store in a dark bottle or foil wrapped bottle

Sample Buffer Stock Solution: 5μl 10mg/ml EtBr
250μl Formamide
50μl **10 x MOPS Buffer**
80μl Formaldehyde
RNA Dye Solution
40% sucrose
0.025% Bromophenol blue (6mg/25ml)

Prehybe Solution
75 ml **PB buffer**
0.3 ml 0.5 M EDTA
52.5 ml 20% SDS
1.5g BSA (good quality BSA)
1.5 ml 10mg/ml denatured salmon sperm DNA
20.7 ml H_2O

Hybe Solution
Prehybe solution + Denatured probe

PB Buffer
67g Na_2HPO_4-$7H_2O$
~2 ml 85% H_3PO_4
pH 7.0
H_2O to 500 ml, autoclave

Wash Buffer 1 (used at 65°C)
40mM **PB Buffer**
1% SDS
1mM EDTA
0.5% BSA (fraction V, junk grade)
Wash Buffer 2 (used at 65 °C)
Wash Buffer 1 except no BSA

CHAPTER **34**

PCR-Sequencing

Sanger Method for DNA Sequencing

One necessity first know a little about the structure of DNA:

1. A segment of DNA, which is ordinarily double stranded, has a specific orientation, as it has a 5' (read as "5 prime") and a 3' ("3 prime") end. This can be simply thought of as a front and tail end to the DNA segment.
2. When DNA is synthesized in the lab, the two strands are separated and new bases are added to the 3' end-thus DNA is assembled from the 5' to 3' end.
3. DNA cannot be synthesized from scratch. A short piece of DNA, called a primer, is required for the reaction to begin.
4. Primers are designed such that they are able to bind to the target DNA, the binding of which is the initiator for DNA synthesis.

DNA sequencing is accomplished by the Fredrick Sanger method for which he won his second Nobel Prize in 1980. become a powerful technique in molecular biology, allowing analysis of genes at the nucleotide level. For this reason, this tool has been applied to many areas of research. For example, the polymerase chain reaction (PCR), a method which rapidly produces numerous copies of a desired piece of DNA, requires first knowing the flanking sequences of this piece. Another important use of DNA sequencing is identifying restriction sites in plasmids. Knowing these restriction sites is useful in cloning a foreign gene into the plasmid. Before the advent of DNA sequencing, molecular biologists had to sequence proteins directly; now amino acid sequences can be determined more easily by sequencing a piece of cDNA and finding an open reading frame. In eukaryotic gene expression, sequencing has allowed researchers to identify conserved sequence motifs and determine their importance in the promoter region. Furthermore, a molecular biologist can utilize sequencing to identify the site of a point mutation. These are only a few examples illustrating the way in which DNA sequencing has revolutionized molecular biology.

Dideoxynucleotide sequencing represents only one method of sequencing DNA. It is commonly called Sanger sequencing since Sanger devised the method. This technique utilizes 2',3'-dideoxynucleotide triphospates (ddNTPs), molecules that differ from deoxynucleotides by the having a hydrogen atom attached to the 3' carbon rather than an OH group. (Fig. 1). These molecules terminate DNA chain elongation because they cannot form a phosphodiester bond with the next deoxynucleotide.

In order to perform the sequencing, one must first convert double stranded DNA into single stranded DNA. This can be done by denaturing the double stranded DNA with NaOH. A Sanger reaction consists of the following: a strand to be sequenced (one of the single strands which was denatured using NaOH), DNA primers (short pieces of DNA that are both complementary to the strand which is to be sequenced and radioactively labelled at the 5' end), a mixture of a particular ddNTP (such as ddATP) with its normal dNTP (dATP in this case), and the other three dNTPs (dCTP, dGTP, and dTTP). The concentration of ddATP should be 1% of the concentration of dATP. The logic behind this ratio is that after DNA polymerase is added, the polymerization will take place and will terminate whenever a ddATP is incorporated into the growing strand. If the ddATP is only 1% of the total concentration of dATP, a whole series of labeled strands will result (Fig. 1). Note that the lengths of these strands are dependent on the location of the base relative to the 5' end.

This reaction is performed four times using a different ddNTP for each reaction. When these reactions are completed, a polyacrylamide gel electrophoresis (PAGE) is performed. One reaction is loaded into one lane for a total of four lanes. The gel is transferred to a nitrocellulose filter and autoradiography is performed so that only the bands with the radioactive label on the 5' end will appear. In PAGE, the shortest fragments will migrate the farthest. Therefore, the bottom-most band indicates that its particular dideoxynucleotide was added first to the labeled primer. In Fig. 2, for example, the band that migrated the farthest was in the ddATP reaction mixture. Therefore, ddATP must have been added first to the primer, and its complementary base, thymine, must have been the base present on the 3' end of the sequenced strand. One can continue reading in this fashion. Note in Fig. 2 that if one reads the bases from the bottom up, one is reading the 5' to 3' sequence of the strand complementary to the sequenced strand. The sequenced strand can be read 5' to 3' by reading top to bottom the bases complementary to the those on the gel.

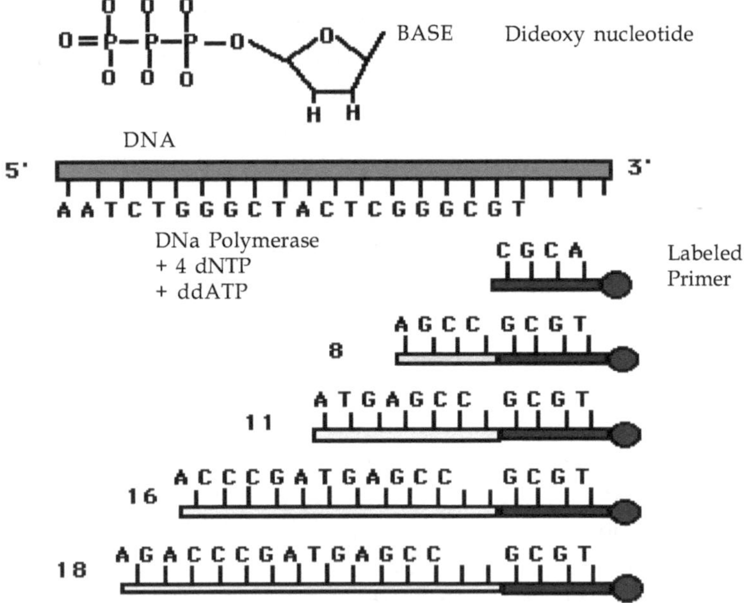

Fig.1: This figure shows the structure of a dideoxynucleotide (notice the H atom attached to the 3' carbon). Also depicted in this figure are the ingredients for a Sanger reaction. Notice the different lengths of labeled strands produced in this reaction.

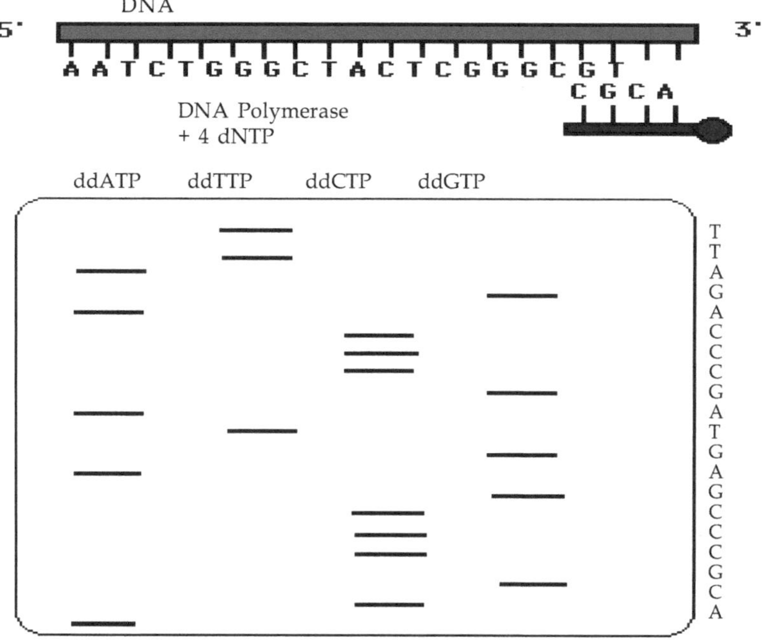

Fig.2: This figure is a representation of an acrylamide sequencing gel. Notice that the sequence of the strand of DNA complementary to the sequenced strand is 5' to 3' ACGCCCGAGTAGCCCAGATT while the sequence of the sequenced strand, 5' to 3', is AATCTGGGCT-ACTCGGGCGT.

The Sequencing Reaction: How the Nucleotide Composition of DNA is Determined

This method essentially involves amplifying a single stranded piece of DNA many a times. Normally, when DNA is amplified, new deoxy-nucleotides (dNTPs) are added as the strand of DNA grows. The Sanger method employs special bases called dideoxy-nucleotides (ddNTPs). These are similar to dNTPs, except for two important differences: they have fluorescent tags attached to them (a different tag for each of the 4 ddNTPs) and are missing a crucial atom that prevents new bases from being added to a DNA strand after a ddNTPs has been added. Thus, once a ddNTP is inserted into a growing DNA strand, synthesis of that strand is stopped. After many repeated cycles of amplification this will result in all the possible lengths of DNA being represented and every piece of synthesized DNA containing a fluorescent label at its terminus.

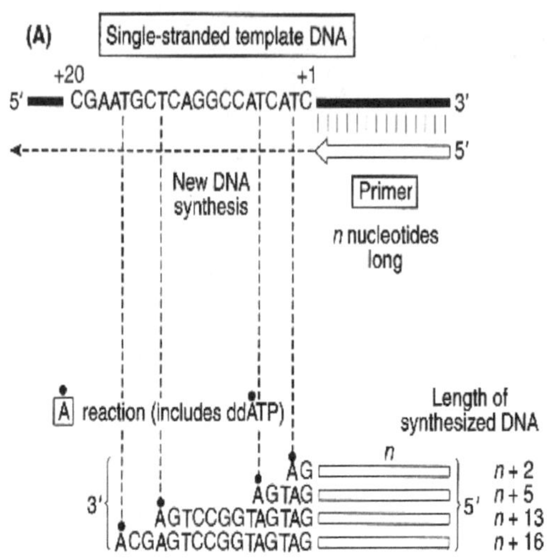

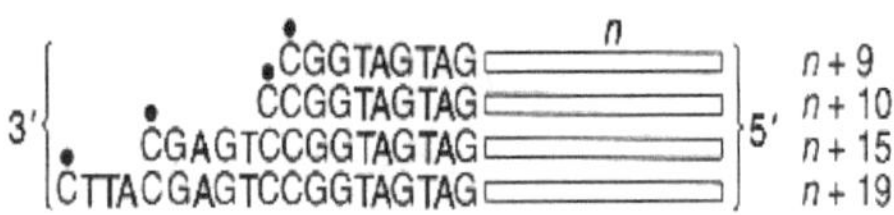

C reaction (includes ddCTP)

G reaction (includes ddGTP)

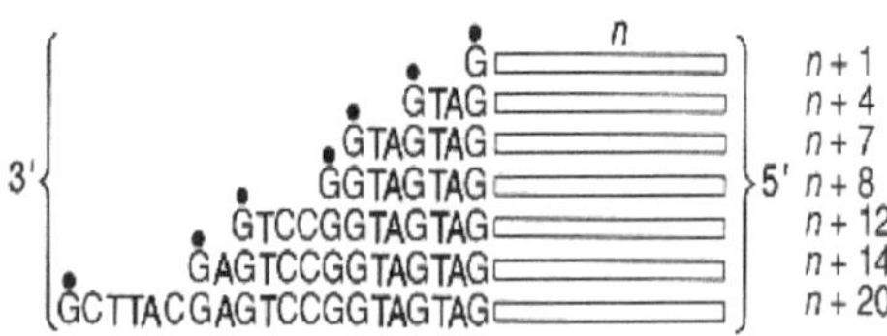

T reaction (includes ddTTP)

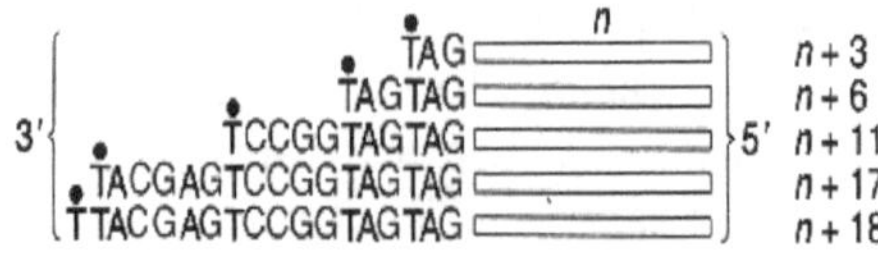

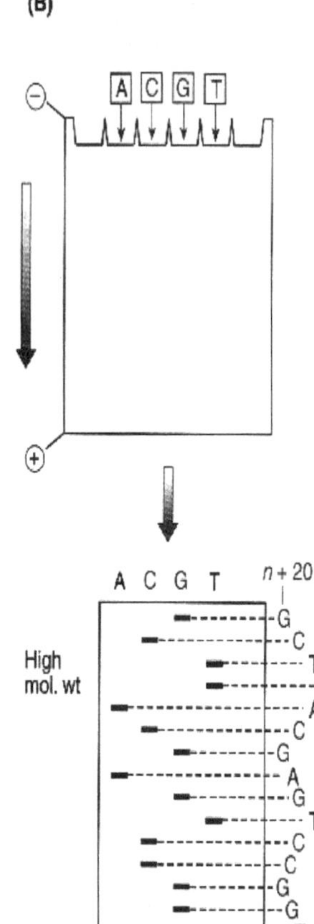

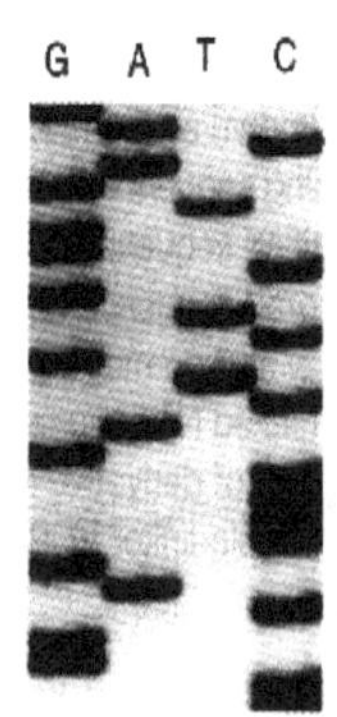

Amplified DNA can then be separated according to size via gel electrophoresis. As the fluorescent DNA reaches the bottom of the gel (now separated from smallest to largest), a laser can pick up the fluorescence of each piece of DNA. The trick to the Sanger method lies in the fact that each ddNTP emits a different fluorescent signal, so that the presence of a ddNTP at the terminus can be recorded on a computer.

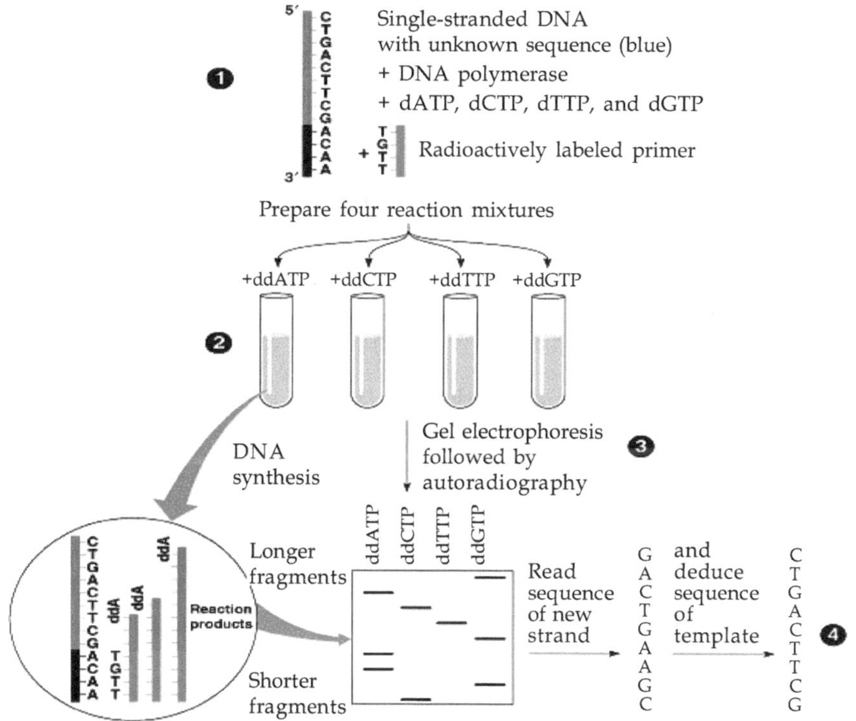

The reaction is set up so that a fluorescent ddNTP is present at every position in the DNA strand (i.e. every possible size of DNA strand is present) so that every nucleotide in the strand can be determined. A computer program can then compile the data into a coloured graph showing the determined sequence.

PCR-sequencing involves determination of the nucleotide sequence within a DNA fragment amplified by the Polymerase Chain Reaction (PCR), using primers (15-35 bp) specific for a particular genomic site. The method most commonly used to determine nucleotide sequences is based on the termination of *in vitro* DNA replication. The procedure starts with annealing a primer to the amplified DNA fragment, followed by dividing the mixture into four subsamples. Subsequently, DNA is replicated *in vitro* by adding the four deoxynucleotides constituting the DNA (dA, dC, dG and dT), a single dideoxynucleotide (ddA, ddC, ddG or ddT) and the enzyme DNA polymerase to each reaction. Sequence extension occurs as long as deoxynucleotides are incorporated in the newly synthesized DNA strand. However, when a dideoxynucleotide is incorporated, DNA replication is terminated. Because each reaction contains many DNA molecules and incorporation of dideoxynucleotides may occur at different positions in the strand, each of the four subsamples contains fragments of varying length terminated at a particular base. Finally, the size fragments in each of the four sub samples are separated by gel-electrophoresis. Because all possible sequence differences within the amplified fragment can be resolved between individuals, PCR-sequencing provides the ultimate measurement of genetic variation. Universal primer pairs to target specific sequences in a wide range of species are currently available for the chloroplast, mitochondrial and ribosomal genome. Because sequencing is costly and time-consuming, most studies have focused on only one or a few loci. Due to this restricted genome coverage and the fact that different genes may evolve at different rates, the extent to which the gene diversity estimated reflects overall genetic diversity is questionable. Because in general insufficient nucleotide variation is detected below the species level, PCR-sequencing is particular useful to address questions at higher taxonomic levels, e.g. phylogeny reconstruction.

Analytical Procedures

- Extraction of DNA
- Amplification of DNA fragments by PCR
- Selective termination of DNA replication in four separate reactions
- Separation of the fragments by polyacrylamide gel-electrophoresis
- Visualization of the fragments by autoradiography or fluorescence.

Main Requirements

- Thermocycler
  Laboratory setup for gel-electrophoresis
- (Laboratory facilities for radioactive labeling)

Advantages

- Low quantities of template DNA required (10-100 ng per reaction)
- High reproducibility
- Fragments studied are of known identity
- Amenable to automation

Disadvantages

- Laborious and technically demanding
- High costs involved
- Technical problems of contamination by foreign DNA when conserved primers are used
- High development costs in case primers for a specific region of interest are unavailable
- Low genome coverage
- Low levels of variation below the species level

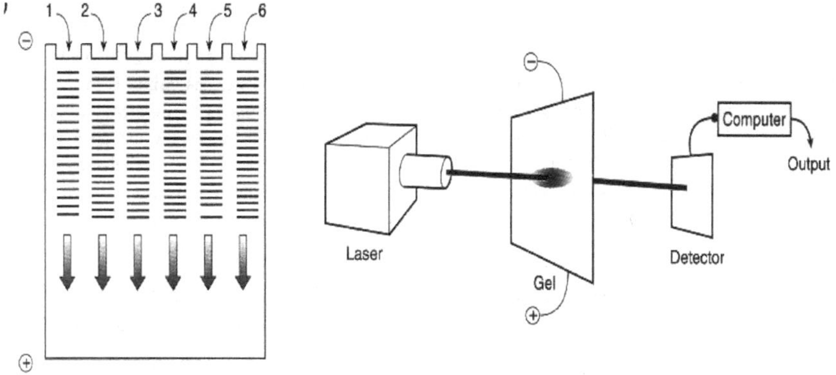

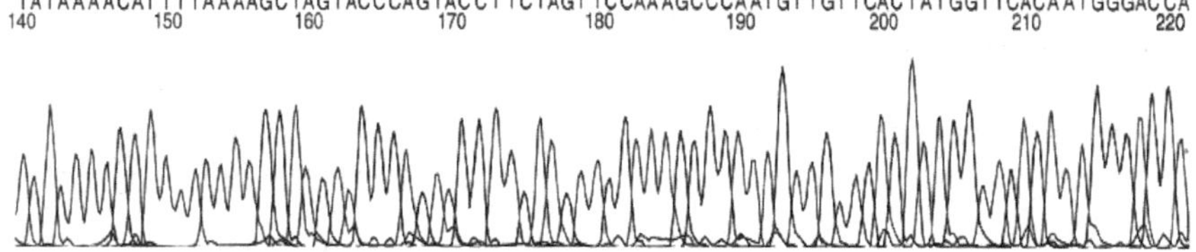

CHAPTER 35

Overview of Metagenomics

Metagenomics is a relatively new field that combines molecular biology and genetics in an attempt to identify, and characterize the genetic material from environmental samples and apply that knowledge. The genetic diversity is assessed by isolation of DNA followed by direct cloning of functional genes from the environmental sample. But more importantly, it is extremely beneficial in the study of synthetically made genomes. Metagenomics is described as "the comprehensive study of nucleotide sequence, structure, regulation, and function". Scientists can study the smallest component of an environmental system by extracting DNA from organisms in the system and inserting it into a synthetic model organism. The model organism then expresses this DNA where it can be studied using standard laboratory techniques. The development of metagenomics stemmed from the ineluctable evidence that as-yet-uncultured microorganisms represent the vast majority of organisms in most environments on earth. This evidence was derived from analyses of 16S rRNA gene sequences amplified directly from the environment, an approach that avoided the bias imposed by culturing and led to the discovery of vast new lineages of microbial life.

Metagenomics is employed as a means of systematically investigating, classifying, and manipulating the entire genetic material isolated from environmental samples. This is a multi-step process that relies on the efficiency of four general steps as shown in the diagram. The procedure consists of: the isolation of genetic material, the manipulation of the genetic material, library construction, and the analysis of genetic material in the metagenomic library. Metagenomics provides a second tier of technical innovation that facilitates study of the physiology and ecology of environmental microorganisms. Novel genes and gene products discovered through metagenomics include the first bacteriorhodopsin of bacterial origin; novel small molecules with antimicrobial activity; and new members of families of known proteins, such as an $Na^{+}(Li^{+})/H^{+}$ ant porter, RecA, DNA polymerase, and antibiotic resistance determinants. Reassembly of multiple genomes has provided insight into energy and nutrient cycling within the community, genome structure, gene function, population genetics and micro heterogeneity, and lateral gene transfer among members of an uncultured community. The application of metagenomic sequence information will facilitate the design of better culturing strategies to link genomic analysis with pure culture studies.

An Overview of the Process of Metagenomics

The first step of the procedure is the isolation of the DNA. Once the DNA from the cells is free, it must be separated from the rest of the sample. This is accomplished by taking advantage of the physical and chemical properties of DNA. Some methods of DNA isolation include density centrifugation, affinity binding, and solubility/precipitation. Once the DNA is collected, it is manipulated so that it can be used in the model organism.

Genomic DNA is relatively large so it is cut up into smaller fragments using enzymes called restriction endonucleases. This results in the smaller, linear fragments of DNA. The fragments are then combined with vectors. Vectors are small units of DNA that can be inserted into cells where they can replicate and produce the proteins encoded on the DNA using the machinery that the cells use to express normal genes. The vectors also contain a selectable marker. Selectable markers provide a growth advantage that the model organism would not normally have (such as resistance to a particular antibiotic) and are used to identify which organisms contain vectors and which ones do not.

The third step is to introduce the vectors with the metagenomic DNA fragments into the model organism. This allows the DNA from organisms that would not grow under laboratory conditions to be grown, expressed, and studied. The DNA inserted in the vector is transformed into cells of a model organism, typically *Escherichia coli*. It can be done by chemical, electrical, or biotechnological methods. The method of transformation is determined based on the type of sample used and the required efficiency of the reaction. The metagenomic DNA in the vectors are all in the same sample initially but the vectors are designed so that only one kind of DNA fragment from the sample will be maintained in each individual cell. The transformed cells are then grown on selective media so that only the cells carrying vectors will survive. Each group of cells that grows is called a colony. Each colony consists of many cloned cells that originated from one single cell. These samples of cells containing all of the metagenomic DNA samples on vectors are called metagenomic libraries. Each colony can be used to create a stock of cells for future study of a single fragment of the DNA from the environmental sample.

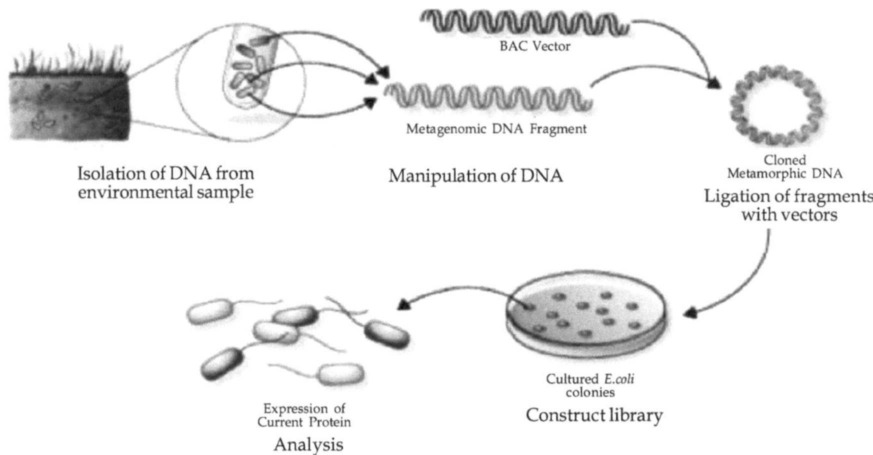

The fourth and final step in the procedure is the analysis of the DNA from the metagenomic libraries. The expression of DNA determines the physical and chemical properties of organisms so there are many potential methods of analysis. A phenotype is the physical attribute associated with expression of a gene. An example of metagenomic analysis would be to look for an unusual color or shape in the model organism. An aspect of the phenotype that is not readily observed is chemical reaction. The chemical properties of the expressed metagenomic DNA can be examined by performing chemical assay on products created by the model organism. This would investigate whether the model organism gained an enzymatic function that it was previously lacking such as use of an unusual nutrient source for growth under conditions that limit normal nutrient availability.

Metagenomic libraries are typically used to search for new forms of a known gene. First, the metagenomic DNA is inserted into a model organism that lacks a specific gene function. Restoration of a physical or chemical phenotype can then be used to detect genes of interest. A genotype is the specific sequence of the DNA and provides another means of analyzing the metagenomic DNA fragment. The sequence of the bases in the DNA can be compared to databases of known DNA to get information regarding the structure and organization of the metagenomic DNA. Comparisons of these sequences can provide insight into how the gene products (proteins) function.

Applications of Metagenomics

Many microorganisms have the ability to degrade waste products, make new drugs for medicine, make environmentally friendly plastics, or even make some of the ingredients of food we eat. By isolating the DNA from these organisms, it provides us with the opportunity to optimize these processes and adapt them for use in our society. As a result of ineffective standard laboratory culture techniques, the potential wealth of biological resources in nature (like microbes) is relatively untapped, unknown, and uncharacterized. Metagenomics represents a powerful tool to access the abounding biodiversity of native environmental samples. The valuable property of metagenomics is that it provides the capacity to effectively characterize the genetic diversity present in samples regardless of the availability of laboratory culturing techniques. Information from metagenomic libraries has the ability to enrich the knowledge and applications of many aspects of industry, therapeutics, and environmental sustainability. This information can then be applied to society in an effort to create a healthy human population that lives in balance with the environment. Metagenomics is a new and exciting field of molecular biology that is likely to grow into a standard technique for understanding biological diversity.

Metabolic Engineering

Metabolic engineering is a new frontier of applied biology. Understanding cellular function and manipulating cells/tissues to perform in a particular manner is the basis for many ventures in the biomedical, biotechnology and bioelectricity, including drug production from cell culture, the creation of artificial organs for replacement of diseased tissues, design of bioremediation processes for waste water clean-up, and creation of energy efficient micro organisms in an effort to develop clean energy. Many independent private companies in the energy sectors have taken the lead on metabolic engineering research, performing cutting-edge work in cell and drug delivery, metabolic engineering, and protein engineering. But today we will focus on the innovative effects of genetically modifying and constructing structures on the molecular level as it has opened a new door in renewable energy.

Metabolic engineering is a field that merges genetic engineering, physiology and molecular systems engineering with the rapidly advancing technologies in synthetic genomics. It aims to design and construct organisms for the production of a desired product, in this case, energy, by genetically engineering the organism's metabolic network and its regulatory vital life functions in a targeted fashion. The related term synthetic biology is also used to describe such activities, but the two are not the same, however easily confused. Generally speaking, metabolic engineering is centered on what is produced by the cell directly, while synthetic biology aims to study cell behavior. To understand metabolic engineering processes, it is important to know that each reaction in the cell is catalyzed by a unique enzyme. That enzyme's concentration can be increased or decreased by changing the number of copies of the gene that simultaneously synthesizes the enzyme or the rate at which the gene is converted into the enzyme. If a specific reaction is deemed counterproductive as par to the goal, the gene that encodes for its enzyme catalyst can be deleted or interrupted by a process known as gene knockout.

Although the metabolic networks within cells have been studied for many years, there is still much to be uncovered, but we can correctly assume that they are vital to the production of biofuels. Unlike pharmaceutical biofuel production, the success of biofuel production is largely determined by the yield of the fuel produced per unit and its rate of productivity. Undesirable byproduct formation, inefficient utilization of carbon reservoirs, and excess production of cell quantities all reduce the overall yield and productivity of a metabolic system.

The study of gene expression levels using high-density DNA has greatly contributed to metabolic engineering. This information provides context for interpreting genomic sequence data. A staggering amount of data now exists for hundreds of different microbial and non-microbial systems, providing cell data that records toxin submission, genetic mutations, and growth conditions. These has been extremely useful in combination with various mutation studies, and have been a crucial part of the identification of what is going on. High-density DNA array technology has also been used to enable scientists with the ability to quickly profile microbial diversity and gene functions from artificial and natural samples. Attempts to process and use the tremendous amount of data generated from transcriptional array studies have given rise to entirely new disciplines within the fields of bioinformatics and computational biology, and necessitated the development of specific protocols. Some scientists speculate that this very experimentation with the interaction between high density DNA and gene structure will unlock a much more efficient method to derive carbon-based energy. This is yet another example of the rapid expansion of microbial synthetic genomics today.

Microbe Cultivation

The past 25 years have yielded unprecedented advances in our appreciation for the breadth and depth of microbial diversity. These advances are mainly derived from the development of advanced nucleic acid-based technologies that have aided in quantifying the striking abundance and genetic diversity of microbes in all of lifeís recognized reaches. For anyone truly interested in microbial diversity and function, the field of molecular ecology can only be embraced with excitement and continued anticipation because of the newly unraveling technologies and discoveries being made that could possible change multiple aspects of our world. Certainly, many aspects of microbial diversity have now been revealed definitively, through gene inventories and related approaches. But despite the triumphs of molecular systematics and ecology, the suggestion that in vitro growth studies of organisms are passed or irrelevant to reaching an understanding of the true diversity and function of microbes in nature is unjustified. Almost all involved in attempting to cultivate new organisms recognize it is a daunting task; one that is constantly reinforced with each publication of a gene inventory. Reassuringly, however, the past 25 years of cultivation-based studies have led to as many significant advances in our understanding of microbes as has modern molecular ecology over the same extremely successful period of time.

In order to produce synthetic and organic microbes capable of being used for sources of energy, the correct atmosphere, or median, must be created. The medium may be purely of chemical makeup and contain organic materials, or it may consist of living organisms such as fertilized eggs or bacteria. Microorganisms growing in or on such a medium form a culture. Acting as a single body of cellular organisms or as individual cells, laboratory cultures are efficient and quick to form and study. When first used, the culture medium should be sterile, meaning that no form of life is present before inoculation with the microorganism.

There are various mediums for the cultivation of microbes, as listed below:

1. *General microbial media* : The most basic medium and process in which bacterium flourish, For the cultivation of bacteria, a combination medium is nutrient broth, a liquid containing proteins, salts, and growth enhancers, that will support many bacteria. To solidify the medium, Non-nutritional substances are added. Polysaccharides that adds no nutrients to a medium, but merely solidifies it, are ideal for this.

2. *Special microbial media* : Special mediums can be utilized for the cultivation of certain microorganisms are cultivated in selective media. These various media impede the augmentation of unwanted organisms while encouraging the growth of the desired cultures.

3. *Isolation methods* : To obtain separated colonies from a mixed culture, various isolation methods can be used. The most common is the streak plate method, in which a sample of mixed bacteria is ìstreakedî several times along the far edge of a Petri dish containing an active medium such as nutrient agar. A loop is flamed and then touched to the first area to retrieve a sample of bacteria from the culture. This sample is then streaked several times in the second area of the medium. The loop is then reflamed, touched to the second area, and streaked once again in the third

area. The process can be repeated in a fourth and fifth area if desired. During incubation, the bacteria will multiply rapidly and form colonies. This is the most effective method in observing and studying synthetic microbial behavior because of its unknown characteristics, and in singularity, scientists are able to directly pinpoint the effects of various DNA alterations.

Microbe Cultivation Methods

For anaerobic microorganisms to be genetically modified and studied, the atmosphere must be oxygen free. To evict surrounding oxygen, the culture media is usually placed within containers with an ample supply of carbon dioxide and hydrogen gas. Anaerobic chambers can also be used within closed compartments, which allows scientists the ability to manipulate and observe genetic makeup.

The illustration demonstrates two process of the isolation methods: the streak plate technique (a) and the pour plate technique (b) Microbial communities are characterized through the works of whole-genome and metagenome sequencing. But what is DNA sequencing? DNA sequencing is the technique of arranging the nucleotide bases of DNA, adenine, guanine, thymine, and cytosine. The discovery of DNA sequencing has dramatically advanced the quality and research involving biotechnology. The intense speed of sequencing acquired with modern DNA sequencing technology has been a significant mechanism in the sequencing of the human genome. The main purpose of this method is to identify and evaluate the features and functions of the approximate 20,000 to 25,000 genes in a human genome. Efficient DNA sequencing is a very intricate process that utilizes the methodical preparation and organization as well as extremely advanced appliances, ranging from various template robots to complex sequencing machines.

The company, Synthetic Genomics, supports DNA sequencing at the J. Craig Venter Institute's Joint Technology Center, one of the world's leading DNA sequencing organizations. The Joint Technology Center is equipped with approximately 100 of the latest and most advanced DNA sequencing appliances at its 60,000 square-foot facility. The Joint Technology Center currently produces 40 million sequence reads per year with the potential ability to its maximum capacity to more than 120 million lanes per year. The center has sequenced the genomes of more than 100 organisms since the first genome of a free-living organism was sequenced in 1995.

DNA Sequencing

Sequence Analysis

The analysis of the data retrieved from DNA sequencing from a range of highly technological appliances, is extremely significant in the world of the advancing principles of biotechnology.

Synthetic Genomics is utilizing a suite of novel and public domain computational mechanisms to aid with the determination and annotation of genes, accumulate sequences and closely compare related genomes. For instance, the Synthetic Genomics Company administers an "interdisciplinary team approach" assimilating the skills of mathematics and biology to obtain different aspects of the most challenging issues related to computational biology. This bioinformatics team collects, classifies, stores and evaluates all of the potential clean and renewable energy sources that can be produced through the analysis of genome sequencing.

Genotypic analysis is usually performed after phenotypic analysis. A typical metagenomic analysis involves several subsequent rounds of the procedure in order to definitively isolate target genes from environmental samples and to effectively characterize the information encoded by the DNA sequence. The information gained from the metagenomic procedure provides information regarding the structure, organization, evolution, and origin of the DNA and can be used in scientific applications for the benefit of society and the environment.

CHAPTER 36

Calculation of Similarity Index Values & Construction of Dendrogram Using NTSYSpc. 2.0

NTSYSpc 2.0 software is used for various types of analysis including calculation of similarity index values and construction of dendrogram. Construction of dendrogram using gel data in the form of bands. This software accepts zero-one data.

STEP 1:

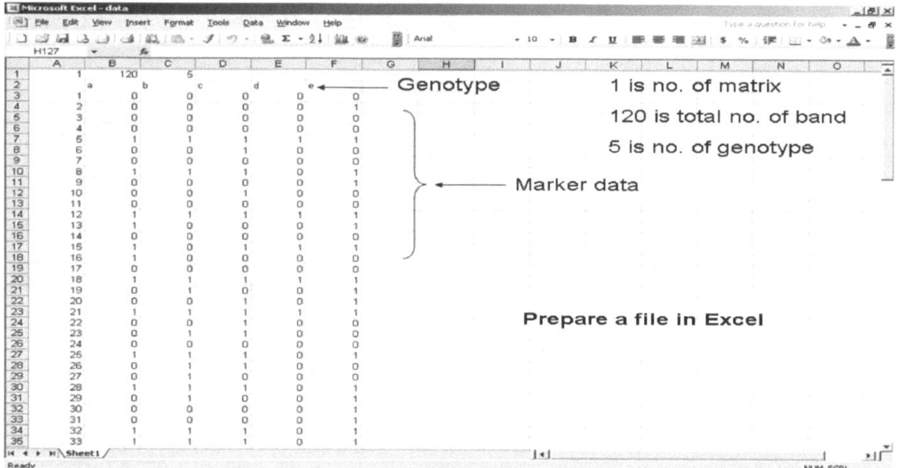

STEP 2:

Microsoft Excel - data
Open the programme NTedit
Programs
NTSYSpc2
NTedit
NTSYSpc 2.0
NTSYSpc help

STEP 3:

Microsoft Excel - data
NTedit 1.07c
Import Excel
using DDE ...
using OLE ...
Import the excel file using DDE

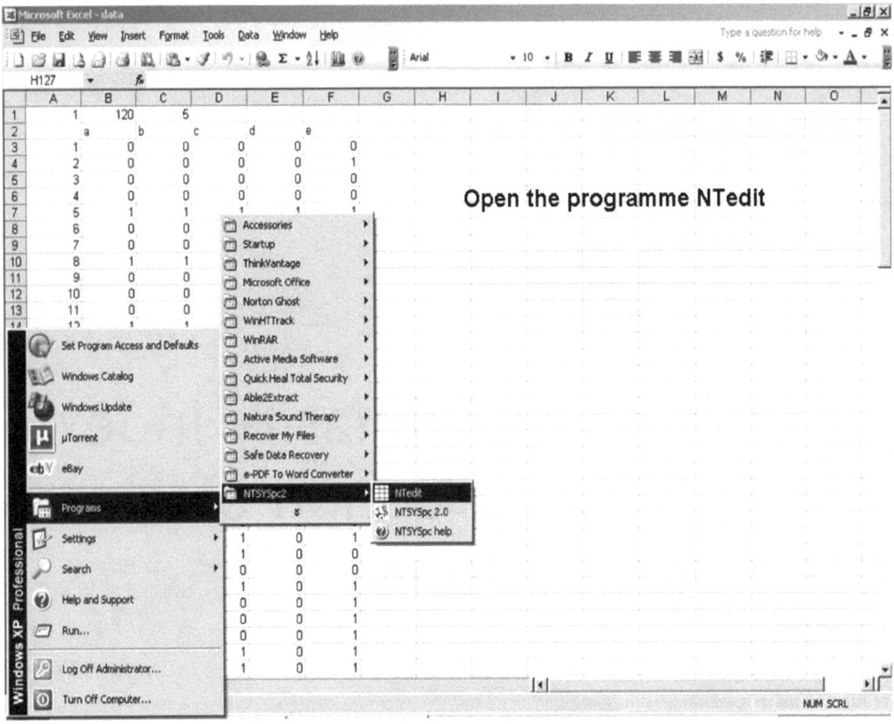

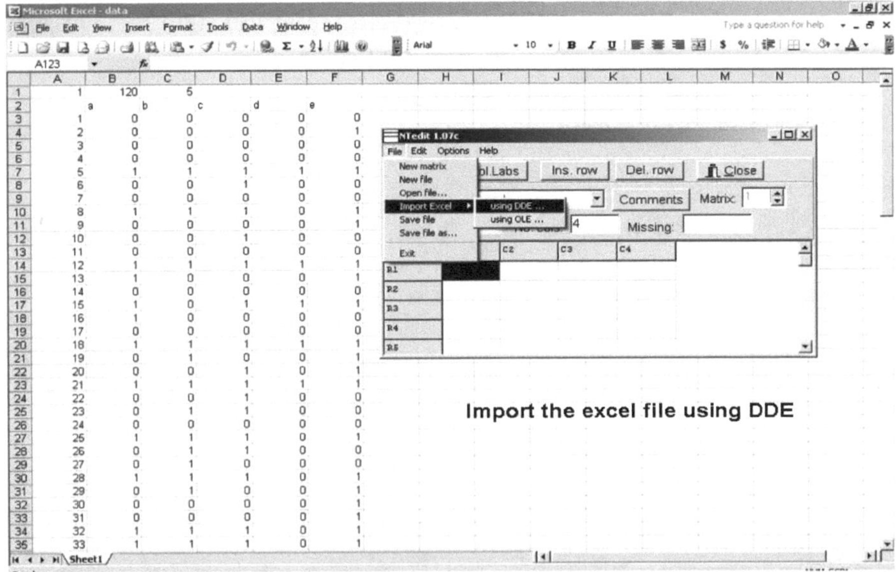

STEP 4:

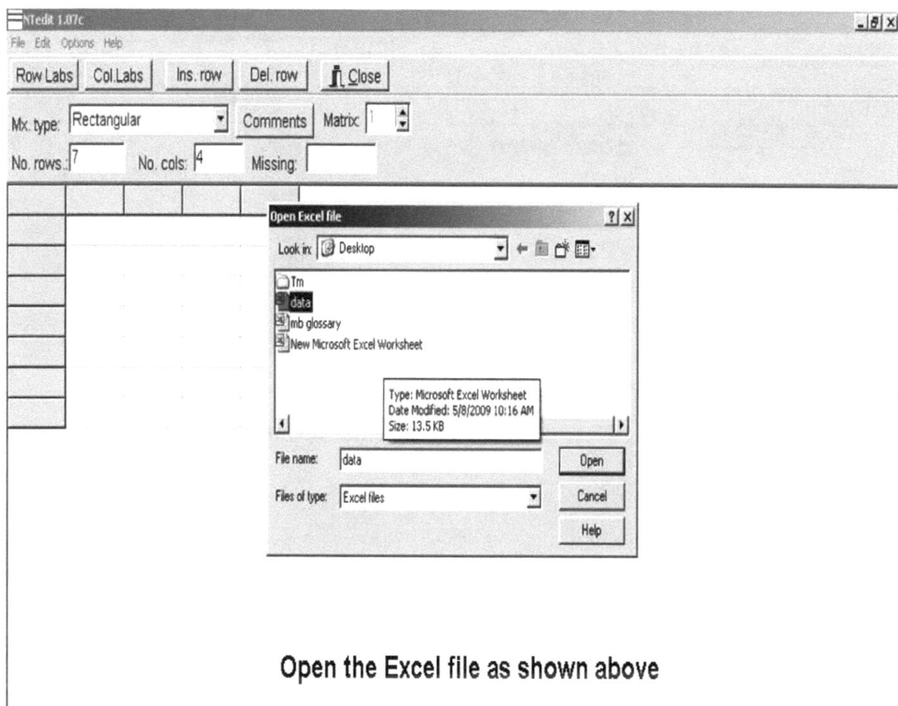

Open the Excel file as shown above

STEP 5:

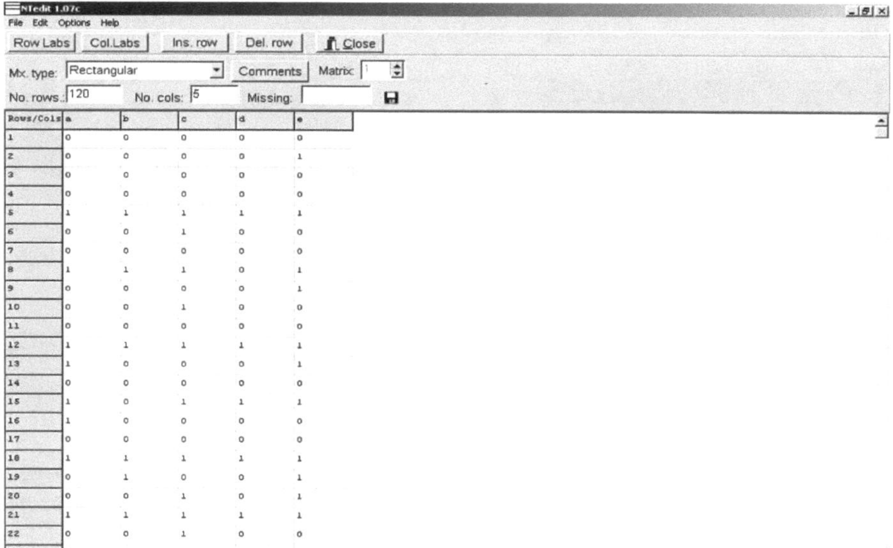

Rows/Cols	a	b	c	d	e
1	0	0	0	0	0
2	0	0	0	0	1
3	0	0	0	0	0
4	0	0	0	0	0
5	1	1	1	1	1
6	0	0	1	0	0
7	0	0	0	0	0
8	1	1	1	0	1
9	0	0	0	0	1
10	0	0	1	0	0
11	0	0	0	0	0
12	1	1	1	1	1
13	1	0	0	0	1
14	0	0	0	0	0
15	1	0	1	1	1
16	1	0	0	0	0
17	0	0	0	0	0
18	1	1	1	1	1
19	0	1	0	0	1
20	0	0	1	0	1
21	1	1	1	1	1
22	0	0	1	0	0

STEP 6:

NTedit 1.07c
Row Labs
Col.Labs
Ins. row
Del. row
Close
Mx. type: Rectangular
Comments
Matrix 1
No. rows 120
No. cols: 5
Missing:
Save data file as...
Save in: Desktop
My Documents
My Computer
My Network Places
avg update 6-05
Bis-PB_JFST-revised
DNA club
Nature
New Folder
New Folder
rec
RNA
File name: data.NTS
Save as type: NTSYS files
Save
Cancel
Help
Save the opened file as data.NTS

STEP 7:

Row Labs
Col.Labs
Ins. row
Del. row
Close
Mx. type: Rectangular
Comments
Matrix 1
No. rows 120
No. cols: 5
Missing:
Set Program Access and Defaults
Windows Catalog
Windows Update
µTorrent
eBay
Programs
Settings
Search
Help and Support
Run...
Log Off Administrator...
Turn Off Computer...
Windows XP Professional
Accessories
Startup
ThinkVantage
Microsoft Office
Norton Ghost
WinHTTrack
WinRAR
Active Media Software
Quick Heal Total Security
Able2Extract
Natura Sound Therapy
Recover My Files
Safe Data Recovery
e-PDF To Word Converter
NTSYSpc2
NTedit
NTSYSpc 2.0
Location: C:\Program Files\NTSYSpc
Open the programme NTSYSpc 2.0 for analysis of data

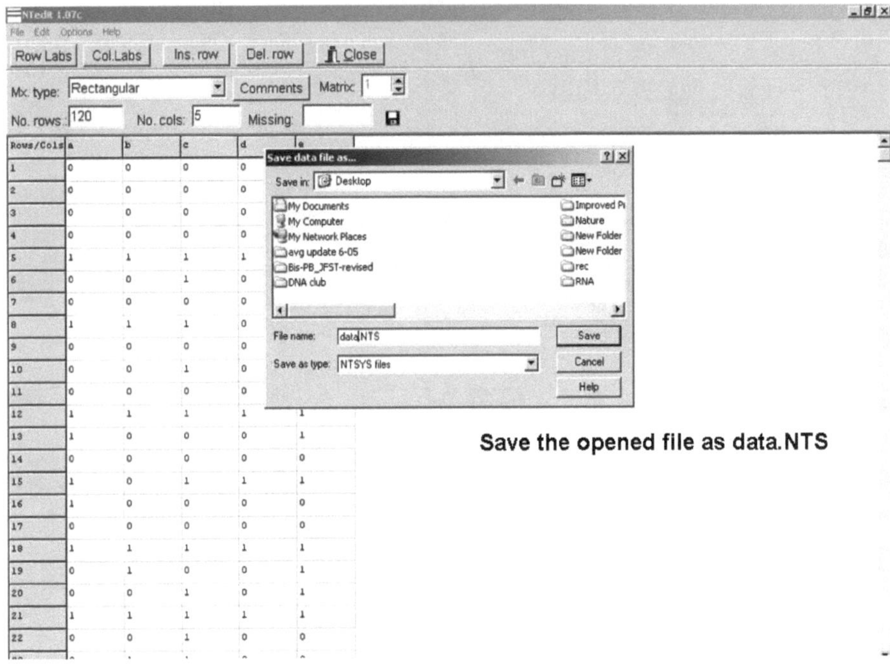

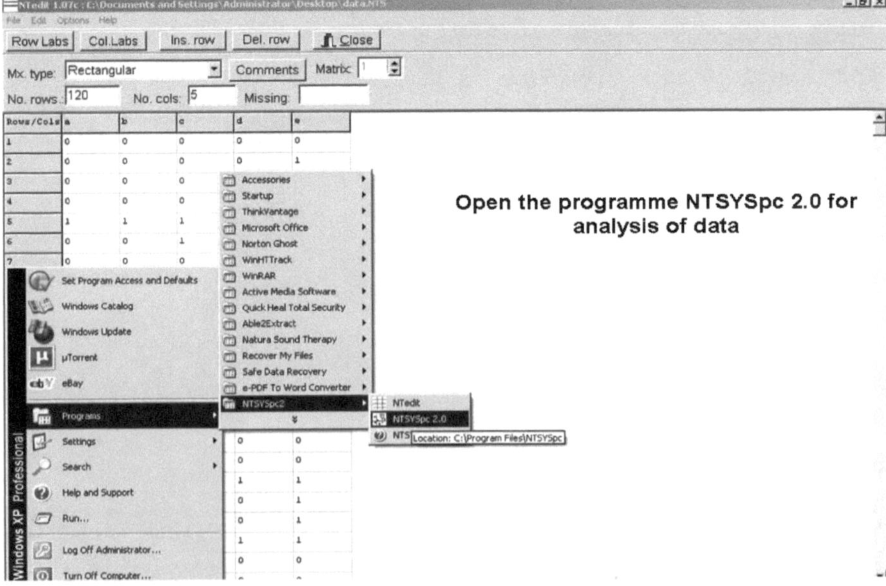

STEP 8:

NTSYSpc
File Options Help
Graphics
General
Similarity
Clustering
Ordination
CanCor/PLS
CVA
Projection
Common PCA
Eigen
SVD
Corresp. Anal.
MD Scale
Open similarity and go to SimQual

STEP 9:

NTSYSpc
File Options Help
Clustering
Ordination
Graphics
General
Similarity
PoolVCV
SimInt
SimGend
SimQual
SimQual
Parameters:
Arguments:
Input file
By rows?
Coefficient
SM
Output file
Positive code
1.00000
Compute
Close
Help
Name of input file
Select input file
Look in:
Desktop
Tm
data.NTS
File name:
data
Files of type:
NTS files
Open
Cancel
Help
Enter the input file, which was created earlier using NTedit

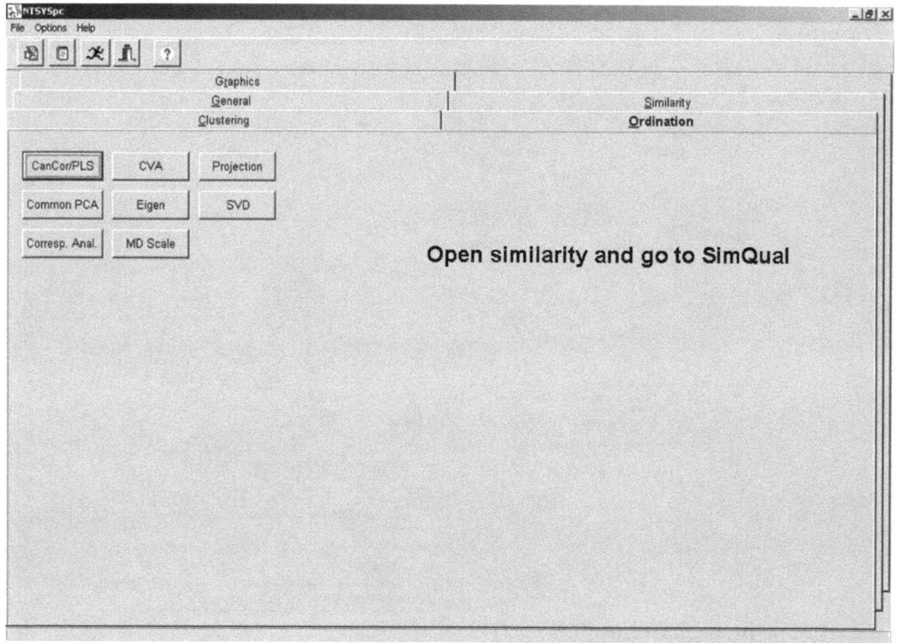

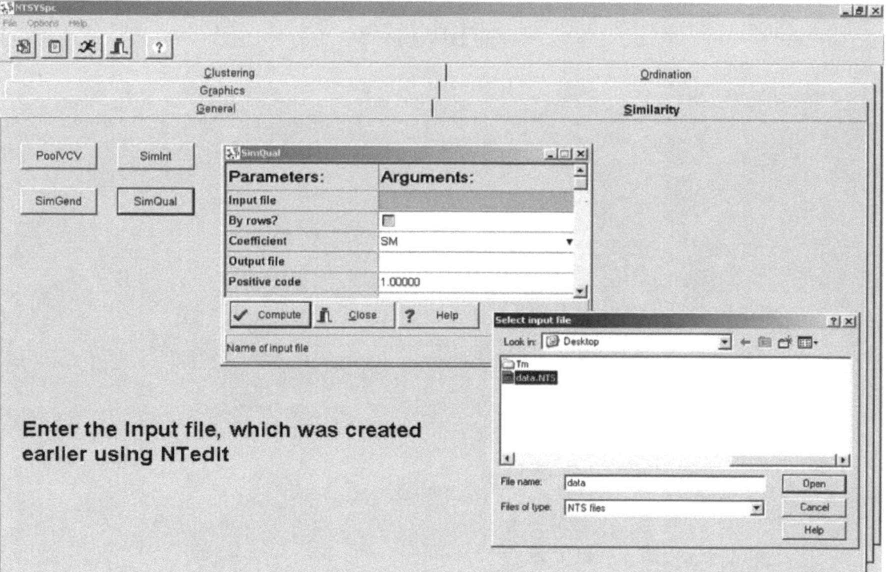

STEP 10:

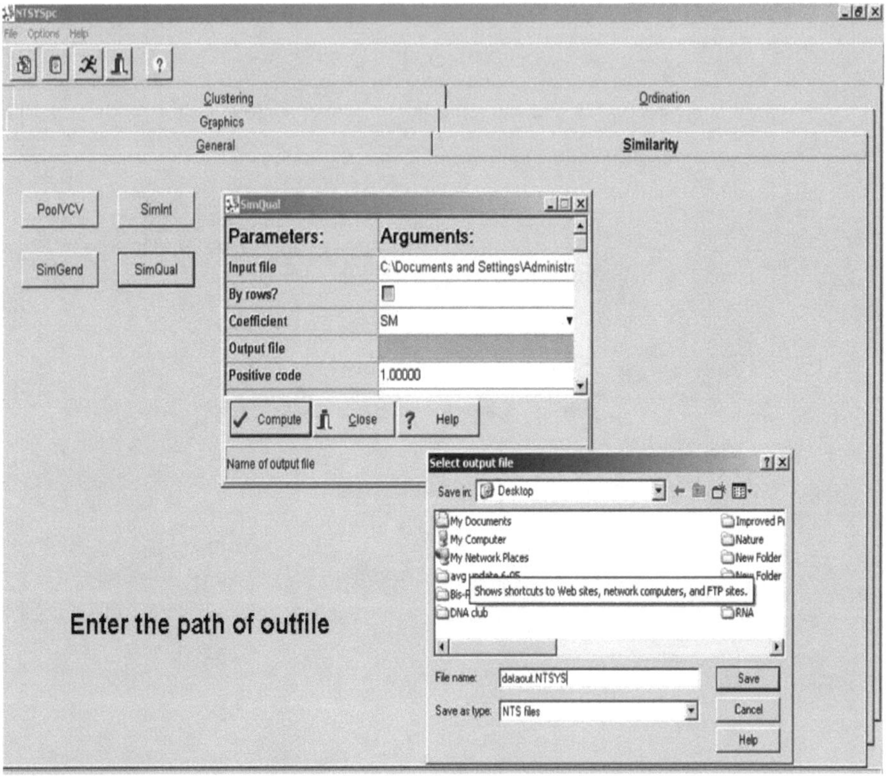

NTSYSpc
File Options Help
Clustering
Ordination
Graphics
General
Similarity
PoolVCV
SimInt
SimGend
SimQual
SimQual

Parameters:	Arguments:
Input file	C:\Documents and Settings\Administr
By rows?	
Coefficient	SM
Output file	
Positive code	1.00000

Compute
Close
Help
Name of output file
Select output file
Save in: Desktop
My Documents
My Computer
My Network Places
avg update 6-05
Bis-P
DNA club
Improved Pr
Nature
New Folder
New Folder
RNA
Shows shortcuts to Web sites, network computers, and FTP sites.
File name: dataout.NTSYS
Save as type: NTS files
Save
Cancel
Help

Enter the path of outfile

STEP 11:

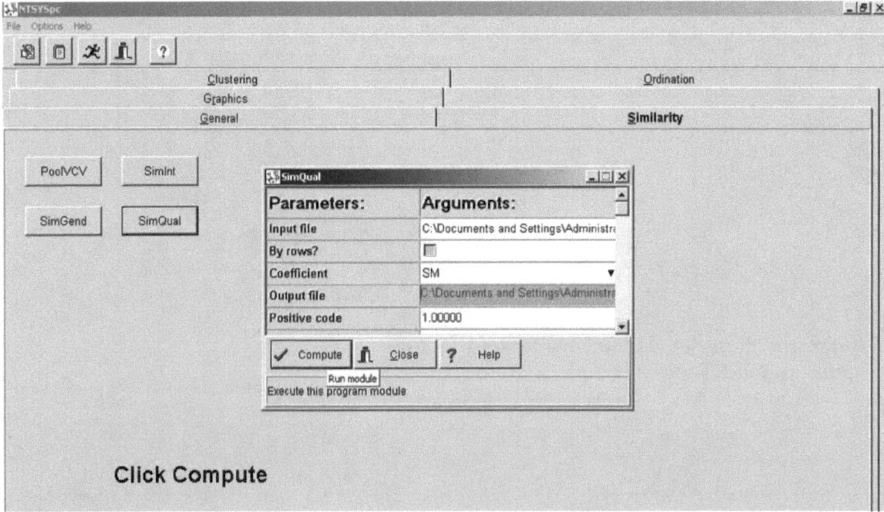

STEP 12:

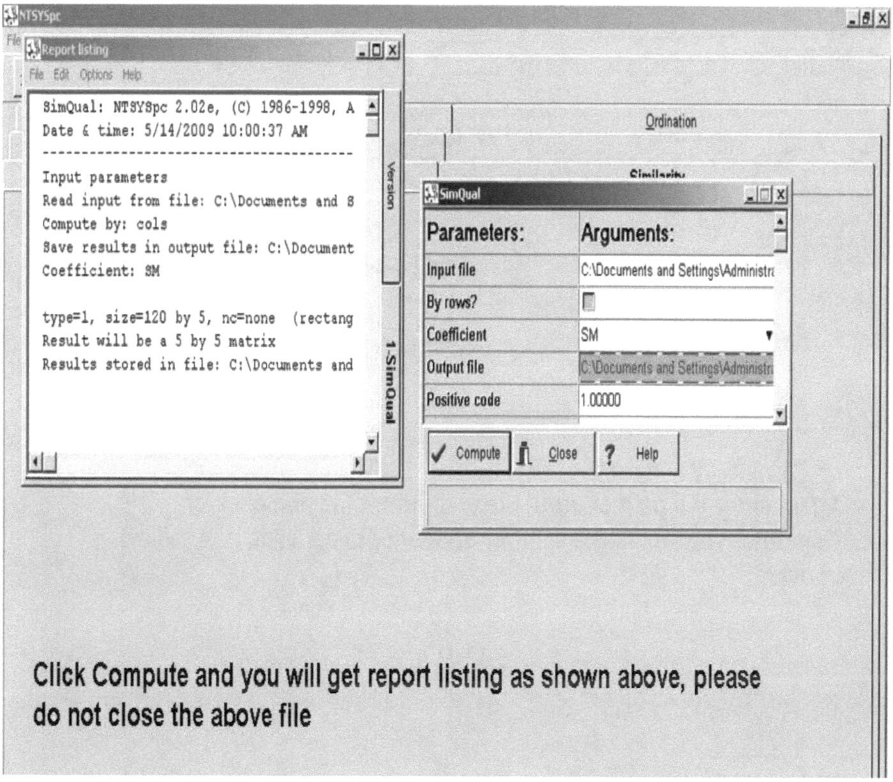

NTSYSpc

Report listing

File Edit Options Help

```
SimQual: NTSYSpc 2.02e, (C) 1986-1998, A
Date & time: 5/14/2009 10:00:37 AM
---------------------------------------
Input parameters
Read input from file: C:\Documents and S
Compute by: cols
Save results in output file: C:\Document
Coefficient: SM

type=1, size=120 by 5, nc=none  (rectang
Result will be a 5 by 5 matrix
Results stored in file: C:\Documents and
```

Version

1-SimQual

Ordination

Similarity

SimQual

Parameters:	Arguments:
Input file	C:\Documents and Settings\Administra
By rows?	
Coefficient	SM
Output file	C:\Documents and Settings\Administra
Positive code	1.00000

Compute Close Help

Click Compute and you will get report listing as shown above, please do not close the above file

STEP 13:

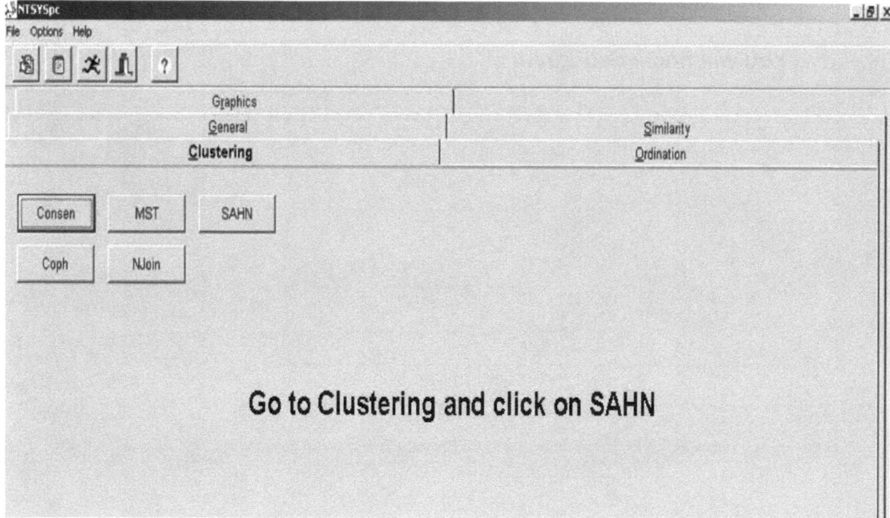

STEP 14:

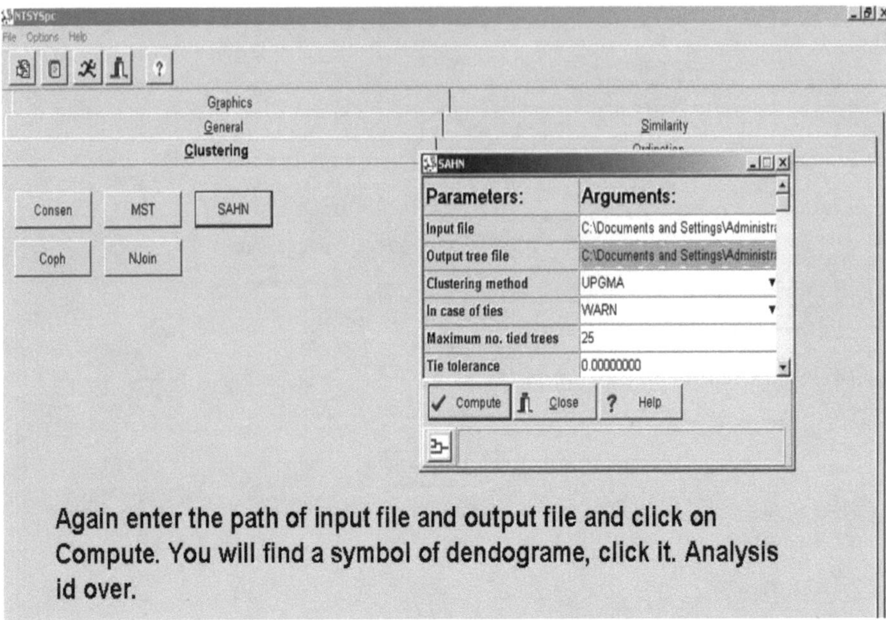

STEP 15:

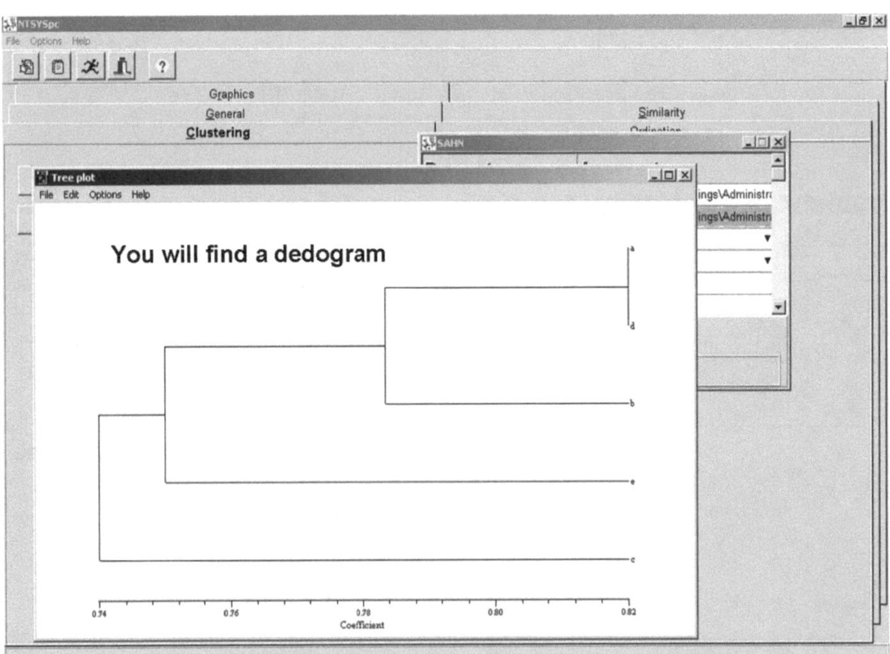

Glossary

Abiotic factors : Non living; moisture, soil, nutrients, fire, wind, temperature, climate

Absorption : Movement of ions and water into as organism as a result of metabolic processes, frequently against an electrochemical potential gradient (active) or as a result of diffusion along an activity gradient (passive).

Absorption field : A system of properly sized and constructed narrow trenches partially filled with a bed of washed gravel or crushed stone into which perforated or open joint pipe is placed. The discharge from the septic tank is distributed through these pipes into trenches and surrounding soil. While seepage pits normally require less land area to install, they should be used only where absorption fields are not suitable and wellwater supplies are not endangered.

Acetogenic bacterium : Prokaryotic organism that uses carbonate as a terminal electron acceptor and produces acetic acid as a waste product.

Acetyleneblock assay : Estimates denitrification by determining release of nitrous oxide (N_2O) from acetylenetreated soil.

Acid soil : Soil with a pH value < 6.6.

Acidophile : Organism that grows best under acid conditions (down to a pH of 1).

Actinomycete : Nontaxonomic term applied to a group of high G + C base composition, Grampositive bacteria that have a superficial resemblance to fungi. Includes many but not all organisms belonging to the order Actinomycetales.

Activated sludge : Sludge particles produced in raw or settled wastewater (primary effluent) by the growth of organisms (including zoogleal bacteria) in aeration tanks in the presence of dissolved oxygen. The term "activated" comes from the fact that the particles are teeming with fungi, bacteria, and protozoa. Activated sludge is different from primary sludge in that the sludge particles contain many living organisms which can feed on the incoming wastewater.

Activation energy : Amount of energy required to bring all molecules in one mole of a substance to their reactive state at a given temperature

Active site : Region of an enzyme where substrates bind.

Adaptive radiation : The evolution of new species or sub-species to fill unoccupied ecological niches.

Adenine : (A) the purine base found attached to the 1´ carbon of deoxyribose in DNA, where it is capable of forming 2 hydrogen bond interactions with the pyrimidine thymine on an opposing nucleic acid strand, and attached to the 1' carbon of ribose in RNA, where it is capable of forming 2 hydrogen bond interactions with the pyrimidine uracil on an opposing nucleic acid strand.

Adenosine triphosphate : A ribose molecule with the purine adenine attached to the 1´ carbon and a chain of three oxygenated phosphorus atoms attached to the 5' carbon. This is a high energy molecule that serves as a precursor for the assembly of RNA and provides energy for many other cellular reactions.

Aflatoxin : A poly ketide secondary fungal metabolite that can cause cancer.

Agar : Complex polysaccharide derived from certain marine algae that is a gelling agent for solid or semisolid microbiological media. Agar consists of about 70% agarose and 30% agaropectin. Agar can be melted at temperature above 100°C; gelling temperature is 40-50°C.

Agarose : A highly refined form of the polysaccharide agar, a polymer purified from seaweed, that can be mixed as a dry powder with a buffer solution, melted with heating, then poured into a mold and allowed to cool to form a solid gel matrix for the electrophoretic separation of macromolecules like DNA and RNA.

Agarose gel electrophoresis : A matrix composed of a highly purified form of agar that is used to separate larger DNA and RNA molecules ranging 20,000 nucleotides.

Agglutination reaction : The formation of an insoluble immune complex by the crosslinking of cells or particles.

Agrobacterium tumefaciens **:**The bacterial organism responsible for the formation of crown gall tumors on susceptible host plants. The gall formation is caused by the transfer of the T-DNA, a portion of the large Ti plasmid resident in the bacterium, to the host plant, where chromosomal integration and expression of T-DNA genes causes alterations in plant hormone levels.

Airborne transmission : The type of infectious organism transmission in which the pathogen is truly suspended in the air and travels over a meter or more from the source to the host.

Akinetes : Specialized, nonmotile, dormant, thickwalled resting cells formed by some cyanobacteria.

Alcoholic fermentation : A fermentation process that produces ethanol and CO2 from sugars.

Alga (plural, algae) : Phototrophic eukaryotic microorganism. Algae could be unicellular or multicellular. Bluegreen algae are not true algae; they belong to a group of bacteria called cyanobacteria.

Aliphatic : Organic compound in which the main carbon structure is a straight chain.

Alkaline conditions : High ph (typically >ph10). Ph is of particular concern during the purification, handling, and storage of nucleic acids. Alkaline conditions, are a ph of greater than 10, will destabilize the base pairing

Alkaline soil : Soil having a pH value >7.3.

Alkalophile : Organism that grows best under alkaline conditions (up to a pH of 10.5).

Alkene : Straight chain or branched organic structure that contains at least one double bond.

Alleles : Different copies of a gene sequence that can occur at the same chromosomal locus, or location, on different copies of the same chromosome. Different alleles are generally detected by the different phenotypic properties associated with allelic variants.

Allelic variations : Different nucleotide sequence versions of the same genetic locus.

Allochthonous flora : Organisms that are not indigenous to the soil but that enter soil by precipitation, diseased tissues, manure, and sewage. They may persist for some time but do not contribute in a significant way to ecologically significant transformations or interactions.

Allosteric site : Site on the enzyme other than the active site to which a nonsubstate compound binds. This may result in a conformational change at the active site so that the normal substrate cannot bind to it.

Allotype : Allelic variants of antigenic determinant(s) found on antibody chains of some, but not all, members of a species, which are inherited as simple Mendelian traits.

Alpha hemolysis : A greenish zone of partial clearing around a bacterial colony growing on blood agar.

Alpha subunit : The smallest of the four subunit types present in most bacterial RNA polymerases, generally present in two copies per RNA polymerase complex.

Alphaproteobacteria : One of the five subgroups of proteobacteria, each with distinctive 16S rRNA sequences. This group contains most of the oligotrophic proteobacteria; some have unusual metabolic modes such as methylotrophy, chemolithotrophy, and nitrogen fixing ability. Many have distinctive morphological feature

Alternative mRNA splicing : The inclusion or exclusion of different exons to form different mRNA transcripts.

Ames test : A test that uses a special Salmonella strain to test chemicals for mutagenicity and potential carcinogenicity.

Amino acid : Any of 20 basic building blocks of proteins– composed of a free amino (NH2) end, a free carboxyl (COOH) end, and a side group (R).

Amino acid activation : The initial stage of protein synthesis in which amino acids are attached to transfer RNA molecules.

Amino group : An NH_2 group attached to a carbon skeleton as in the amines and amino acids.

Aminoacyl or acceptor site (A site) : The site on the ribosome that contains an aminoacyltRNA at the beginning of the elongation cycle during protein synthesis; the growing peptide chain is transferred to the aminoacyltRNA and lengthens by an amino acid.

Aminoglycoside antibiotics : A group of antibiotics synthesized by Streptomyces and Micromonospora, which contain a cyclohexane ring and amino sugars; all aminoglycoside antibiotics bind to the small ribosomal subunit and inhibit protein synthesis.

Amphotericin B : An antibiotic from a strain of Streptomyces nodosus that is used to treat systemic fungal infections; it also is used topically to treat candidiasis.

Ampicillin (β-lactamase) : An antibiotic derived from penicillin that prevents bacterial growth by interfering with cell wall synthesis.

Amplification : DNA cloning methods were developed in part as a response to the need to overcome the low yield of a given DNA fragment from a large genome, amplifying the yield of a desired DNA fragment during the purification process. Chloramphenicol amplification is the process of increasing plasmid yield from a bacterial culture by the addition of the protein synthesis inhibitor chloramphenicol, which blocks bacterial chromosomal replication while allowing continued replication of plasmids containing certain origins of replication. This procedure is most commonly used with plasmids containing an origin derived from a cole1 plasmid, although it can also be used to obtain increased yields of bacteriophage M13 replicative form. PCR amplification is the use of the polymerase chain reaction to increase the amount of a specific DNA fragment.

Anaerobe : An organism that grows in the absence of oxygen.

Anaerobic respiration : Metabolic process whereby electrons are transferred from an organic, or in some cases, inorganic compounds to an inorganic acceptor molecule other than oxygen. The most common acceptors are nitrate, sulfate, and carbonate.

Anaplerotic reactions : Reactions that replenish depleted tricarboxylic acid cycle intermediates.

Anergy : A state of unresponsiveness to antigens. Absence of the ability to generate a sensitivity reaction to substances that are expected to be antigenic.

Anion exchange capacity : Sum total of exchangeable anions that a soil can adsorb. Expressed as centimoles of negative charge per kilogram of soil.

Anneal : The pairing of complementary DNA or RNA sequences, via hydrogen bonding, to form a double-stranded polynucleotide. Most often used to describe the binding of a short primer or probe

Annotation : The process of determining the location of specific genes in a genome map after it has been produced by nucleic acid sequencing.

Anoxic : Literally "without oxygen." An adjective describing a microbial habitat devoid of oxygen.

Anoxygenic photosynthesis : Type of photosynthesis in green and purple bacteria in which oxygen is not produced.

Antagonist : Biological agent that reduces the number or diseaseproducing activities of a pathogen.

Antheridium : Male gametangium found in the phylum Oomycota (Kingdom Stramenopila) and phylum Ascomycota (Kingdom Fungi).

Anthrax : An infectious disease of animals caused by ingesting *Bacillus anthracis* spores. Can also occur in humans and is sometimes called woolsorter's disease.

Anthropogenic : Derived from human activities.

Antibiosis : Inhibition or lysis of an organism mediated by metabolic products of the antagonist; these products include lytic agents, enzymes, volatile compounds, and other toxic substances.

Antibiotic : Organic substance produced by one species of organism that in low concentrations will kill or inhibit growth of certain other organisms.

Antibiotic resistance : The ability of a microorganism to produce a protein that disables an antibiotic or prevents transport of the antibiotic into the cell.

Antibody : Protein that is produced by animals in response to the presence of an antigen and that can combine specifically with that antigen.A glycoprotein produced in response to the introduction of an antigen; it has the ability to combine with the antigen that stimulated its production. Also known as an immunoglobulin (Ig).

Anti codon : A nucleotide base triplet in a transfer RNA molecule that pairs with a complementary base triplet, or codon, in a messenger RNA molecule. See Codon, Messenger RNA, RNA.

Antigen : Substance that can incite the production of a specific antibody and that can combine with that antibody. .A foreign (nonself) substance (such as a protein, nucleoprotein, polysaccharide, or sometimes a glycolipid) to which lymphocytes respond; also known as an immunogen because it induces the immune response.

Antigenic determinant : A surface feature of a microorganism or macromolecule, such as a glycoprotein, that elicits an immune response.

Antigenic drift : A small change in the antigenic character of an organism that allows it to avoid attack by the immune system.

Antigenic shift : A major change in the antigenic character of an organism that alters it to an antigenic strain unrecognized by host immune mechanisms.

Antigenic switching : The altering of a microorganism's surface antigens through genetic rearrangement, to elude detection by the host's immune system.

Antigenpresenting cells : Antigenpresenting cells (APCs) are cells that take in protein antigens, process them, and present antigen fragments to B cells and T cells in conjunction with class II MHC molecules so that the cells are activated. Macrophages, B cells, dendritic cells, and Langerhans cells may act as APCs.

Antimetabolite : A compound that blocks metabolic pathway function by competitively inhibiting a key enzyme's use of a metabolite because it closely resembles the normal enzyme substrate.

Antimicrobial agent : An agent that kills microorganisms or inhibits their growth

Antiparallel : Aligned with the two sugar phosphate backbones in opposite orientations, with the 5′ end of each polynucleotide aligned with the 3′ end of the complementary polynucleotide.

Antisense RNA : A complementary RNA sequence that binds to a naturally occurring (sense) mRNA molecule, thus blocking its translation.

Antiseptic : Agent that kills or inhibits microbial growth but is not harmful to human tissue.

Apoenzyme : The protein part of an enzyme that also has a non protein component.

Apoptosis : Programmed cell death. The fragmentation of a cell into membrane bound particles that are eliminated by phagocytosis. Apoptosis is a physiological suicide mechanism that preserves homeostasis and occurs during normal tissue turnover. It is responsible for cell death in pathological circumstances, such as exposure to low concentrations of xenobiotics and infections by HIV and various other viruses.

Aporepressor : An inactive form of the repressor protein, which becomes the active repressor when the co repressor binds to it

Arbuscule : Special "treeshaped" structure formed within root cortical cells by arbuscular mycorrhizal fungi.

Archaea : Evolutionarily distinct group (domain) of prokaryotes consisting of the methanogens, most extreme halophiles and hyperthermophiles, and *Thermoplasma*.

Aromatic : Organic compounds which contain a benzene ring, or a ring with similar chemical characteristics.

Aseptic technique : Manipulating sterile instruments or culture media in such a way as to maintain sterility.

Asexual reproduction : Nonsexual means of reproduction which can include grafting and budding.

Assimilatory nitrate reduction : Conversion of nitrate to reduced forms of nitrogen, generally ammonium, for the synthesis of amino acids and proteins.

Associative dinitrogen fixation : Close interaction between a freeliving diazotrophic organism and a higher plant that results in an enhanced rate of dinitrogen fixation.

Associative symbiosis : Close but relatively casual interaction between two dissimilar organisms or biological systems. The association may be mutually beneficial but is not required for accomplishment of a particular function.

Asymmetric : An object with one side, end or aspect different from another. A DNA strand is polar due to the asymmetric nature of the sugar phosphate backbone. The 3' carbon of one sugar moiety is connected to the 5' carbon of the next, giving the overall strand distinguishable 5' and 3' ends.

Autogenous infection : An infection that results from a patient's own microbiota, regardless of whether the infecting organism became part of the patient's microbiota subsequent to admission to a clinical care facility

Autolysins : Enzymes that partially digest peptidoglycan in growing bacteria so that the peptidoglycan can be enlarged

Autolysis : Spontaneous lysis.

Autoradiography : Detecting radioactivity in a sample, such as a cell or gel, by placing it in contact with a photographic film.

Autosome : A chromosome that is not involved in sex determination

Autotroph : Organism which uses carbon dioxide as the sole carbon source.

Autotrophic nitrification : Oxidation of ammonium to nitrate through the combined action of two chemoautotrophic organisms, one forming nitrite from ammonium and the other oxidizing nitrite to nitrate.

Auxotroph : A mutated prototroph that lacks the ability to synthesize an essential nutrient and therefore must obtain it or a precursor from its surroundings.

Axenic : Literally "without strangers." A system in which all biological populations are defined, such as a pure culture.

B cell, also known as a B lymphocyte : A type of lymphocyte derived from bone marrow stem cells that matures into an immunologically competent cell under the influence of the bursa of Fabricius in the chicken and bone marrow in nonavian species. Following interaction with antigen, it becomes a plasma cell, which synthesizes and secretes antibody molecules involved in humoral immunity.

Bacillus : Bacterium with an elongated, rod shape.

***Bacillus thuringiensis* (Bt) :** A bacterium that kills insects; a major component of the microbial pesticide industry

Backcross : Crossing an organism with one of its parent organisms

Bacteremia : The presence of viable bacteria in the blood.

Bacteria : All prokaryotes that are not members of the domain Archaea.The domain that contains procaryotic cells with primarily diacyl glycerol diesters in their membranes and with bacterial rRNA. Bacteria also is a general term for organisms that are composed of procaryotic cells and are not multicellular.

Bacterial artificial chromosome (BAC) : A cloning vector constructed from the *E. coli* Ffactor plasmid that is used to clone foreign DNA fragments in *E. coli*.

Bacterial photosynthesis : A lightdependent, anaerobic mode of metabolism. Carbon dioxide is reduced to glucose, which is used for both biosynthesis and energy production. Depending on the hydrogen source used to reduce CO2, both photolithotrophic and photoorganotrophic reactions exist in bacteria.

Bacteriochlorophyll : Light absorbing pigment found in green sulfur and purple sulfur bacteria.

Bacteriocide : A class of antibiotics that kills bacterial cells

Bacteriocin : Agent produced by certain bacteria that inhibits or kills closely related isolates and species.

Bacteriophage : A virus that infects bacteria. Bacteriophages have proven very useful in molecular biology as models of control of gene expression, as sources of enzymes, and as cloning vectors. Some that are particularly important in molecular biology include lambda, m13, and t4.

Bacteriorhodopsin : A protein containing retinal found in the membranes of certain extremely halophilic Archaea and which is involved in lightmediated ATP synthesis.

Bacteriostat : A class of antibiotics that prevents growth of bacterial cells

Bacteroid : Altered form of cells of certain bacteria. Refers particularly to the swollen, irregular vacuolated cells of rhizobia in nodules of legumes.A modified, often pleomorphic, bacterial cell within the root nodule cells of legumes; after transformation into a symbiosome it carries out nitrogen fixation.

Baeocytes : Small, spherical, reproductive cells produced by pleurocapsalean cyanobacteria through multiple fission.

Balanced growth : Microbial growth in which all cellular constituents are synthesized at constant rates relative to each other

Barophile : An organism able to live optimally at high hydrostatic pressure.

Barotolerant : An organism able to tolerate high hydrostatic pressure, although growing better at normal pressures.

Basal body : The cylindrical structure at the base of procaryotic and eucaryotic flagella that attaches them to the cell.

Basal medium : A medium which allows the growth of many types of microorganisms which do not require any special nutrient supplements, e.g. nutrient broth.

Base : In a nucleic acid, the term "base" refers to a nucleoside. The bases in two RNA or DNA strands are the portions that interact in base pairing. In general chemistry, a base is hydrogen ion acceptor. Chemical bases give a low ph ("basic") solution when dissolved in water. A basic molecule is positively charged in aqueous solution at neutral ph. Despite the presence of the nucleosides, the overall charge on a nucleic acid polymer is negative at neutral ph, due to the highly negative charge on the phosphates of the backbone.

Base composition : Proportion of the total bases consisting of guanine plus cytosine or thymine plus adenine base pairs. Usually expressed as a guanine + cytosine (G + C) value, e.g. 60% G+C.

Base pairing : Nucleotide bases form the core of the double stranded DNA helix. To maintain a constant distance between the two opposing sugar phosphate backbones, it is necessary for the bases to align in combinations with similar dimensions, a requirement that is fulfilled by always aligning a purine (A or G) with a pyrimidine (T or C). When present in the core of the DNA helix, A and T are each capable of two hydrogen bond interactions in their most stable configuration, while G and C are capable of three hydrogen bond interactions. The AT/GC (or WatsonCrick) base pairing rules dictate that the most stable condition for a DNA double helix occurs when A is aligned and hydrogen bonded with T and G is aligned and hydrogen bonded with C.

Batch culture : A culture of microorganisms produced by inoculating a closed culture vessel containing a single batch of medium

Batch process : A treatment process in which a tank or reactor is filled, the wastewater (or solution) is treated or a chemical solution is prepared and the tank is emptied. The tank may then be filled and the process repeated. Batch processes are also used to cleanse, stabilize or condition chemical solutions for use in industrial manufacturing and treatment processes.

B cell antigen receptor (BCR) : A trans membrane immunoglobulin complex on the surface of a B cell that binds an antigen and stimulates the B cell. It is composed of a membrane bound immunoglobulin, usually IgD or a modified IgM, complexed with another membrane protein (the Iga/Igb heterodimer).

Benthic : Zone the sediment

Beta hemolysis : A zone of complete clearing around a bacterial colony growing on blood agar. The zone does not change significantly in color.

Beta prime subunit : One member of the pair of large subunits of many bacterial RNA polymerases.

Beta subunit : One member of the pair of large subunits of many bacterial RNA polymerases.

Beta-DNA : The normal form of DNA found in biological systems, which exists as a right-handed helix

Beta-Lactamase : Ampicillin resistance gene

Beta-proteobacteria : One of the five subgroups of proteobacteria, each with distinctive 16S rRNA sequences. Members of this subgroup are similar to the alpha proteobacteria metabolically, but tend to use substances that diffuse from organic matter decomposition in anaerobic zones.

Binal symmetry : The symmetry of some virus capsids (e.g., those of complex phages) that is a combination of icosahedral and helical symmetry.

Binary fission : Division of one cell into two cells by the formation of a septum. It is the most common form of cell division in bacteria.

Binomial nomenclature : System of having two names, genus and specific epithet, for each organism.

Bioaccumulation : Intracellular accumulation of environmental pollutants such as organic materials by living organisms. Accumulation of a chemical substance in living tissue.

Bioaugmentation : The addition to the environment of microorganisms that can metabolize and grow on specific organic compounds.

Bioavailability : The availability of chemicals to potentially biodegradative microorganisms.

Biochemical oxygen demand (BOD) : Amount of dissolved oxygen consumed in five days by biological processes breaking down organic matter. In particular: The requirement for molecular oxygen by microbes during oxidation of biological substances in sewage. The BOD test measures the oxygen consumed (in mg/L) over 5 days at 20 degrees C

Biodegradable : Substance capable of being decomposed by biological processes.

Biodegradation : The breakdown of organic substances by microorganisms.

Biofilm : Microbial cells encased in an adhesive, usually a polysaccharide material, and attached to a surface. Organized microbial systems consisting of layers of microbial cells associated with surfaces, often with complex structural and functional characteristics. Biofilms have physical/chemical gradients that influence microbial metabolic processes. They can form on inanimate devices (catheters, medical prosthetic devices) and also cause fouling (e.g., of ships' hulls, water pipes, cooling towers).

Biogeochemistry : Study of microbial mediated chemical transformations of geochemical interest, such as nitrogen or sulfur cycling.

Bio insecticide : A pathogen that is used to kill or disable unwanted insect pests. Bacteria, fungi, or viruses are used, either directly or after manipulation, to control insect populations.

Bioluminescence : The production of light by living cells, often through the oxidation of molecules by the enzyme luciferase

Biomagnification : Increase in the concentration of a chemical substance as it is progresses to higher trophic levels of a food chain.

Bioremediation : The process by which living organisms act to degrade or transform hazardous organic contaminants. Use of microorganisms to remove or detoxify toxic or unwanted chemicals from an environment.

Biosolid : The resides of wastewater treatment. Formerly called sewage sludge.

Biostimulation : A process that increases activity of microorganisms biodegrading contaminants. For example, addition of nutrients, oxygen, or other electron donors and acceptors.

Biosynthesis : Production of needed cellular constituents from other, usually simpler, molecules.

Biotechnology : Use of living organisms to carry out defined physiochemical processes having industrial or other practical application.

Bio tower : An attached culture system. A tower filled with a media similar to rachet or plastic rings in which air and water are forced up a counterflow movement in the tower.

Biotransformation : Alteration of the structure of a compound by a living organism or enzyme.

Biotrophic : Nutritional relationship between two organisms in which one or both must associate with the other to obtain nutrients and grow.

Bioventing : The process of supplying oxygen in situ to oxygen deprived soil microbes by forcing air through unsaturated contaminated soil at low flow rates. This stimulates biodegradation and minimizes stripping volatiles into the atmosphere. Frequently used to remediate soil under structures since it is relatively noninvasive.

Black smoker : Thermal vent emitting very hot water and minerals.

Brown rot fungus : Fungus that attacks cellulose and hemicellulose in wood, leaving darkcolored lignin and phenolic materials behind.

Brownfield : An abandoned, idled, or underused industrial or commercial facility where expansion or redevelopment is complicated by a real or perceived environmental contamination.

Bulk density, soil : Mass of dry soil per unit bulk volume (combined volume of soil solids and pore space).

Burst size : The number of phages released by a host cell during the lytic life cycle.

Capsid : Protein coat of a virus.

Capsomere : An individual protein subunit of the virus capsid.

Capsule : Compact layer of polysaccharide exterior to the cell wall in some bacteria.

Carbohydrate : Any chemical compound which consists of only carbon (C), oxygen (O), and hydrogen (H) elements, for examples, sugars, starches, and cellulose are carbohydrates. Also the ratio of hydrogen to oxygen atoms in carbohydrates is usually 2:1.

Carbon cycle : Sequence where carbon dioxide is converted to organic forms by photosynthesis or chemo synthesis recycled through the biosphere, with partial incorporation into sediments, and ultimately returned to its original state through respiration or combustion.

Carboxyl group : A COOH group attached to a carbon skeleton as in the carboxylic acids and fatty acids.

Carboxysomes : Polyhedral cellular inclusions of crystline ribulose bisphosphate carboxylase (RubisCO), the key enzyme of the Calvin cycle

Carcinogen : Substance which causes the initiation of tumor formation. Frequently a mutagen.

Catabolism : Biochemical processes involved in the breakdown of organic compounds, usually leading to the production of energy.

Catabolite repression : Transcriptionlevel inhibition of a variety of inducible enzymes by glucose or other readily used carbon source.

Catalyst : A compound that increases the rate of a chemical reaction by acting to decrease the activation energy necessary for the reaction to begin. Catalysts can be chemical compounds, such as the ammonium sulphate added to initiate polymerization of an acrylamide solution, or biological, such as the enzymes that mediate virtually all chemical reactions in a cell.

Catalytic antibody (abzyme) : An antibody selected for its ability to catalyze a chemical reaction by binding to and stabilizing the transition state intermediate.

Catalytic RNA (ribozyme) : A natural or synthetic RNA molecule that cuts an RNA substrate.

Cation exchange capacity (CEC) : Sum of exchangeable cations that a soil can adsorb at a specific pH. Expressed as centimoles of positive charge per kilogram of soil (cmolc kg^{1}).

Cation : A positively charged ion

CD95 pathway : The CD95 receptor is found on many nucleated eucaryotic cells. When the receptor is bound to a specific ligand (CD95L), the CD95CD95L complex activates several cytoplasmic proteins that initiate a cellular suicide cascade leading to apoptosis

cDNA : Complementary DNA; a DNA molecule generated by reverse transcription from an RNA (generally mRNA) template.

cDNA library : A collection recombinant vector molecules containing inserts of DNA fragments synthesized from a mixture of mRNA molecules, generally by reverse transcriptasemediated elongation of an oligodt primer annealed to the 3' polyA tail of the mRNA molecules purified from a eukaryotic cell under a specific growth condition. These collections are tissuespecific in that they contain cDNA inserts that represent only the genes that were expressed at the time the mRNA was purified.

Cell : Fundamental unit of living matter.

Cell wall : Layer or structure that lies outside the cytoplasmic membrane; it supports and protects the membrane and gives the cell shape.

Centers of origin : Usually the location in the world where the oldest cultivation of a particular crop has been identified.

Central dogma : Francis Crick's seminal concept that in nature genetic information generally flows from DNA to RNA to protein

Centrifugation : Separating molecules by size or density using centrifugal forces generated by a spinning rotor. G forces of several hundred thousand times gravity are generated in ultracentrifugation

Centromere : The central portion of the chromosome to which the spindle fibers attach during mitotic and meiotic division

Cephalosporin : A group of beta-lactam antibiotics derived from the fungus Cephalosporium, which share the 7aminocephalosporanic acid nucleus.

CFU : Colony forming units. Viable microorganisms (bacteria, yeasts & mould) capable of growth under the prescribed conditions (medium, atmosphere, time and temperature) develop into visible colonies (colony forming units) which are counted. The term colony forming unit (CFU) is used because a colony may result from a single microorganism or from a clump / cluster of microorganisms.

Chaperonin : A protein that aids in the correct folding of other proteins and the assembly of multi subunit structures.

Chargetomass ratio : This ratio significantly affects the mobility of a macromolecule through a solution when driven by an electric field (two molecules of identical mass but different charge will move at different rates in an electric field). Since at neutral ph, the majority of the net charge on DNA is derived from the negatively charged phosphate groups in the DNA backbone, as DNA increases in size, the total charge increases at the same rate. The resulting chargetomass ratio therefore remains constant, and DNA fragments of different sizes all move at about the same rate in an electric field. For separation of the fragments according to size, it is necessary to force the fragments to migrate through a molecular sieve or matrix of many small pores that allows the smaller fragments to move faster than the larger fragments. Agarose and polyacrylamide gels are most commonly used as a matrix for size separation of DNA and RNA fragments.

Chelate (chelator) : Organic chemical that forms ring compound in which a metal is held between two or more atoms strongly enough to diminish the rate at which it becomes fixed by soil, thereby making it more available for plant and microbial uptake.

Chemical cleavage : DNA can be chemically cleaved by one of two common methods acid depurination followed by hydrolysis, or methylation followed by incubation in pyridine. Acid depurination and hydrolysis introduces nicks into DNA, reducing the size of DNA fragments on denaturation, a step frequently employed in Southern blot transfer protocols and in the preparation of hybridization probes. Methylation/piperidine cleavage is an important step in MaxamGilbert sequencing and in chemical footprinting reactions.

Chemical method : (of nucleotide sequence determination) determination of nucleotide sequence by the MaxamGilbert method.

Chemotaxis : Oriented movement of a motile organism with reference to a chemical agent. May be positive (toward) or negative (away) with respect to the chemical gradient.

Chemotrophs : Organisms that obtain energy from the oxidation of chemical compounds

Chitin : A tough, resistant, nitrogencontaining polysaccharide forming the walls of certain fungi, the exoskeleton of arthropods, and the epidermal cuticle of other surface structures of certain protists and animals.

Chloramphenicol : A broadspectrum antibiotic that is produced by Streptomyces venezuelae or synthetically; it binds to the large ribosomal subunit and inhibits the peptidyl transferase reaction

Chloroplast : Chlorophyllcontaining organelle of photosynthetic eukaryotes.

Chromatid : Each of the two daughter strands of a duplicated chromosome joined at the centromere during mitosis and meiosis.

Chromatin : The DNAcontaining portion of the eucaryotic nucleus; the DNA is almost always complexed with histones. It can be very condensed (heterochromatin) or more loosely organized and genetically active (euchromatin).

Chromatography : Any technique used to separate different species of molecules (or ions) by subjecting them to two different carrier phases: mobile and stationary phases.

Chromogen : A colorless substrate that is acted on by an enzyme to produce a colored end product.

Chromogenic : Producing color; a chromogenic colony is a pigmented colony.

Chromophore group : A chemical group with double bonds that absorbs visible light and gives a dye its color.

Chromosomal attachment site : An area of a cell or nuclear membrane where a chromosome appears to be anchored to assist in replication and segregation of new chromosomes to progeny cells.

Chromosome : Genetic element carrying information essential to cellular metabolism. Prokaryotes have a single chromosome, consisting of a circular DNA molecule. Eukaryotes contain more than one chromosome, each containing a linear DNA molecule complexed with specific proteins.

Chromosome walking : Working from a flanking DNA marker, overlapping clones are successively identified that span a chromosomal region of interest

Chytrid : Fungal organism in the phylum Chytridiomycota that consists of a spherical cell from which short thin filamentous branches (rhizoids) grow that resemble fine roots.

Cistron : A DNA sequence that codes for a specific polypeptide; a gene.

Clamp connection : Small branch of a fungal hypha that connects two compartments separated by a septum and helps to maintain a dikaryon in each hyphal compartment; characteristic of fungi in the phylum Basidiomycota.

Clarification : A process in which suspended material is removed from a wastewater. This may be accomplished by sedimentation, with or without chemicals, or filtration.

Classification : (i) Arrangement of organisms into groups based on mutual similarity or evolutionary relatedness. (ii) Systematic arrangement of soils into groups or categories on the basis of their characteristics.

Clay : Soil particle < 0.002 mm in diameter.

Clone : (i) Population of cells all descended from a single cell. (ii) Number of copies of a DNA fragment to be replicated by a phage or plasmid.

Cloning : A colony that grows from a single bacterium is said to be clonal. Molecular cloning is the process of inserting DNA fragments into a vector and generating recombinants, each of which contains only a few of the DNA fragments present in the original DNA sample.

Cloning vector : DNA molecule that is able to bring about the replication of foreign DNA fragments

Coagulants : Chemicals which cause very fine particles to clump (collect) together into larger particles. This makes it easier to separate the solids from the water by settling, skimming, draining, or filtering.

Coat protein (capsid) : The coating of a protein that enclosed the nucleic acid core of a virus

COD : Chemical oxygen demand the amount of oxygen in mg/l required to oxidize both organic and oxidizable inorganic compounds

Codon : A sequence of three nucleotides in mRNA that directs the incorporation of an amino acid during protein synthesis or signals the start or stop of translation.

Cohesion : Force holding a solid or liquid together, owing to attraction between like molecules.

Colicin : A plasmid encoded protein that is produced by enteric bacteria and binds to specific receptors on the cell envelope of sensitive target bacteria, where it may cause lysis or attack specific intracellular sites such as ribosomes.

Coliform : Gramnegative, nonsporeforming facultative rod that ferments lactose with gas formation with 48 hours at 35°C. Often an indicator organism for fecal contamination of water supplies. *Escherichia coli* and *Enterobacter* are important members.

Colloid fraction : Organic and inorganic matter with very small particle size and a correspondingly large surface area per unit of mass.

Colloids : Very small, finely divided solids (particles that do not dissolve) that remain dispersed in a liquid for a long time due to their small size and electrical charge.

Colonization : Establishment of a community of microorganisms at a specific site or ecosystem.

Colony : Clone of bacterial cells on a solid medium that is visible to the naked eye.

Colony blot : A procedure in which an imprint of the colonies grown on the surface of a solid nutrient medium is transferred to a membrane and the cells lysed or disrupted, allowing the DNA or protein to bind to the membrane. Following appropriate treatment, the membrane can be exposed to either a hybridization probe for the identification of colonies containing a specific nucleic acid sequence or to an antibody for the identification of colonies containing a specific protein.

Colony forming units (CFU) : The number of microorganisms that can form colonies when cultured using spread plates or pour plates, an indication of the number of viable microorganisms in a sample.

Cometabolism : Transformation of a substrate by a microorganism without deriving energy, carbon, or nutrients from the substrate. The organism can transform the substrate into intermediate degradation products but fails to multiply at its expense.

Common vehicle transmission : The transmission of a pathogen to a host by means of an inanimate medium or vehicle.

Community : All organisms that occupy a common habitat and interact with one another.

Competent : In a genetic sense, the ability to take up DNA.

Competition : Rivalry between two or more species for a limiting factor in the environment that usually results in reduced growth of participating organisms.

Competitive exclusion principle : Two competing organisms overlap in resource use, which leads to the exclusion of one of the organisms.

Complement system : A group of plasma proteins that plays a major role in an animal's defensive immune response. Complementary In reference to base pairing, the ability of two polynucleotide sequences to form a double stranded helix by hydrogen bonding between bases in the two sequences.

Complementary DNA or RNA : The matching strand of a DNA or RNA molecule to which its bases pair.

Complementary nucleotides : Members of the pairs adenine-thymine, adenine-uracil, and guanine cytosine that have the ability to hydrogen bond to one another.

Complementation : Restoration of a damaged physical property or capability. Genetic complementation involves introduction of nucleotide sequences that ultimately, through one of several different mechanisms, restore gene function. Biochemical complementation generally involves the introduction of a protein component that restores function to an inactive protein complex.

Complementation analysis : Generally refers to the mixture of genes derived from two or more different mutants in an attempt to establish a chromosomal order or location of mutations.

Complex medium : Medium whose precise chemical composition is unknown. Also called undefined medium.

Complex viruses : Viruses with capsids having a complex symmetry that is neither icosahedral nor helical.

Complexity : The total number of base pair combinations in a DNA fragment or DNA sample. For example, although both sequences contain 3 a's, 3 c's, 3 g's, and 3 t's, the sequence AATCTGCCAGTG is more complex than the sequence ACTGACTGACTG, which is a linear repeat of the sequence ACTG.

Compost : Organic residues which have been mixed, piled, and moistened, with or without addition of fertilizer and lime, and generally allowed to undergo thermophilic decomposition until the original organic materials are substantially altered or decomposed.

Concatemer : A long DNA molecule consisting of several genomes linked together in a row.

Conditional mutations : Mutations that are expressed only under certain environmental conditions.

Conjugants : Complementary mating types that participate in a form of protozoan sexual reproduction called conjugation.

Conjugation : In prokaryotes, transfer of genetic information from a donor cell to a recipient cell by celltocell contact.

Conjugation : The joining of two bacteria cells when genetic material is transferred from one bacterium to another

Conjugative plasmid : Selftransmissible plasmid; a plasmid that encodes all the functions needed for its own intercellular transmission by conjugation.

Constitutive enzyme : Enzyme always synthesized by the cell regardless of environmental conditions.

Constitutive promoter : An unregulated promoter that allows for continual transcription of its associated gene

Contiguous (contig) map : The alignment of sequence data from large, adjacent regions of the genome to produce a continuous nucleotide sequence across a chromosomal region

Cosmid : A plasmid vector with lambda phage cos sites that can be packaged in a phage capsid; it is useful for cloning large DNA fragments.

Covalent : Nonionic chemical bond formed by a sharing of electrons between two atoms.

Covalently closed circular DNA : A doublestranded circular DNA molecule containing no nicks in the sugarphosphate backbone. These molecules can contain various degrees of twisting stress around the axis of the DNA helix, which can cause the appearance of turns or superhelical twists to relieve the stress.

Crista : Inner membrane in a mitochondrion, site of respiration.

Cross feeding : (i) Specific type of syntrophy where two populations cooperate to metabolize a compound. (ii) One organism consuming products excreted by another organism.

Cross-hybridization : The hydrogen bonding of a single- stranded DNA sequence that is partially but not entirely complementary to a singlestranded substrate. Often, this involves hybridizing a DNA probe for a specific DNA sequence to the homologous sequences of different species

Crossing-over : The exchange of DNA sequences between chromatids of homologous chromosomes during meiosis

Cross-pollination : Fertilization of a plant from a plant with a different genetic makeup

Cryptins : Peptides produced by Paneth cells in the intestines. Cryptins are toxic for some bacteria, although their mode of action is not known.

Cryptococcosis : An infection caused by the basidiomycete, Cryptococcus neoformans, which may involve the skin, lungs, brain, or meninges

Crystallizable fragment : The stem of the Y portion of an antibody molecule. Cells such as macrophages bind to the Fc region, and it also is involved in complement activation

Culture : Population of microorganisms cultivated in an artificial growth medium. A pure culture is grown from a single cell; a mixed culture consists of two or more microbial species or strains growing together.

Current : (Used during electrophoresis) the amount of electricity passing through a circuit per unit time. Current (i) is measured in amperes ("amps", a), which are coulombs per second. The amount of current passing through a circuit (e.g., through an electrophoresis gel) depends on the voltage (v) and the resistance (r) according to the formula v=ir. The resistance of a gel depends on several parameters, including the crosssectional area of the current path (usually corresponding to gel thickness) and the salt concentration of the buffer. Higher resistance (e.g., using lower salt concentrations) gives less current at a given voltage, and thus leads to less heat generation. Excessive heat can distort sample mobility or cause melting of an agarose gel matrix.

Cyst : Resting stage formed by some bacteria, nematodes, and protozoa in which the whole cell is surrounded by a protective layer; not the same as endospore.

Cytogenetics : Study that relates the appearance and behavior of chromosomes to genetic phenomenon

Cytokine : A general term for nonantibody proteins, released by a cell in response to inducing stimuli, which are mediators that influence other cells. Are produced by lymphocytes, monocytes, macrophages, and other cells.

Cytoplasm : Cellular contents inside the cell membrane, excluding the nucleus.

Cytoplasmic membrane : Selectively permeable membrane surrounding the cell's cytoplasm.

Cytosine(C) : A base found attached to the 1´ carbon of deoxyribose in DNA and RNA, where it is capable of forming 3 hydrogen bond interactions with the purine guanine on an opposing nucleic acid strand.

Dalton : A unit of measurement equal to the mass of a hydrogen atom, 1.67 x 10E-24 gram/L (Avogadro's number).

Degradation : Process whereby a compound is usually transformed into simpler compounds.

Denaturation : Process where doublestranded DNA unwinds and dissociates into two single strands. The reverse of DNADNA hybridization.

Density gradient centrifugation : High-speed centrifugation in which molecules "float" at a point where their density equals that in a gradient of cesium chloride or sucrose.

Deoxyadenosine triphosphate : A deoxyribose molecule with the purine adenine attached to the 1´ carbon and a chain of three oxygenated phosphorus atoms attached to the 5' carbon. This is a high energy molecule that serves as a precursor for the assembly of dna.

Deoxycytidine triphosphate : A deoxyribose molecule with the pyrimidine cytosine attached to the 1´ carbon and a chain of three oxygenated phosphorus atoms attached to the 5' carbon. This is a high energy molecule that serves as a precursor for the assembly of dna.

Deoxyguanosine triphosphate : A deoxyribose molecule with the purine guanine attached to the 1´ carbon and a chain of three oxygenated phosphorus atoms attached to the 5' carbon. This is a high energy molecule that serves as a precursor for the assembly of dna.

Deoxynucleoside triphosphate : (dntp = dATP, dCTP, dGTP, or dTTP) any of the precursors required for enzymatic sythesis of DNA.

Deoxyribonucleic acid (DNA) : Polymer of nucleotides connected via a phosphatedeoxyribose sugar backbone; the genetic material of the cell.

Deoxyribose : (2 deoxyribose) the 5carbon sugar that forms an integral part of the sugarphosphate backbone of DNA and to which the nucleotide bases are attached at the 1', or first, carbon. This compound is structurally the same as ribose except for the absence of a hydroxyl residue at the 2', or second, carbon of deoxyribose, a feature that makes deoxyribosephosphate polymers more stable than ribosephosphate polymers.

Derepressible enzyme : Enzyme that is produced in the absence of a specific inhibitory compound acting at the transcriptional level.

Detritus : Dead plant and animal matter, usually consumed by bacteria, but some remains

Dew point : The temperature to which air with a given quantity of water vapor must be cooled to cause condensation of the vapor in the air.

Diatom : Alga with siliceous cell walls that persist as a skeleton after death. Any of the microscopic unicellular or colonial alga constituting the class Bacillariophyceae.

Diatomaceous earth : Geologic deposit of fine, grayish siliceous material composed chiefly or wholly of the remains of diatoms. It may occur as a powder or as a porous, rigid material.

Diazotroph : Organism that can use dinitrogen as its sole nitrogen source, i.e. capable of N_2 fixation.

Dideoxyadenosine triphosphate : A nucleoside triphosphate analog (a dideoxynucleoside triphosphate) that causes termination of an elongating DNA molecule when incorporated by DNA polymerase in place of adenosine triphosphate.

Dideoxycytosine triphosphate : A nucleoside triphosphate analog (a dideoxynucleoside triphosphate) that causes termination of an elongating DNA molecule when incorporated by DNA polymerase in place of cytosine triphosphate.

Dideoxyguanosine triphosphate : A nucleoside triphosphate analog (a dideoxynucleoside triphosphate) that causes termination of an elongating DNA molecule when incorporated by DNA polymerase in place of guanosine triphosphate.

Dideoxynucleoside triphosphate : Ddntp.GIF a nucleoside triphosphate analog that causes termination of an elongating DNA molecule when incorporated by DNA polymerase because of the absence of a 3´hydroxyl residue on the sugar portion of the compound, which is necessary for the formation of the 3' to 5' phosphate linkage in an elongating DNA chain.

Dideoxynucleotide (didN) : A deoxynucleotide that lacks a 3' hydroxyl group, and is thus unable to form a 3'-5' phosphodiester bond necessary for chain elongation. Dideoxynucleotides are used in DNA sequencing and the treatment of viral diseases.

Dideoxynucleotide sequencing : A method of nucleotide sequence determination that is dependent on the enzymatic elongation of a specific sequencing primer annealed to a template DNA with the elongation reaction performed in the presence of a basespecific chainterminating nucleoside triphosphate analogs, such as a dideoxynucleoside triphosphate. Denaturing polyacrylamide gel electrophoresis of the chaintermination products generated from extension of the sequence primer resolves the DNA fragments in order of increasing size to generate a nested set of DNA fragments that reveals the nucleotide sequence of the DNA fragment.

Dideoxythymidine triphosphate : A nucleoside triphosphate analog (a dideoxynucleoside triphosphate) that causes termination of an elongating DNA molecule when incorporated by DNA polymerase in place of thymidine triphosphate.

Differential medium : Cultural medium with an indicator, such as a dye, which allows various chemical reactions to be distinguished during growth.

Diffused air aeration : A diffused air activated sludge plant takes air, compresses it, and then discharges the air below the water surface of the aerator through some type of air diffusion device.

Diffusion (nutrient) : Movement of nutrients in soil that results from a concentration gradient.

Digest : To cut DNA molecules with one or more restriction endonucleases

Dikaryon : Two nuclei present in the same hyphal compartment; they constitute a homokaryon when both nuclei are genetically the same or a heterokaryon when each nucleus is genetically different from the other.

Dilution plate count method : Method for estimating the viable numbers of microorganisms in a sample. The sample is diluted serially and then transferred to agar plates to permit growth and quantification of colonyforming units.

Dinitrogen fixation : Conversion of molecular dinitrogen (N_2) to ammonia and subsequently to organic combinations or to forms useful in biological processes.

Diploid : In eukaryotes, an organism or cell with two chromosome complements, one derived from each haploid gamete.

Diploid cell : A cell which contains two copies of each chromosome

Direct count : Method of estimating the total number of microorganisms in a given mass of soil by direct microscopic examination.

Directional cloning : DNA insert and vector molecules are digested with two different restriction enzymes to create noncomplementary sticky ends at either end of each restriction fragment. This allows the insert to be ligated to the vector in a specific orientation and prevents the vector from recircularizing

Disinfectant : Agent that kills microorganisms.

Dissolved solids : Chemical substances either organic or inorganic that are dissolved in a waste stream and constitute the residue when a sample is evaporated to dryness.

DNA diagnosis : The use of DNA polymorphisms to detect the presence of a disease gene.

DNA fingerprint : The unique pattern of DNA fragments identified by Southern hybridization (using a probe that binds to a polymorphic region of DNA) or by polymerase chain reaction (using primers flanking the polymorphic region).

DNA fingerprinting : Molecular genetic techniques to assess possible differences among DNA in a samples.

DNA fragment size standard : A DNA fragment or mixture of fragments of known size subjected to gel electrophoresis along with unknown DNA samples to enable preparation of a standard curve for estimation of the sizes of unknown fragments.

DNA library : Collection of cloned DNA fragments which in total contain genes from the entire genome of an organism; also called a gene library.

DNA ligase : An enzyme that catalyzes sealing of breaks or nicks in the backbone of a DNA molecule. The *E. Coli* DNA ligase, encoded by a gene in the Escherichia coli chromosome, requires DPN rather than ATP as a cofactor, and cannot ligate blunt or flush ends. The ATPdependent DNA ligase encoded by bacteriophage T4 and synthesized during infection of Escherichia coli is capable of sealing breaks in association with 5' cohesive termini, 3' cohesive termini, or blunt ends that have no cohesive overlap.

DNA polymerase : An enzyme capable of extending the 3' hydroxyl residue of an oligo or polynucleotide primer annealed to a singlestranded DNA template, resulting in net synthesis of a complementary DNA strand. This is the only orientation in which DNA polymerization has been found to occur.

DNA polymorphism : One of two or more alternate forms (alleles) of a chromosomal locus that differ in nucleotide sequence or have variable numbers of repeated nucleotide units

DNA sequencing : Procedures for determining the nucleotide sequence of a DNA fragment

Do : Dissolved Oxygen a measure of the oxygen dissolved in water expressed in milligrams per liter.

Dominant : An allele is said to be dominant if it expresses its phenotype even in the presence of a recessive allele. See Allele, Phenotype, Recessive

Dominant gene : A gene whose phenotype is when it is present in a single copy.

Double helix : Describes the coiling of the antiparallel strands of the DNA molecule, resembling a spiral staircase in which the paired bases form the steps and the sugar-phosphate backbones form the rails

Double digestion : Treatment of a nucleic acid sample, usually DNA, with either simultaneous or successive restriction enzymes with differing recognition cleavage sites to establish a restriction map of the DNA fragment.

Double-stranded complementary DNA : A duplex DNA molecule copied from a cDNA template.

Doubling time : Time needed for a population to double in number or biomass.

Downstream : The region extending in a 3' direction from a gene.

Drug resistance gene : A gene conferring resistance to an antibiotic or class of antibiotics. These genes form the selective basis for distinguishing cells containing a vector from cells that do not, a key aspect of cloning technology.

Duplex DNA : Double-stranded DNA.

Ecology : The study of the interactions of organisms with their environment and with each other

EcoRI: Restriction endonuclease purified from bacterial strain Escherichia coli RY13 recognizing the sequence GAATTC and cleaving after the first G to generate the fourbase 5' cohesive terminus AATT.

Ecosystem : Community of organisms and the environment in which they live.

EGL : Energy grade line a line that represents the elevation of energy head in feet of water flowing in a pipe, conduit, or channel.

E_h : Potential generated between an oxidation or reduction halfreaction and the H electrode in the standard state.

Electrolytic process : A process that causes the decomposition of a chemical compound by the use of electricity.

Electron acceptor : Substance that accepts electrons during an oxidationreduction reaction. An electron acceptor is an oxidant.

Electron donor : Substance that donates electrons in an oxidationreduction reaction. An electron donor is a reductant.

Electrophilic compounds : Chemicals that attack or are drawn to regions in other chemicals in which electrons are readily available; oxidizing agents act as electrophilic compounds.

Electrophoresis : The technique of separating charged molecules in a matrix to which is applied an electrical field

Electroporation : A method for transforrning DNA, especially useful for plant cells, in which high voltage pulses of electricity are used to open pores in cell membranes, through which foreign DNA can pass.

ELISA : Enzymelinked immunosorbent assay. An immunoassay that uses specific antibodies to detect antigens or antibodies in body fluids. The antibodycontaining complexes are visualized through enzyme coupled to the antibody. Addition of substrate to the enzymeantibodyantigen complex results in a colored product.

Eluviation : Removal of soil material from a layer of soil as a suspension .

Emulsion : A liquid mixture of two or more liquid substances not normally dissolved in one another, one liquid held in suspension in the other

Encapsidation : Process by which a virus' nucleic acid is enclosed in a capsid.

Endonuclease : Endoenzyme that cleaves phosphodiester bonds within a nucleic acid molecule.

Endophyte : Organism growing within a plant. The association may be symbiotic or parasitic.

Enrichment culture : Technique in which environmental (including nutritional) conditions are controlled to favor the development of a specific organism or group of organisms.

Enteric bacteria : General term for a group of bacteria that inhabit the intestinal tract of humans and other animals. Among this group are pathogenic bacteria such as *Salmonella* and *Shigella.*

Entnerdoudoroff pathway (ED pathway) : A pathway that converts glucose to pyruvate and glyceraldehydephosphate by producing 6phosphogluconate and then dehydrating it.

Enzyme : A biological molecule that catylizes a specific chemical reaction. Most enzymes are proteins; a ribozyme is an enzyme composed of rna.

Enzyme : Protein within or derived from a living organism that functions as a catalyst to promote specific reactions.

Enzymelinked immunosorbent assay (ELISA) : Immunoassay that uses specific antibodies to detect antigens or antibodies. The antibodycontaining complexes are visualized through an enzyme coupled to the antibody. Addition of substrate to the enzymeantibodyantigen complex results in a colored product.

Episome : Plasmid that replicates by inserting itself into the bacterial chromosome.

Episomes : DNA elements that are capable of either replicating extrachromosomally or integrating into the host chromosome. Bacteriophage lambda, for example, is an episome that replicates extrachromosomally during the lytic, or progenyproducing, phase of life, but can integrate into the bacterial chromosome during the lysogenic phase.

Epitope : The region of an antigen to which the variable region of an antibody binds.

Ethidium bromide : A fluorescent dye used to stain DNA and RNA. The dye fluoresces when exposed to UV light.

Eubacteria : Old term for the Bacteria.

Eukarya : Phylogenetic domain containing all eukaryotic organisms.

Eukaryote : Organism having a unit membrane bound nucleus and usually other organelles.

Eurythermal : Bodies of water which are located at the lower end of a river and are subject to tidal fluctuations.

Eutrophic : Having high concentrations of nutrients optimal, or nearly so, for plant or animal growth. Can be applied to nutrient or soil solutions and bodies of water.

Eutrophic : lakes have high nutrient status and high productivity

Exergonic reaction : Chemical reaction that proceeds with the liberation of energy.

Exobiology : Branch of biology concerned with the effects of extraterrestrial environments on living organisms.

Exo enzyme : Enzyme that acts at the end of a polymer cleaving off monomers and dimers and sometimes larger chain fragments

Exon : A portion of a genomic DNA sequence that is transcribed and assembled as part of a mature mRNA molecule.

Exo nuclease : A protein capable of removing nucleotides from the terminal end of a polynucleotide. A 5'>3' exonuclease removes nucleotides from the 5´phosphate terminus, and a 3'>5' exonuclease removes them from the 3´hydroxyl terminus.

Exponential growth : Period of sustained growth of a microorganism in which the cell number constantly doubles within a fixed time period.

Express : To translate a gene's message into a molecular product

Expression vector : A DNA cloning vector designed to accomplish the expression of protein from DNA inserts, usually by synthesizing the expressed protein as an extension of a carrier protein encoded by the vector.

Extra cellular : Outside the cell.

Extra chromosomal : In addition to the chromosomes of a cell. Extra chromosomal propagation of DNA molecules is replication of a DNA molecule separate from the host chromosome. This replication may be completely independent of the chromosomal replication apparatus, or it may be strongly dependent on chromosomal replication protein factors. A bacterial plasmid is an example of an extra chromosomal replicative element. Many viruses replicate as extra chromosomal elements.

Exudate : Low molecular weight metabolites that leak from plant roots into soil.

Facultative organism : Organism that can carry out both options of a mutually exclusive process (e.g., aerobic and anaerobic metabolism).

Feedback inhibition : Inhibition by an end product of the biosynthetic pathway involved in its synthesis.

Fermentation : Metabolic process in which organic compounds serve as both electron donors and electron acceptors.

Fertilizer : Any organic or inorganic material of natural or synthetic origin (other than liming materials) added to a soil to supply one or more elements essential to plant growth.

Field capacity : Content of water, on a mass or volume basis, remaining in a soil after being saturated with water and after free drainage is negligible.

Fingerprint : The term "DNA fingerprint" refers to the DNA patterns generated by any of several methods used to detect sequence variations between different species or different individuals of the same species. A PCR fingerprint is the pattern of DNA bands that results when a set of oligonucleotide primers is used to PCR amplify a subset of DNA bands from a genome. The primers may be specific and designed to amplify DNA fragments corresponding to a particular combination of genes, or may be rather nonspecific in that the identity of the amplified DNA fragments remains unknown. The intent of a PCR fingerprint is generally to generate a DNA fragment pattern characteristic of a particular species or individual. RFLP patterns can also be used for DNA fingerprinting.

Flanking region : The DNA sequences extending on either side of a specific locus or gene

Fluorescent : Able to emit light of a certain wavelength when activated by light of a shorter wavelength.

Fluorescent antibody : Antiserum conjugated with a fluorescent dye, such as fluorescein or rhodamine.

Fluxes : Rate of emission, sorption, or deposition of a material from one pool to another. For example, the exchange of methane between the land and the atmosphere is a flux, while the production of methane within the soil is not.

Footprint : The region where a protein interacts with a polynucleotide, generally a doublestranded DNA fragment. The region is detected by the altered access of the DNA molecule to either enzymes, such as the endonuclease dnasei, or to chemical agents, such as methylating compounds.

Free energy : Intrinsic energy contained in a given substance that is available to do work.

Frequency of enzyme cleavage : The statistical frequency or probability of occurrence of a site of N bases in length and containing no ambiguous bases can be estimated by the formula frequency = 1/(4)N. When ambiguous bases are present, such as in the five base sequence aperients, the calculation must to take into account the allowance of either purine in the central position. Frequency = 1/4 X 1/4 X 2/4 X 1/4 X 1/4.

Gamete : A haploid sex cell, egg or sperm, that contains a single copy of each chromosome

Gas chromatography : Chromatographic technique in which the stationary phase is a solid or an immobile liquid and the mobile phase is gaseous. The gaseous samples are separated based on their differential adsorption to the stationary phase.

GC : Because GC base pairing has three hyrogen bonds compared to an AT pair's two bonds, the relative amount of G plus C (GC composition) in a DNA sample often correlates with the relative resistance of doublestranded DNA to denaturation.

Gel : Inert polymer, usually made of agarose or polyacrylamide, that separates macromolecules such as nucleic acids or proteins during electrophoresis.

Gel electrophoresis : The process of using an electric field to force charged macromolecules through a resolving matrix, causes the formation of bands contining molecules with similar mobility.

GEM : A genetically engineered microorganism

Gene : The heritable units of expression that encodes a physical characteristic of an organism and is passed from parents to progeny. A gene is the region of a chromosome where mutations all influence a single phenotypic trait that can be passed from parents to progeny.

Gene amplification : The presence of multiple genes. Amplification is one mechanism through which proto-oncogenes are activated in malignant cells

Gene cloning : Isolation of a desired gene from one organism and its incorporation into a suitable vector for the production of large amounts of the gene.

Gene expression : The process of producing a protein from its DNA- and mRNA-coding sequences

Gene family : A collection of similar genes at different chromosomal loci.

Gene flow : The exchange of genes between different but (usually) related populations

Gene frequency : The percentage of a given allele in a population of organisms.

Gene insertion The addition of one or more copies of a normal gene into a defective chromosome.

Gene library : A collection of recombinants of sufficient number and size of DNA insert to ensure a high probability that all regions of chromosomal DNA from the organism will be present in the collection.

Gene linkage : The hereditary association of genes located on the same chromosome.

Gene modification : The chemical repair of a gene's defective DNA sequence

Gene pool : The totality of all alleles of all genes of all individuals in a particular population.

Gene probe : A strand of nucleic acid which can be labeled and hybridized to a complementary molecule from a mixture of other nucleic acids.

Gene splicing : Combining genes from different organisms into one organism.

Gene translocation : The movement of a gene fragment from one chromosomal location to another, which often alters or abolishes expression

Generation time : Time needed for a population to double in number or biomass.

Genetic assimilation : Eventual extinction of a natural species as massive pollen flow occurs from another related species and the older crop becomes more like the new crop.

Genetic code : The three-letter code that translates nucleic acid sequence into protein sequence. The relationships between the nucleotide base-pair triplets of a messenger RNA molecule and the 20 amino acids that are the building blocks of proteins. See Base pair, Nucleic acid, Nucleotide.

Genetic disease : A disease that has its origin in changes to the genetic material, DNA. Usually refers to diseases that are inherited in a Mendelian fashion, although noninherited forms of cancer also result from DNA mutation.

Genetic drift : Random variation in gene frequency from one generation to another.

Genetic engineering : *In vitro* techniques for the isolation, manipulation, recombination, and expression of DNA.

Genetic information : Information stored in genes and transmissible to an organism's offspring.

Genetic linkage map : A linear map of the relative positions of genes along a chromosome. Distances are established by linkage analysis, which determines the frequency at which two gene loci become separated during chromosomal recombination

Genetic marker : A gene or group of genes used to "mark" or track the action of microbes.

Genetic material : Maintaining a high degree of fidelity when storing genetic information is crucially important to the viability of an organism, since the rate at which the genetic information accumulates damage directly directly influences longterm survival. The primary genetic material of most organisms is DNA; organisms that can tolerate higher mutation rates (including certain viruses) may use RNA.

Genome : Complete set of genes present in an organism.

Genomic equivalent : The amount of DNA necessary to be present in a purified sample to guarantee that all genes will be present. This number increases with the total genome size of an organism and can be calculated by converting the size of a genome in base pairs to micrograms of DNA.

Genomic library : A library composed of fragments of genomic DNA.

Genotype : Precise genetic constitution of an organism.

Genotype : The structure of DNA that determines the expression of a trait

Genus : A category including closely related species. Interbreeding between organisms within the same category can occur.

Genus (plural, genera) : The first name of the scientific name (binomial); the taxon between family and species.

GEO : Genetically engineered organism.

Germ cell : Reproductive cell.

Germ cell (germ line) gene therapy : The repair or re- placement of a defective gene within the gamete-forming tissues, which produces a heritable change in an organism's genetic constitution

GMO : Genetically modified organism

Gram stain : Differential stain that divides bacteria into two groups, Grampositive and Gramnegative, based on the ability to retain crystal violet when decolorized with an organic solvent such as ethanol. The cell wall of Grampositive bacteria consists chiefly of peptidoglycan and lacks the outer membrane of Gramnegative cells.

Gravitational water : Portion of total soil water potential due to differences in elevation.

Green (sulfur) bacteria : Anoxygenic phototrophs containing chlorosomes and bacteriochlorophyll c, c_s, d or e and light harvesting chlorophyll

Growth factor : Organic compound necessary for growth because it is an essential cell component or precursor of such components and cannot be synthesized by the organism itself. Usually required in trace amounts.

Growth rate : The rate at which growth occurs, usually expressed as the generation time.

Haploid : In eukaryotes, an organism or cell containing one chromosome complement and the same number of chromosomes as the gametes.

Heavy metals : Those metals which have densities > 5.0 Mg m^3. These include the metallic elements Cu, Fe, Mn, Mo, Co, Zn, Cd, Hg, Ni, and Pb. Al and Se have densities < 5 but are also considered heavy metals.

Heterochromatin : Dark-stained regions of chromosomes thought to be for the most part genetically inactive

Heterocyst : Differentiated cyanobacterial cell that carries out dinitrogen fixation.

Heteroduplex : A double-stranded DNA molecule or DNA-RNA hybrid, where each strand is of a different origin

Hetero fermentation : Any fermentation in which there is more than one major end product. Synonym of hetero lactic fermentation

Heterogeneous nuclear RNA (hnRNA) : The name originally given to large RNA molecules found in the nucleus, which are now known to be unedited mRNA transcripts, or pre-mRNAs.

Heterokaryon : Hypha that contains at least two genetically dissimilar nuclei.

HGH : Human growth hormone.

High fidelity replication : The production of copies of a DNA molecule with the introduction of very few errors, a necessity for ensuring that progeny receive a full set of the functional genes from the parent organism.

Hind II : Restriction endonuclease purified from bacterial strain Hemophilus inlfuenzae d, recognizing the sequence AAGCTT and cleaving after the first A to generate the fourbase 5' cohesive terminus AGCT.

Histone : Any of a group of small, basic proteins associated with maintaining the structure of chromosomes.

Holomictic : In these lakes there is a periodic (usually once a year in temperate zone lakes) mixing of the zones

Homologous : A term used to describe nucleotide sequences that are very similar in structure or function, often encoding closely related gene products or serving similar regulatory roles during gene expression. Homologous genes produce very similar gene products in different organisms. The term is sometimes used with a more precise definition to mean genes that share a common evolutionary ancestor; saying two genes are "similar" does not imply any judgement about evolutionary relationship.

Homologous chromosomes : Chromosomes that have the same linear arrangement of genes – a pair of matching chromosomes in a diploid organism.

Homologous recombination : The exchange of DNA fragments between two DNA molecules or chromatids of paired chromosomes (during crossing over) at the site of identical nucleotide sequences

Homopolymer tail : When provided with only one of the four nucleoside triphosphates, terminal transferase will add a homopolymer extension of the provided nucleoside triphosphate (such as CCCCCCCCC when provided with only dctp) to a DNA fragment, known as a homopolymer tail. This feature has been used to add complementary cohesive homopolymer termini to allow annealing and joining of two different DNA fragments with otherwise incompatible termini. The polyA tail found in polyadenylated mrna is an example of a naturally occurring homopolymer tail.

Homothallic : Hyphae that are selfcompatible in that sexual reproduction occurs in the same organism by meiosis and genetic recombination; fusion of hypha results in a dikaryon or diploid.

Host : An organism that will serve to maintain and propagate a DNA molecule. The vast majority of routine recombinant DNA work has been accomplished using the bacterium *Escherichia coli* as a host cell.

Host range : The variety of organisms into which a gene can be inserted and retain functionality. Plasmids with a very narrow host range, like cole1 and F factor, will only function in *Escherichia coli* and closely related enteric bacteria, while plasmids with a very broad host range, like rp4, will function in a wide variety of distantly related gram negative bacteria.

Human Genome Project : A project coordinated by the National Institutes of Health (NIH) and the Department of Energy (DOE) to determine the entire nucleotide sequence of the human chromosomes.

Hybrid : The offspring of two parents differing in at least one genetic characteristic (trait). Also, a heteroduplex DNA or DNA-RNA molecule

Hybridization : The hydrogen bonding of complementary DNA and/or RNA sequences to form a duplex molecule

Hybridoma : A hybrid cell, composed of a B Iymphocyte fused to a tumor cell, which grows indefinitely in tissue culture and is selected for the secretion of a specific antibody of interest

Hydrocarbon : Any chemical compound containing only carbon and hydrogen elements. Some simple examples of hydrobarbons are: methane (CH_4), ethylene (C_2H_4), ethane (C_2H_6), etc

Hydrogen bond : A noncovalent chemical interaction in which a hydrogen atom is "shared" between two electron donors.

Hydrogen oxidizing bacterium : Facultative lithotrophs that, in the absence of an oxidizable organic source, oxidize H_2 for energy and synthesize carbohydrates with carbon dioxide as their source of carbon.

Immobilization : Conversion of an element from the inorganic to the organic form in microbial or plant biomass.

Immunity : The ability of a human or animal body to resist infection by microorganisms or their harmful products such as toxins.

Immunoblot : Western blot.

Immunoblot (western blot) : Detection of proteins immobilized on a filter by complementary reaction with specific antibody.

Immunofluorescence : Technique to visualize specific antibodies and any attached homologous antigens by means of conjugating the antibodies to a fluorescent dye.

Immunogen : Substance which is capable of eliciting immune response. An immunogen usually has a fairly high molecular weight (usually greater than 10,000), thus, a variety of macromolecules such as proteins, lipoprotein, polysaccharides, and some nucleic acids can act as immunogens.

Immunoglobulin : Antibody.

***In situ* :** Refers to performing assays or manipulations with intact tissues

***In vitro* :** Literally "in glass"; it describes whatever happens in a test tube or other receptacle, as opposed to *in vivo*. When a study or an experiment is done outside the living organism, in test tube, it is done *in vitro*.

***In vivo* :** In the body, in a living organism, as opposed to *in vitro*; when a study or an experiment is done in the living organism, it is done *in vivo*.

Inducible enzyme : Enzyme synthesized (induced) in response to the presence of an external substance (the inducer).

Infection Infrared (IR) : Growth of an organism within another living organism. The portion of the electromagnetic spectrum with wavelengths from about 0.75 μm to 1 mm.

Inhibition : Prevention of growth or function.

Initiation codon : The mRNA sequence AUG, coding for methionine, which initiates translation of mRNA.

Inoculate : To treat with microorganisms for the purpose of creating a favorable response. For example, treatment of legume seeds with rhizobia to stimulate N_2 fixation.

Inoculum : Material used to introduce a microorganism into a suitable situation for growth.

Insertion : Genetic mutation in which one or more nucleotides are added to DNA.

Insertion mutations : Changes in the base sequence of a DNA molecule resulting from the random integration of DNA from another source

Insertion sequence (IS element) : Simplest type of transposable element. Has an only gene involved in transposition.

Insertional inactivation : The inactivation of a phenotype by insertion of a DNA fragment

Instability : Some DNA fragments tend to undergo recombination processes when cloned and transferred from their native host to a different organism. Sensitivity to host recombination processes, production of detrimental gene products, difficulty in replication of certain DNA sequences, and a variety of other factors can contribute to DNA instability in a new host.

Integration : Process by which a DNA molecule becomes incorporated into another genome.

Intergenic regions : DNA sequences located between genes that comprise a large percentage of the human genome with no known function.

Interspecies hydrogen transfer : The process in which organic matter is degraded anaerobically by the interaction of several groups of microorganisms in which hydrogen production and hydrogen consumption are closely coupled among species.

Intracellular : Inside the cell.

Introgression : Backcrossing of hybrids of two plant populations to introduce new genes into a wild population.

Intron : A noncoding DNA sequence within a gene that is initially transcribed into messenger RNA but is later snipped out. See Coding, DNA, Messenger RNA, Transcription

Ions : Atoms, groups of atoms, or compounds, which are electrically charged as a result of the loss of electrons (cations) or the gain of electrons (anions).

Isolation : Any procedure, in which an organism presents in a particular sample or environment, is obtained in pure culture.

Isotope : Different form of the same element containing the same number of protons and electrons, but differing in the number of neutrons.

Jaccard coefficient (S_J) : An association coefficient used in numerical taxonomy; it is the proportion of characters that match, excluding those that both organisms lack.

Joining (J) segment : A small DNA segment that links genes to yield a functional gene encoding an immunogobulin

Kanamycin : An antibiotic of the aminoglycoside family that poisons translation by binding to the ribosomes

kanr : Kanamycin resistance gene

Karyogamy : Fusion in a cell of haploid (N) nuclei to form a diploid (2N).

Karyotype : All of the chromosomes in a cell or an individual organism, visible through a microsope during cell division

Koch's postulates : Set of laws formulated by Robert Koch to prove that an organism is the causal agent of disease.

Lag phase : Period after inoculation of fresh growth medium during which population numbers do not increase.

Lambda : A naturally occurring double stranded Bacteriophage with a linear genome of about 50,000 base pairs. This bacteriophage has been manipulated in vitro for the construction of many specialized cloning vectors that take advantage of the viral DNA packaging system to selectively clone specific size classes of DNA fragments or to facilitate screening for desired recombinants.

Ligand : Molecule, ion, or group bound to the central atom in a chelate or a coordination compound.

Ligase (DNA ligase) : An enzyme that catalyzes a condensation reaction that links two DNA molecules via the formation of a phosphodiester bond between the 3' hydroxyl and 5' phosphate of adjacent nucleotides

Ligate : The process of joining two or more DNA fragments

Ligation : A nicksealing reaction that requires ATP as an energy source. This reaction is mediated by the enzyme ligase, commonly the ligase from the bacetiophage T4 (T4 DNA ligase).

Light compensation point : where the rate of photosynthesis is lower than the rate of respiration usually about 1% of the light intensity of sunlight

Lime, agricultural : Soil amendment containing calcium carbonate, magnesium carbonate or other materials to neutralize soil acidity and furnish calcium or magnesium or both for plant growth.

Lineage : A chart that traces the flow of genetic information from generation to generation

Linkage : The frequency of coinheritance of a pair of genes and/or genetic markers, which provides a measure of their physical proximity to one another on a chromosome.

Linked genes/markers : Genes and/or markers those are so closely associated on the chromosome that they are co inherited in 80% or more of cases.

Linker : A short, double-stranded oligonucleotide containing a restriction endonuclease recognition site, which is ligated to the ends of a DNA fragment.

Litter : Surface layer of the forest floor consisting of freshly fallen leaves, needles, twigs, stems, bark, and fruits.

Locus : A chromosomal site involved in the production of a specific gene product. Loci are generally defined by analysis of the location of mutations.

Locus (plural = loci) : A specific location or site on a chromosome

Luminescence : Production of light.

Luxury uptake : The absorption by plants of nutrients in excess of their need for growth. Luxury contents accumulated during early growth may be used for later growth.

Lysis : Rupture of a cell, resulting in loss of cell contents.

Lysogeny : An association where a prokaryote contains a prophage and the virus genome is replicated in synchrony with the host chromosome.

M13 : A naturally occurring bacteriophage that contains a singlestranded circular genome of about 6400 bases that infects only enteric bacteria containing the F factor or producing the F pilus protein. The infecting virus is converted to a double stranded circular replicative form that can be purified and digested with restriction enzymes similar to bacterial plasmids. This virus has been manipulated in vitro for the construction of specialized cloning vectors that exist in a double stranded form in the bacterial cell bit produce an infectious bacteriophage particle containing a single stranded circular form, an excellent template for nucleotide sequence analysis.

Macronutrient : A substance required in large amounts for growth, usually attaining a concentration of > 500 mg kg^1 in mature plants. Usually refers to N, P, K, Ca, Mg, and S.

Magnetosome : Small particle of Fe_3O_4 present in cells that exhibit magnetotaxis.

Magnetotactic bacteria : Bacteria that can orient themselves in the earth's magnetic field due to the presence of magnetosomes.

Mapping : Determining the physical location of a gene or genetic marker on a chromosome

MaxamGilbert : Inventors of a method of nucleotide sequence determination that is dependent on the sensitivity of methylated DNA nucleotides to chemical cleavage by piperidine. With this approach, an endlabeled DNA fragment is subjected to partial chemical methylation under four or more conditions to preferentially methylate A, C, G, and T, followed by chemical cleavage of the methylated bases with piperidine. Denaturing polyacrylamide gel electrophoresis of the cleavage products resolves the DNA fragments in order of increasing size to generate an endlabeled nested set of DNA fragments that reveals the nucleotide sequence of the labeled DNA fragment.

Medium (plural, media) : Any liquid or solid material prepared for the growth, maintenance, or storage of microorganisms.

Megabase cloning : The cloning of very large DNA fragments.

Megaprimer : A term used to describe large polynucleotides, often themselves the initial product of a PCR reaction, that serve as primers for DNA synthesis, typically during the later stages of a multistep PCR reaction designed to assemble a PCR DNA fragment of fairly large size from smaller DNA fragments.

Meiosis : In eukaryotes, reduction division, the process by which the change from diploid to haploid occurs.

Messenger RNA (mRNA) : The class of RNA molecules that copies the genetic information from DNA, in the nucleus, and carries it to ribosomes, in the cytoplasm, where it is translated into protein

Methylation : DNA in a cell can be subject to the action of various enzymes that modify DNA bases, most often by the addition of a methyl group to the base. This modification may be very specific, such as the sequence specific methylation of restriction endonuclease cleavage sites by the corresponding modification methylase, or quite general, such as a routine methylation of cytosine by a particular organism. Treatment with a chemical methylating agent like dimethyl sulfate under different ionic conditions can cause the basespecific methylation of the different bases in DNA.

Microbiology : Study of microorganisms.

Microcosm : A community or other unit that is representative of a larger unity.

Microenvironment : Immediate physical and chemical surroundings of a microorganism.

Microfauna : Protozoa, nematodes and arthropods generally < 200 micrometers long.

Microhabitat : Clusters of micro aggregates with associated water within which microbes function. May be composed of several microsites (e.g., aerobic and anaerobic).

Microinjection : A means to introduce a solution of DNA, protein, or other soluble material into a cell using a fine microcapillary pipet.

Micrometer : Onemillionth of a meter, or 10^6 meter, the unit usually used for measuring microorganisms.

Micronutrient : Chemical element necessary for growth found in small amounts, usually < 100 mg kg^1 in a plant. These elements consist of B, Cl, Cu, Fe, Mn, Mo, and Zn.

Microorganism (microbe) : Living organism too small to be seen with the naked eye (< 0.1 mm); includes bacteria, fungi, protozoans, microscopic algae, and viruses.

Micropore : Relatively small soil pore, generally found within structural aggregates and having a diameter < 30 micrometers.

Microsequence variation : The slight differences in nucleotide sequence observed when the same chromosomal region is compared between two or more individuals, particularly when the individuals are of the same species.

Miniprep : Rapid preparation of a small amount of partially purified DNA from a culture of cells, usually sufficient for a few

Mitochondrion (plural, mitochondria) : Eukaryotic organelle responsible for processes of respiration and oxidative phosphorylation.

Mitosis : Highly ordered process by which the nucleus divides in eukaryotes.

Mixotroph : Organism able to assimilate organic compounds as carbon sources while using inorganic compounds as electron donors. Compare with autotroph and heterotroph.

Modification methylase : An enzyme that recognizes a restriction endonuclease cleavage site and adds a methyl group to one or more of the bases within the site to prevent cleavage by the restriction enzyme.

Molecular biology : The study of the biochemical and mo- lecular interactions within living cells.

Molecular cloning : The biological amplification of a specific DNA sequence through mitotic division of a host cell into which it has been transformed or transfected

Molecular genetics : The study of the flow and regulation of genetic information between DNA, RNA, and protein molecules.

Molecule : Result of two or more atoms combining by chemical bonding.

Motility : Movement of a cell under its own power.

mRNA : Messenger RNA, the RNA molecule that is used by the ribosome as a template for the assembly of protein.

Mucigel : Gelatinous material at the surface of roots grown in normal nonsterile soil. It includes natural and modified plant mucilages, bacterial cells, and their metabolic products (e.g., capsules and slimes), and colloidal mineral and organic matter from the soil.

Multi-locus probe : probe that hybridizes to a number of different sites in the genome of an organism.

Multiple cloning site : A genetically engineered collection of several different restriction enzyme cleavage sites that allow convenient insertion of DNA fragments into a cloning vector.

Mutagen : Substance that causes the mutation of genes.

Mutant : Organism, population, gene, or chromosome that differs from the corresponding wild type by one or more base pairs.

Mutation : Heritable change in the base sequence of the DNA of an organism.

Mycovirus : Virus that infects fungi.

Nanopore : Soil pore having dimensions measured in nanometers. Materials encased in nano pores are beyond the reach of microorganisms and enzymes.

NAPL : Nonaqueous phase liquid. This can be lighter than water (LNAPL), or more dense than water (DNAPL).

Nested primers : Two oligonucleotide primers of different, usually nonoverlapping, sequences that anneal to sites close together and in the same orientation on a particular DNA target sequence. When used in two successive PCR reactions, these primers can greatly enhance the specificity of the final amplification product.

Nested set : A collection of DNA or RNA fragments of different sizes derived from a larger polynucleotide, with all fragments sharing one common end. Of great importance to nucleotide sequence determination, the nested set in the MaxamGilbert method is defined by the endlabel used to visualize the products of the chemical modification and cleavage reactions and in the Sanger method by the primer used for the enzymatic elongation and chaintermination reactions.

Nick translation : A procedure for making a DNA probe in which a DNA fragment is treated with DNase to produce single-stranded nicks, followed by incorporation of radioactive nucleotides from the nicked sites by DNA polymerase I.

Nitrocellulose : A membrane used to immobilize DNA, RNA, or protein, which can then be probed with a labeled sequence or antibody

Nomenclature : System of naming organisms.

Nonphosphorylated termini : Although most DNA contains a phosphate residue at the 5′ terminus, the enzyme phosphorylase can remove that terminal phosphate, leaving a 5' hydroxyl residue. This reaction is frequently employed to block ligation, since the enzyme DNA ligase is unable to ligate a non phosphorylated 5' terminus.

Nonpolar : Possessing hydrophobic (water repelling) characteristics and not easily dissolved in water.

Northern blot : The process in which the RNA bands present in a gel are transferred to a membrane and incubated with a hybridization probe to reveal the location of RNA bands complementary to the probe.

Northern hybridization : A procedure in which RNA fragments are transferred from an agarose gel to a nitrocellulose filter, where the RNA is then hybridized to a radioactive probe.

Nuclease : A class of enzymes that degrades DNA and/or RNA molecules by cleaving the phosphodiester bonds that link adjacent nucleotides. In deoxyribonuclease (DNase), the substrate is DNA. In endonuclease, it cleaves at internal sites in the substrate molecule. Exonuclease progressively cleaves from the end of the substrate molecule. In ribonuclease (RNase), the substrate is RNA. In the S1 nuclease, the substrate is single-stranded DNA or RNA

Nucleic acids : The two nucleic acids, deoxyribonucleic acid (DNA) and ribonucleic acid (RNA), are made up of long chains of molecules called nucleotides. See DNA, RNA, Nucleotides

Nuclein : The term used by Friedrich Miescher to describe the nuclear material he discovered in 1869, which today is known as DNA.

Nucleoid : Aggregated mass of DNA that makes up the chromosome of prokaryotic cells.

Nucleophilic compound : Chemical that attracks or is drawn to electrondeficient regions in other chemicals; reducing agents act as nucleophilic compounds.

Nucleoside : An Nglycosyl derivative of a heterocyclic base, typically deoxyribose attached to adenine (deoxyadenosine, da), cytosine (deoxycytidine, dc), guanine (deoxyguanosine, dg), or thymine (thymidine, T or dt) in DNA and ribose attached to adenine (adenosine, A), cytosine (cytidine, C), guanine (guanosine, G), or uracil (uridine, U).

Nucleoside : A building block of DNA and RNA, consisting of a nitrogenous base linked to a five carbon sugar.

Nucleoside analog : A synthetic molecule that resembles a naturally occuring nucleoside, but that lacks a bond site needed to link it to an adjacent nucleotide.

Nucleotide : A building block of DNA and RNA, consisting of a nitrogenous base, a five-carbon sugar, and a phosphate group. Together, the nucleotides form codons, which when strung together form genes, which in turn link to form chromosomes

Nucleotide sequence map : A map that shows the position of restriction sites and important genetic features based on analysis of the nucleotide sequence of a DNA fragment.

Nucleus : Membrane enclosed structure containing the genetic material (DNA) organized in chromosomes.

Nutrient : Substance taken by a cell from its environment and used in catabolic or anabolic reactions.

Obligate : (i) Adjective referring to an environmental factor (for example, oxygen) that is always required for growth. (ii) Organism that can grow and reproduce only by obtaining carbon and other nutrients from a living host, such as obligate symbiont.

Offspring : Progeny. Genetic inheritance is the transmission of genetic information, and thus passage of genetically determined traits, to offspring. It is crucial to the viability of a species that the offspring receive functional copies of all of the genetic information of the parent organism. Processes of cell division and sexual reproduction must function in a manner that ensures an adequate degree of integrity of genetic information transmitted to progeny.

Oligonucleotide : A short, usually chemically synthesized polynucleotide. An oligonucleotide is typically 10 to 30 bases in size, but may be over 100 bases long. Synthetic oligonucleotides are frequently used as primers in sanger dideoxy sequencing or pcr, or as probes to identify a gene sequence of interest.

Open circular DNA : A double stranded circular DNA molecule that has been nicked in one of the strands to allow the release of any superhelical turns present in the molecule. The open circular form migrates more slowly during gel electrophoresis than a covalently closed circular, or supercoiled, molecule of the same size due to the associated differences in conformation, or shape, of the molecules.

Open reading frame : A long DNA sequence that is unin- terrupted by a stop codon and encodes part or all of a protein

Operator : A prokaryotic regulatory element that interacts with a repressor to control the transcription of adjacent structural genes

Operon : Cluster of genes whose expression is controlled by a single operator; typical in prokaryotic cells.

Overlapping DNA termini : The ends of two DNA molecules which are complementary and capable of hybridization to one another. Such structures include the "sticky ends" left by compatible restriction enzymes. See cohesive terminus.

Overlapping reading frames : Start codons in different reading frames generate different polypeptides from the same DNA sequence.

Paleontology : The study of the fossil record of past geo- logical periods and of the phylogenetic relationships between ancient and contemporary plant and animal species

Palindromic sequence : A DNA locus whose 5'-to-3' sequence is identical on each DNA strand. The sequence is the same when one strand is read left to right and the other strand is read right to left. Recognition sites of many restriction enzymes are palindromic. See DNA

pAMP : Ampicillin-resistant plasmid developed for this laboratory course

Particle size : Effective diameter of a particle measured by sedimentation, sieving or micrometric methods.

pBR322 : A derivation of ColE1, one of the first plasmid vectors widely used

PCR : See polymerase chain reaction.

Pedigree : A diagram mapping the genetic history of a par- ticular family

Pellicle : Relatively rigid layer of proteinaceous elements just beneath the cell membrane in many protozoa and algae.

Peptidoglycan : Rigid layer cell walls of bacteria, a thin sheet composed of N acetylglucosamine, N acetylmuramic acid, and a few amino acids. Also called murein.

Persistence : Ability of an organism to remain in a particular setting for a period of time after it is introduced

PH : Negative logarithm of the hydrogen ion activity. The degree of acidity (or alkalinity) of a soil as determined by means of a glass or other suitable electrode or indicator at a specified moisture content or soilwater ratio, and expressed in terms of the pH scale.

Phagotrophic : Form of feeding where animals, such as protozoans, engulf particulate nutrients, such as bacterial cells or detritus.

Phenotype : Observable properties of an organism.

Phenotypic screening : Examination of the physical characteristics or properties of collection of recombinant transformants in the attempt to find a transformant cell containing a recombinant molecule that endcodes a specifc phenotypic trait.

Phosphatase : An enzyme that hydrolyzes esters of phosphoric acid, removing a phosphate group

Phosphobacterium : Bacterium that is especially good at solubilizing the insoluble inorganic phosphate in soil.

Phosphodiester bond : Type of covalent bond linking nucleotides together in a polynucleotide.

Phosphorylated sugar backbone : The fundamental chemical composition of a nucleic acid is a polymer backbone composed of the structure sugaroxygen-phosphateoxygensugar where deoxyribose is the sugar present in DNA and ribose the sugar present in RNA. The sugar molecules are modified by the presence of a purine or pyrimidine base attached to the first, or 1', carbon and the oxygenphosphorusoxygen bond links the third, or 3', carbon to the fifth, or 5', carbon of the adjacent sugar molecule.

Phosphorylated termini : Nucleic acid ends that have an attached phosphate group, most commonly existing in the form of a 5´phosphate.

Phototaxis : Movement toward light.

Phylogeny : Ordering of species into higher taxa and the construction of evolutionary trees based on evolutionary (genetic) relationships.

Physical map : A map showing physical locations on a DNA molecule, such as restriction sites, and sequence-tagged sites

Piperidine : Incubation of methylated DNA with the chemical piperidine causes elimination of the methylated base and cleavage of the DNA strand at that point. This reaction is important to MaxamGilbert sequence analysis and to chemical footprinting studies that examine the interaction of DNA binding proteins with a target DNA sequence.

Plant growthpromoting rhizobacteria (PGPR) : Broad group of soil bacteria that exert beneficial effects on plant growth usually as root colonizers. Many members of the genus *Pseudomonas.*

Plaque : Localized area of lysis or cell inhibition caused by virus infection on a lawn of cells.

Plaque lift : A procedure in which an imprint of the plaques that appear in a surface layer of phagesensitive indicator bacteria grown on the surface of a solid nutrient medium is transferred to a membrane, allowing the DNA or protein to bind to the membrane. Following appropriate treatment, the membrane can be exposed to either a hybridization probe for the identification of colonies containing a specific nucleic acid sequence or to an antibody for the identification of colonies containing a specific protein.

Plaques of infected cells : The zones of either lysed cells, as caused by bacteriophage lambda, or inhibited cell growth, as caused by bacteriophage M13, that occur in the surface of an indicator lawn of bacteriophagesensitive cells in the presence of the bacteriophage particles released by infected cells.

Plasmid : Covalently closed, circular piece of DNA which, as an extrachromosomal genetic element, is not essential for growth.

Plasmid vector : A molecular cloning vector made from a plasmid.

Plasmogamy : Fusion of the contents of two cells, including cytoplasm and nuclei.

Point mutation : A change in a single base pair of a DNA sequence in a gene

Polar : Possessing hydrophilic characteristics and generally water soluble.

Polar flagellation : Condition of having flagella attached at one end or both ends of the cell.

Poly(A) polymerase : Catalyzes the addition of adenine residues to the 3' end of pre-mRNAs to form the poly(A) tail.

polyA tail : A stretch of adenosine residues that is posttranscriptionally attached to the 3' terminus of many eukaryotic mrna molecules during the maturation process.

Polyacrylamide : A mixture of acrylamide monomer and bisacrylamide polymerized together by the addition of a catalyst to form a solid matrix used for gel electrophoresis of macromolecules.

Polyacrylamide gel electrophoresis : Electrophoresis through a matrix composed of a synthetic polymer, used to separate proteins, small DNA, or RNA molecules of up to 1000 nucleotides. Used in DNA sequencing

Polybetahydroxy-butyrate (PHB) : Common storage material of prokaryotic cells consisting of betahydroxybutyrate or other betaalkanoic acids.

Polylinker : A short DNA sequence containing several re- striction enzyme recognition sites that is contained in cloning vectors

Polymerase (DNA) : Synthesizes a double-stranded DNA molecule using a primer and DNA as a template.

polymerase chain reaction (PCR) : A procedure that en- zymatically amplifies a DNA polymerase

Polymerase chain reaction (PCR) : Method for amplifying DNA *in vitro*, involving the use of oligonucleotide primers complementary to nucleotide sequences in target genes and the copying of the target sequences by the action of DNA polymerase.

Polymorphisms : Variant forms of a particular gene that occur simultaneously in a population

Polyploid : Having multiple copies of a chromosome.

Polysome : Strings of ribosomes attached by strands of mRNA.

Pore space : Portion of soil bulk volume occupied by soil pores.

Porin : A protein channel in the lipopolysaccharide layer of Gramnegative bacteria.

Primer : A short DNA or RNA fragment annealed to single-stranded DNA, from which DNA polymerase extends a new DNA strand to produce a duplex molecule

Primer dependent : Requiring the presence of an oligonucleotide to serve as an initiation point for a particular reaction, typically DNA synthesis. Primer dependent synthesis of DNA on an RNA template is mediated by the enzyme reverse transcriptase, and enzyme normally associated with the conversion of the RNA genome of a retrovirus to a DNA copy during the viral life cycle.

Primer independent : Not requiring the presence of an oligonucleotide to serve as an initiation point for a particular reaction, such as the process of RNA synthesis. Primer independent synthesis of DNA has not been routinely demonstrated with the DNA polymerizing enzymes that have been examined.

Probe : A sequence of DNA or RNA, labeled or marked with a radioactive isotope, used to detect the presence of complementary nucleotide sequences

Processing : Before mRNA processing, a newly synthesized RNA molecule is a complementary copy of the DNA strand that served as the template for RNA synthesis. Processing or maturation of the RNA molecule to its final form can involve endo or exonucleolytic cleavage or cleavage/resealing steps (splicing), that remove specific regions of the RNA molecule before the RNA becomes biologically functional. Since a cDNA molecule is generally synthesized using an mRNA molecule as a template, the cDNA sequence corresponds to that of the mature, or fully processed RNA molecule. Regions such as the extreme 5′ terminus and initially transcribed intervening sequences are generally absent from an isolated cDNA molecule. Certain characteristic processing site sequence features, such as the junctions associated with removal of intervening sequences, may be retained in a cDNA molecule, but the biological significance of such features must be confirmed by comparison with the genomic DNA sequence.

Prokaryote : Organism lacking a unit membranebound nucleus and other organelles, usually having its DNA in a single circular molecule.

Promoter : Site on DNA where the RNA polymerase binds and begins transcription.

Pronucleus : Either of the two haploid gamete nuclei just prior to their fusion in the fertilized ovum

Propagule : Cell unit capable of developing into a complete organism.

Protein assembly machinery of a cell : The ribosome.

Protein complex : A collection of proteins that aggregate or associate with one another to form the enzymatically or structurally active form.

Protein contaminant : Proteins that are inadvertently purified along with DNA or RNA and often interfere with subsequent nucleic acid manipulation and analysis.

Protoplast : Cell from which the wall has been removed.

pUC : A widely used expression plasmid containing a -galactosidase gene

Pure culture : Population of microorganisms composed of a single strain. Such cultures are obtained through selective laboratory procedures and are rarely found in a natural environment.

Purine : The chemical structure from which adenine and guanine are derived.

Pyrimidine : The chemical structure from which cytosine, thymine, and uracil are derived.

Q_{10} : Relative increase in a reaction rate with temperature. It is expressed as the increase over a 10°C interval.

Radioimmunoassay : An immunological assay employing radioactive antibody or antigen for the detection of certain substances in body fluids.

Radioisotope : An isotope of an element that undergoes spontaneous decay with the release of radioactive particles

Reading frame : A series of triplet codons beginning from a specific nucleotide. Depending on where one begins, each DNA strand contains three different reading frames

Reannealing : Process where two complementary single strands of DNA automatically hybridize back into a single, doublestranded molecule upon cooling.

Recalcitrant : Resistant to microbial attack.

Recognition sequence (site) : A nucleotide sequence – composed typically of 4, 6, or 8 nucleotides – that is recognized by a restriction endonuclease. Type II enzyrnes cut (and their corresponding modification enzymes methylate) within or very near the recognition sequence

Recombinant : A cell that results from recombination of genes

Recombinant DNA : The collection of methods that allow the isolation, specific restriction enzyme fragmentation, rearrangement into new combinations by ligation, and propagation of nucleic acids; also, a DNA molecule generated by application of this methodology.

Recombination : Process by which genetic elements in two separate genomes are brought together in one unit.

Recombination frequency : The frequency at which crossing over occurs between two chromosomal loci – the probability that two loci will become unlinked during meiosis.

Regulatory gene : A gene whose protein controls the activity of other genes or metabolic pathways.

Regulatory region : Although classically defined in terms of the presence of mutations that affect regulation of gene expression, this is now considered to be the region of a gene where RNA polymerase and other accessory transcription modulator proteins bind and interact to control RNA synthesis. Although the promoter is an integral part of the regulatory region, this region may also contain binding sites for proteins that function in either a positive or a negative modulating fashion and various nucleotide sequence features, such as attenuators, may contribute to regulation of transcription.

Relaxed plasmid : A plasmid that replicates independently of the main bacterial chromosome and is present in 10-500 copies per cell.

Renaturation : The process of two singlestranded nucleic acid molecules annealing or base pairing to form a doublestranded molecule.

Repetitive sequence element : Any fairly short nucleotide sequences that is repeated many times, often in tandem arrays, in a genome. Atatatatatatatatatatat and gatgatgatgatgatgatgat are examples of simple repetitive sequence elements.

Replacement vector : A DNA cloning vector that takes advantage of the DNA packaging mechanism of bacteriophage lambda to replace a portion of the vector DNA with the desired insert DNA. This approach optimizes the generation of recombinant bacteriophage that contain DNA inserts larger than routinely possible with plasmid cloning vectors.

Replicase : A nucleic acidsynthesizing enzyme involved in the reproduction of a nucleic acid genome. This term is often applied to the fairly small, specialized polymerases encoded by viral genomes.

Replication : Conversion of one doublestranded DNA molecule into two identical doublestranded DNA molecules.

Replication origin : A DNA sequence, generally fairly hostspecific, whose presence allows the initiation of DNA replication in a host cell.

Replicon : A chromosomal region containing the DNA sequences necessary to initiate DNA replication processes

Repression : Process by which the synthesis of an enzyme is inhibited by the presence of an external substance (the repressor).

Repressor : A DNA-binding protein in prokaryotes that blocks gene transcription by binding to the operator

Restriction endonuclease : Enzyme that recognizes and cleaves specific DNA sequence, generating either blunt or singlestranded (sticky) ends.

Restriction fragment length polymorphism (RFLP) : Method to identify differences between similar genes from different organisms. Digestion of genes with restriction endonucleases followed by separation of the resulting fragments by gel electrophoresis yields banding patterns that are characteristic of the individual gene.

Restriction map : A chart or listing of relative positioning of restriction sites within a DNA molecule.

Restriction site : The nucleotide sequence at which a restriction endonuclease binds and cleaves DNA. More formally called a restriction endonuclease recognition site.

Reverse transcriptase : An RNAdependant DNA polymerase. Reverse transcription is found in retroviruses, whose genome is composed of RNA, but goes through a phase of its life cycle in which its sequence information if converted to DNA. In vitro, reverse transcriptase is commonly used for cdna synthesis.

Reverse transcription : The process of generating a DNA copy from an RNA template. An enzyme capable of mediating this reaction is a reverse transcriptase.

Rho subunit : The subunit of bacterial RNA polymerase that appears to be involved in termination of transcription at a class of sequences known as rhodependent terminators.

Ribonuclease : A enzyme capable of cleaving or degrading RNA.

Ribonucleic acid : (RNA) A nucleic acid using ribose as the sugar moiety in the backbone chain. Some major uses of RNA in the cell are as constituents of the ribosome (rrna), as transfer molecules temporarily linked to amino acids during protein synthesis (trna), and as messenger molecules transferring genetic information from the chromosome to the ribosome (mrna).

Ribonucleic acid (RNA) : Polymer of nucleotides connected via a phosphateribose backbone, involved in protein synthesis.

Ribonucleoside triphosphate : (NTP) any of the four RNA precursor compounds ATP, CTP, GTP, and UTP.

Ribosomal RNA (r RNA) : Types of RNA found in the ribosome; some participate actively in the process of protein synthesis.

Ribosome : A subcellular component which catalyzes translation of RNA sequences into proteins. Ribosomes are composed of proteins and ribosomal RNA (rrna).

RNA : Ribonucleic acid.

RNA degradation : The disassembly of a polyribonucleotide by either chemical or enzymatic mechanisms.

RNA polymerase : An enzyme or enzyme complex capable of synthesizing RNA.

RNA polymerization : The assembly of ribonucleoside triphosphates into a polyribonucleotide.

RNA template : An RNA molecule that serves as the annealing site for an oligonucleotide primer and directs the assembly of a complmentary polynucleotide, such as an mrna template primed and used as a template for the synthesis of cdna by reverse transcriptase.

RNADNA hybrid : A doublestranded molecule in which a strand of DNA has annealed with or base paired with a complementary RNA molecule.

RNAse H : Protein capable of degrading the RNA from an RNA

S1 nuclease : An enzyme capable of degrading singlestranded but not doublestranded nucleic acids.

Sadenosylmethionine : The chemical compound used as a cofactor by many Type I restriction endonucleases and other DNA methylating enzymes.

salt : The ability of two complementary DNA strands to remain associated with each other as a double stranded molecule is dependent on several factors, including the amount of salt in the solution, the ph of the solution, the temperature of the solution, the size of the complementary region, and the percent GC base pairing in the complementary region. Decreasing the concentration of salt in a DNA solution tends to destabilize the base pairing interactions, forcing a doublestranded molecule towards the singlestranded state.

SAM : Sadenosylmethionine.

Sanger : Inventor of the method of determination of nucleotide sequence by the enzymatic chain termination. See dideoxynucleotide sequencing.

Satellite RNA (viroids) : A small, self-splicing RNA molecule that accompanies several plant viruses, including tobacco ring spot virus.

Satellite sequence : Any DNA sequence that is repeated at such high copy number in the genome of an organism that it will form a band isolated from the bulk of the chromosomal DNA during equilibrium buoyant density gradient centrifugation.

Screening : The process of searching through a library or other collection of recombinant transformants in the attempt to identify a clone containing a recombinant vector containing at least a portion of the gene recognized by the probe.

Selectable marker : A gene whose expression allows one to identify cells that have been transforrned or transfected with a vector containing the marker gene.

Selective marker : A gene that encodes a product whose expression allows detection of the presence of the selective marker in a host cell. Resistance to antibiotics, growth in the presence of a specific nutrient, and the formation of plaques of infected cells on a sensitive lawn of host cells are all selective markers that have been employed in the construction of cloning vectors.

Selective medium : Medium that allows the growth of certain types of microorganisms in preference to others. For example, an antibiotic containing medium allows the growth of only those microorganisms resistant to the antibiotic.

Sequence : The order of monomeric units in biological polymer (typically DNA or protein). In DNA and RNA molecules, genetic information is contained in the order of the nucleotide bases attached to the sugar phosphate backbone of the molecule. The order of these bases is consequently probably the single most important biological aspect of DNA structure. The sequence of amino acids in a protein is sometimes referred to as "primary structure".

Sequence identity : A term that has been used in reference to the degree of similarity between two or more nucleotide sequences, generally in the context of "percentage of nucleotide sequence identity".

Sequencespecific cleavage : Introduction of a nick in the sugar phosphate backbone of RNA or DNA only where a particular nucleic acid sequence is present.

Sequence-tagged site (STS) : A unique (single-copy) DNA sequence used as a mapping landmark on a chromosome.

Serial dilution : Series of stepwise dilutions (usually in sterile water) performed to reduce the populations of microorganisms in a sample to manageable numbers.

Shuttle vector : A cloning vector that contains more than one origin of replication, allowing function of the vector in different host cells. Most shuttle vectors contain an origin and selectable marker that allow routine propagation in *Escherichia coli* and an origin and selectable marker that allow transfer to the organism for which the DNA inserts are to be cloned, thereby allowing examination of the function of cloned DNA fragments in the original host environment.

Siderochromes : Compounds produced by microorganisms that are involved with the uptake of iron by those microorganisms (see siderchromes)

Siderophore : Nonporphyrin metabolite secreted by certain microorganisms that forms a highly stable coordination compound (chelate) with iron; a highaffinity ironbinding compound. There are two major types: catecholates and hydroxamates.

Sigma subunit : The subunit of the RNA polymerase complex that is responsible for conferring recognition specificity to the initiation of RNA synthesis on a DNA template.

Silver stain : A process in which silver salts are used to stain and visualize various biological macromolecules, including nucleic acids and proteins.

Single stranded : A polynucleotide molecule or portion of a polynucleotide molecule that is not annealed to a complementary polynucleotide molecule.

Site : (i) In ecology, area described or defined by its biotic, climatic, and soil conditions as related to its capacity to produce vegetation. (ii) Area sufficiently uniform in biotic, climatic, and soil conditions to produce a particular climax vegetation.

Sitedirected mutagenesis : Insertion of a different nucleotide at a specific site in a molecule using recombinant DNA methodology.

Sitespecific ribonucleases : Proteins that recognize a specific nucleotide sequence and introduce a nick in the sugar phosphate backbone of RNA.

Small nuclear RNA (snRNA) : Short RNA transcripts of 100-300 bp that associate with proteins to form small nuclear ribonucleo protein particles (snRNPs), which participate in RNA processing

Soil biochemistry : Branch of soil science concerned with enzymes and the reactions, activities, and products of soil microorganisms.

Soil extract : Solution separated from a soil suspension or from a soil by filtration, centrifugation, suction or pressure.

Soil microbiology : Branch of soil science concerned with soilinhabiting microorganisms and their functions and activities.

Soil organic matter (SOM) : Organic fraction of the soil exclusive of undecayed plant and animal residues. Often synonymous with humus.

Soil salinity : Amount of soluble salts in a soil. The conventional measure of soil salinity is the electrical conductivity of a saturation extract.

Soil science : Science dealing with soils as a natural resource on the surface of the earth including soil formation, classification and mapping, and physical, chemical, biological, and fertility properties of soils per se; and these properties in relation to their use and management.

Soil solution : Aqueous liquid phase of the soil and its solutes.

Southern hybridization (Southern blotting) : A procedure in which DNA restriction fragments are transferred from an agarose gel to a nitrocellulose filter, where the denatured DNA is then hybridized to a radioactive probe (blotting).

Spatial variability : Variation in soil properties (i) laterally across the landscape, at a given depth, or with a given horizon, or (ii) vertically downward through the soil.

Species : A classification of related organisms that can freely interbreed.

Splicing : The intron removal step of mrna processing. The unprocessed RNA molecule must be cleaved to remove the intron, and resealed (ligated). The endonuclease and ligase activities involved is splicing are performed by ribonucleoprotein complexes. In most cases, introns must be removed must be removed before the mrna becomes biologically functional.

Stationary phase : Period during the growth cycle of a population in which growth rate equals the death rate.

Sterilization : Rendering an object or substance free of viable microbes.

Sticky end : A protruding, single-stranded nucleotide se- quence produced when a restriction endonuclease cleaves off center in its recognition sequence.

Stoichiometric : A chemical term meaning that the molecules referred to are present in constant proportions; for nucleic acids this generally equates to equal numbers of molecules. Complete digestion of a DNA fragment with a restriction enzyme dictates that all resulting smaller DNA fragments derived from the original band should be stoichiometric, or present in the same number, with the relative staining intensity of each band proportional to fragment size following gel electrophoresis. Bands that are not present in stoichiometric levels are indicative of digestion artefacts, such as contaminating DNA fragments, incomplete enzyme activity, or contaminating enzyme activity.

Stringency : Reaction conditions – notably temperature, salt, and pH – that dictate the annealing of single-stranded DNA/DNA, DNA/RNA, and RNA/RNA hybrids. At high stringency, duplexes form only between strands with perfect one-to-one complementarity; lower stringency allows annealing between strands with some degree of mismatch between bases.

Stringent plasmid : A plasmid that only replicates along with the main bacterial chromosome and is present as a single copy, or at most several copies, per cell

Sub cloning : Large DNA fragments inserted into a vector for propagation often contain more nucleotide sequence than necessary for particular uses. Restriction enzyme digestion or PCR are often used to isolate and clone smaller DNA fragments from the original large DNA fragment, a process known as subcloning.

Substrate : (i) Substance, base, or nutrient on which an organism grows. (ii) Compounds or substances that are acted upon by enzymes or catalysts and changed to other compounds in the chemical reaction.

Supercoiled DNA : A DNA molecule, containing superhelical twists. Supercoiled plasmid DNA is generally circular; if the circle is nicked in either or both strands, the twists can relax or unwind.

Supergene : A group of neighboring genes on a chromosome that tend to be inherited together and sometimes are functionally related

Synthesis of DNA : The net assembly of a polydeoxynucleotide. In a biological system, this process generally requires the four precursor deoxynucleoside triphosphates, a dna template with an annealed dna synthesis primer, and the enzyme dna polymerase (which may be reverse transcriptase in some viral systems). DNA synthesis can be done enzymatically in vitro (as in PCR), or chemically (as in oligonucleotide synthesis). Assembly of the DNA chain has been found to always occur only in the 5′ to 3′ orientation, extending the 3′hydroxyl of the primer along the template by adding residues complemetary to the template nucleotide sequence.

Systemic : Not localized in a particular place of the body; an infection disseminated widely through the body is said to be systemic.

T4 : A bacteriophage with a fairly complex genome that causes the synthesis of many virusencoded proteins during the infection of the host bacterium Escherichia coli. Of particular interest due to their use in in vitro DNA manipulation are the enzymes T4 DNA ligase, T4 polynucleotide kinase, and T4 DNA polymerase.

Taq DNA polymerase : The thermostable DNA polymerase purified from the thermophilic bacteria *Thermus acquaticus*, frequently used to form PCR.

TATA box : An adenine- and thymine-rich promoter sequence located 25-30 bp upstream of a gene, which is the binding site of RNA polymerase.

T-DNA (transfer DNA, tumor-DNA) : The transforming region of DNA in the Ti plasmid of *Agrobacterium tumefaciens*

Telomere : The end of a chromosome

Temperate virus : Virus which upon infection of a host does not necessarily cause lysis but whose genome may replicate in synchrony with that of the host.

Temperature of DNA solution : The ability of two complementary DNA strands to remain associated with each other as a double stranded molecule is dependent on several factors, including the amount of salt in the solution, the ph of the solution, the temperature of the solution, the size of the complementary region, and the percent GC base pairing in the complementary region. Increasing temperature destabilizes base pairing interactions, forcing a double stranded molecule towards the single stranded state.

Template : An RNA or single-stranded DNA molecule upon which a complementary nucleotide strand is synthesized

Template independent DNA synthesis : Net synthesis of DNA in the absence of a template. A few enzymes have been found to be capable of generating a random polymer in the absence of a template. The most significant of these enzymes, terminal transferase, will extend the 3´hydroxyl terminus of a DNA molecule by the addition of whatever nucleoside triphosphates are present. When provided with only one of the four nucleoside triphosphates, terminal transferase will add a homopolymer extension (such as CCCCCCCCC when provided with only dctp) to a DNA fragment, known as a homopolymer tail, a feature that has been used to add complementary cohesive homopolymer termini to allow annealing and joining of two different DNA fragments with otherwise incompatible termini.

Terminal deletion : Removal of nucleotides from either the 5´phosphate or the 3´hydroxyl terminal end of a DNA molecule. These deletions can be enzymatically generated with restriction enzymes, endonucleases, exonucleases, and a variety of PCRbased strategies. Such deletions are often constructed to facilitate nucleotide sequence analysis or to investigate the role of the deleted region in some aspect of gene structure or function.

Termination codon : Any of three mRNA sequences (UGA, UAG, UAA) that do not code for an amino acid and thus signal the end of protein synthesis. Also known as stop codon.

Termination of transcription : Just as RNA polymerases require specific signals for initiation of RNA synthesis on a DNA template, these enzymes may require specific signals that end transcription and cause the polymerase to release from the DNA template.

Terminator region : A DNA sequence that signals the end of transcription

Test : Hard external covering or shell.

Tetracycline : A protein synthesis inhibiting antibiotic occasionally used in the selection of bacterial transformants containing a cloning vector encoding resistance to this antibiotic, generally because of the presence of a tetracycline resistance gene derived from a transposable element. Resistance is generally accomplished by blocking entry of tetracycline into the host cell, rather than by inactivation of the antibiotic.

Thermostable DNA polymerase : Any DNA polymerase capable of retaining biological activity at very high temperatures (80100OC).

Ti (tumor-inducing) plasmid : A giant plasmid of *Agrobacterium tumefaciens* that is responsible for tumor formation in infected plants. Ti plasmids are used as vectors to introduce foreign DNA into plant cells.

Toxin : Microbial substance able to induce host damage.

Trace gas : Gas other than nitrogen and oxygen in the atmosphere, particularly those gases that are active in the chemistry or radiation balance of the atmosphere.

Trans capsidation : The partial of full coating of the nucleic acid of one virus with a coat protein of a differing virus.

Transcription : Templatedependant synthesis of RNA by RNA polymerase.The template gene is usually DNA, though there are also RNAdependant RNA polymerases used, for example, by certain viruses. This process produces a RNA molecule complementary to one of the template strands.

Transcription initiation : RNA polymerase begins the synthesis of RNA on a DNA template at a specific nucleotide sequence termed a promoter. The initiation process can be very complex, requiring not only the presence of RNA polymerase but also additional accessory transcription factors that may function in both positive and negative roles in modulating RNA synthesis.

Transduction : Transfer of host genetic information via a virus or bacteriophage particle.

Transfection : The introduction of infectious DNA, such as bacteriophage lambda DNA, into a host cell capable of propagating the infectious DNA, usually by the production and dissemination of phage or virus particles.

Transformant : A cell whose phenotypic properties have been altered, generally by the introduction of new DNA sequences.

Transformant : In prokaryotes, a cell that has been genetically altered through the uptake of foreign DNA. In higher eukaryotes, a cultured cell that has acquired a malignant phenotype.

Transformation : In prokaryotes, the natural or induced uptake and expression of a foreign DNA sequence—typically a recombinant plasmid in experimental systems. In higher eukaryotes, the conversion of cultured cells to a malignant phenotype—typically through infection by a tumor virus or transfection with an oncogene.

Transformation efficiency : The number of bacterial cells that uptake and express plasmid DNA divided by the mass of plasmid used (in transformants/ microgram). (See Transformation.)

Transforming oncogene : A gene that upon transfection converts a previously immortalized cell to the malignant phenotype.

Transgenic : Describes genetically modified plants or animals containing foreign genes inserted by means of recombinant DNA techniques.

Transgenic animal : Genetically enginnered animal or offspring of genetically engineered animals. The transgenic animal usually contains material from at lease one unrelated organism, such as from a virus, plant, or other animal. See Transgenic.

Transgenic plant : Genetically engineered plant or offspring of genetically engineered plants. The transgenic plant usually contains material from at least one unrelated organisms, such as from a virus, animal, or other plant. See Transgenic.

Translation : Protein is assembled by ribosomes that use the linear order of nucleotide bases present on a template messenger RNA molecule to direct the order of addition of amino acid residues to a growing polypeptide chain.

Translocation : The movement or reciprocal exchange of large-chromosomal segments, typically between two different chromosomes.

Transposable element : Genetic element that can to move (transpose) from one site on a chromosome to another.

Transposition : Movement of a piece of DNA around the chromosome, usually through the function of a transposable element.

Transposon : Transposable element of which, in addition to genes involved in transposition, carries other genes; often confers selectable phenotypes such as antibiotic resistance.

Transposon mutagenesis : Insertion of a transposon into a gene; this inactivates the host gene leading to a mutant phenotype and also confers the phenotype associated with the transposon gene.

Twofold rotational symmetry : A double stranded nucleic acid where the 5' to 3' nucleotide sequence on one strand is identical to that on the complementary strand.

Type I restriction endonuclease : Typically large proteins that have more than one subunit, that require Sadenosylmethionine, ATP, and magnesium, that are able to both cleave and modify DNA in a sitespecific manner, and have a relatively slow in their rate of action when purified.

Type II restriction endonuclease : Typically small proteins that have a single subunit, require magnesium (or another divalent metal ion) but no other cofactors for activity, and are able to cleave DNA in a sitespecific manner with a relatively high rate when purified. These are accompanied in the cell by a corresponding restriction modification enzyme, generally a methylase, that recognizes and modifies the same sequence to prevent cleavage by the restriction endonuclease.

Type II restriction enzyme : Type II restriction endonuclease.

Ultraviolet light : (UV) light in the wavelength range of 280 to 320 nanometers that is used to illuminate DNA and RNA samples bound wit fluorescent dyes, such as ethidium bromide, allowing visualization of the nucleic acids.

Upstream : The region extending in a 5' direction from a gene.

Uracil : (U) a pyrimidine base found attached to the 1´ carbon of ribose in RNA, where it is capable of forming 2 hydrogen bond interactions with the purine adenine on an opposing nucleic acid strand.

Vector : (i) Plasmid or virus used in genetic engineering to insert genes into a cell. (ii) Agent, usually an insect or other animal, able to carry pathogens from one host to another.

Virus : A subcellular infectious organism. Viruses have highly specialized nucleic acid genomes generally packaged in protein or proteolipid coats or capsules that are capable of inserting the genome into a susceptible host cell, where the genome is able to replicate and produce progeny to be released by the infected cell.

Voltage : The electromotive force (voltage) applied to drive charged molecules through a gel matrix in order to separate the molecules by size and/or charge during electrophoresis. Proper voltage can be critical to the resolution of different species in a mixed sample. A very low voltage may allow diffusion of samples and loss of band resolution, while a very high voltage can distort sample mobility. Excessive power (Watts = voltage times current) can generate excessive heat, and distort sample mobility or melt an agarose gel matrix.

Water content : Water contained in a material expressed as the mass of water per unit mass of ovendry material.

Western blot : The process in which the protein bands present in a polyacrylamide gel are electrophoretically transferred to a membrane, incubated with a primary antibody directed against a specific protein target, then incubated with a labeled secondary antibody (or comparable protein) directed against the primary antibody to reveal the location of any primary antibody

Wild type : An organism as found in nature; the organism before it is genetically engineered.

Xenobiotic : Compound foreign to biological systems. Often refers to humanmade compounds that are resistant or recalcitrant to biodegradation and decomposition.

X-linked disease : A genetic disease caused by a mutation on the X chromosome. In X-linked recessive conditions, a normal female "carrier" passes on the mutated X chromosome to an affected son.

X-ray crystallography : The diffraction pattern of X-rays passing through a pure crystal of a substance.

Xray diffraction : The helical structure of DNA was proposed nearly 40 years ago based on the study of the diffraction of Xrays by crystals of DNA. The decades of subsequent DNA studies fundamentally support the basic structure as originally proposed.

Yield problem : The dilemma that as a genome increases in relative size, an individual DNA fragment becomes a smaller percentage of a DNA sample, making direct isolation of an individual DNA fragment technically impossible from a large genome.

Z-DNA : A region of DNA that is "flipped" into a lefthanded helix, characterized by alternating purines and pyrimidines, and which may be the target of a DNA-binding protein.

1' carbon : By the conventional chemical nomenclature used to refer to the various carbon atoms in an organic molecule, the carbon to which the purine or pyrimidine base is attached to deoxyribose in DNA and to ribose in RNA is designated the 1' carbon, the adjacent carbon in the sugar the 2' carbon, the next carbon the 3', the next the 4', and the last the 5' carbon. Moieties or residues attached to a particular carbon are identified by the number of the carbon to which they are attached, such as the 5´phosphate and the 3´hydroxyl residues of a nucleic acid.

16S rRNA : Large polynucleotide (about 1,500 bases) that functions as a part of the small subunit of the ribosome of prokaryotes and from whose sequence evolutionary information can be obtained; the eukaryotic counterpart is 18s rRNA.

3' cohesive DNA termini : The singlestranded, complementary regions of DNA that are formed when both of the strands of a doublestranded DNA molecule are cleaved in an asymmetric manner. The restriction endonuclease psti, for example, cleaves its recognition sequence "CTGCAG" between the A and G to generate the 4base, 3'cohesive terminus "TGCA". Psti.GIF

3' to 5' orientation : The direction along a polynucleotide molecule as defined by proceeding from the 3´hydroxyl terminus towards the 5´phosphate terminus. This is the direction opposite to that in which dna and rna are synthesized enzymatically. An example of an enzyme which proceeds with this orientation is a 3'5' exonuclease (a "proofreading" exonuclease), which can assist in correcting errors introduced during dna replication.

3'hydroxyl : The hydrogenated oxygen molecule attached to the 3', or third, carbon of either deoxyribose or ribose.

5' cohesive DNA termini : The singlestranded, complementary regions of DNA that are formed when both of the strands of a doublestranded DNA molecule are cleaved in an asymmetric manner. The restriction endonuclease EcoRi for example, cleaves the sequence "GAATTC" between the G and A to generate the 4base, 5′cohesive terminus "AATT". Ecori.GIF

5' end : In mRNA, the 5' end of a newly synthesized molecule corresponds to the transcription initiation point associated with a promoter. RNA processing steps and degradation may remove or modify this terminus so that the transcription initiation point is absent from the mature mRNA molecule and, consequently, from any cDNA made from the mature mRNA. By convention, nucleic acid sequences are written with the 5' end at the left.

5' terminus : The 5′ end of a nucleic acid strand.

5' to 3' orientation : The direction along a polynucleotide molecule as defined by proceeding from the 5′phosphate terminus towards the 3′hydroxyl terminus or by extending a polynucleotide by the addition of residues to the 3′hydroxyl terminus.

5' to 3' phosphate linkage : The oxygen phosphorus oxygen linkage that connects the 5' carbon of deoxy ribose or ribose to the 3' carbon of the adjacent sugar molecule in a nucleic acid.

5′phosphate : The oxygenated phosphorus atom attached through an oxygen molecule to the 5' carbon of deoxyribose or ribose.

Index

C

D

E

F

G

H

R

S

T

U

W

X

Zeitfracht Medien GmbH
Ferdinand-Jühlke-Straße 7
99095 Erfurt, Deutschland
produktsicherheit@kolibri360.de